Ecological Integrity and International Law

This book examines how the principle of ecological integrity in international law and governance can help address our current global crises.

The principle of ecological integrity points to the basic foundations that must exist for life – all life – not only to exist but also to flourish. Ecological integrity, or well-being, cuts across legal and geopolitical spheres, insofar as it is the decisions within our systems of governance that determine the integrity of our various living systems. This book provides a comprehensive examination of how the principle of ecological integrity can help us respond to our current crises of global conflict and insecurity. Including chapters by both academics and practitioners and by both experienced and emerging authors, the book covers a range of contemporary issues – the climate crisis, war, the pandemic, food, and more – at the intersection of peace, public health, and global security.

It will appeal to scholars across a range of fields, including law, sociology, political science, environmental studies, and public health.

Klaus Bosselmann is Professor of Environmental Law and Founding Director of the New Zealand Centre for Environmental Law, University of Auckland, New Zealand.

Kathryn Gwiazdon is Executive Director of the Center for Environmental Ethics and Law and Professor of International Law, Northern Illinois University College of Law, United States.

Virginia Zambrano is Professor of Law at the University of Salerno, Italy.

Ecological Integrity and International Law

Peace, Public Health, and Global Security

Edited by Klaus Bosselmann, Kathryn Gwiazdon, and Virginia Zambrano

Routledge
Taylor & Francis Group
a GlassHouse Book

First published 2025
by Routledge
4 Park Square, Milton Park, Abingdon, Oxon OX14 4RN

and by Routledge
605 Third Avenue, New York, NY 10158

Routledge is an imprint of the Taylor & Francis Group, an informa business

A GlassHouse book

British Library Cataloguing-in-Publication Data
A catalogue record for this book is available from the British Library

Library of Congress Cataloging-in-Publication Data
Names: Bosselmann, Klaus, editor. | Gwiazdon, Kathryn, editor. | Zambrano, Virginia, editor.
Title: Ecological integrity and international law : peace, public health, and global security / edited by Klaus Bosselmann, Kathryn Gwiazdon, and Virginia Zambrano.
Description: Abingdon, Oxon [UK] ; New York, NY : Routledge, 2025. | Includes bibliographical references and index.
Identifiers: LCCN 2024043874 (print) | LCCN 2024043875 (ebook)
Subjects: LCSH: Environmental law, International. | Ecological integrity. | Public health laws, International. | Security, International.
Classification: LCC K3585 .E35 2025 (print) | LCC K3585 (ebook) | DDC 344.04/6—dc23/eng/20240918
LC record available at https://lccn.loc.gov/2024043874
LC ebook record available at https://lccn.loc.gov/2024043875

ISBN: 978-1-032-57763-0 (hbk)
ISBN: 978-1-032-57767-8 (pbk)
ISBN: 978-1-003-44087-1 (ebk)

DOI: 10.4324/9781003440871

Typeset in Sabon
by Apex CoVantage, LLC

In honor of the 30th anniversary of the Global Ecological Integrity Group, this book is dedicated to Laura Westra – our founder, our leader, our friend – and to all participants, past, present, and future. May we forever come together in common cause, united in our hearts and minds and critical inquiry, forever trying to make the world more just – and just a bit better – for those yet to come.

Contents

Contributors

Renata Adami is Professor at the University of Salerno, Department of Physics, and Researcher with the Interdepartmental Centre NANO_MATES (Research Centre for Nanomaterials and Nanotechnology at the University of Salerno) in Fisciano, Italy.

Orsolya Bányai is Professor at the University of Debrecen, Faculty of Law, Environmental and Labor Law Department, in Debrecen, Hungary.

Klaus Bosselmann is Professor of Environmental Law at the University of Auckland, Faculty of Law; Founding Director of the New Zealand Centre for Environmental Law; Chair of the Ecological Law and Governance Association; Co-Chair of the Global Ecological Integrity Group; and Chair of the Earth Trusteeship Initiative.

Roman Buchtele is Assistant Professor at the University of South Bohemia, Faculty of Economics, in České Budějovice, Czech Republic.

Eva Cudlínová is Associate Professor at the University of South Bohemia, Faculty of Economics, in České Budějovice, Czech Republic.

Salvatore De Pasquale is Professor at the University of Salerno, Department of Physics, and Researcher with the Interdepartmental Centre NANO_MATES (Research Centre for Nanomaterials and Nanotechnology at the University of Salerno), in Fisciano, Italy.

Cristiane Derani is Professor at the Federal University of Santa Catarina, Department of Law, in Florianópolis, Brazil; Fellow at the Cambridge Centre for Environment, Energy, and Natural Resource Governance; and Deputy Chair of the Ethics Specialist Group of the IUCN World Commission on Environmental Law.

J. Ronald Engel is Professor *Emeritus* at Meadville/Lombard Theological School in Chicago, Illinois, USA.

Alicia Epstein is Researcher and Lecturer at Brandenburgische Technische Universität Cottbus-Senftenberg in Cottbus, Germany.

Joseph M. Gabris is Associate Attorney at DeHay & Elliston, LLP, and is pursuing his master of laws (LLM) in environmental, energy, and natural resources law at Lewis & Clark Law School in Portland, Oregon, USA.

Geoffrey Garver is Adjunct Professor of Environment and Sustainability at McGill University and Concordia University in Montréal, Canada; Law and Governance Lead of the Leadership for the Ecozoic program at McGill University; and Founding Member of the Ecological Law and Governance Association.

Giada Giacomini is a Postdoctoral researcher at the Geneva Graduate Institute of International and Development Studies, International Law Department, in Geneva, Switzerland.

Janice Gray is Honorary Associate Professor at the University of New South Wales, School of Private and Commercial Law, Faculty of Law and Justice, in Sydney, Australia.

Marco Ettore Grasso is *Cultore* on the subjects of theoretical philosophy and philosophy of sustainable development at the Department of Humanities, University of Macerata, Italy.

Emanuele Fratto Rosi Grippaudo is Assistant Professor of Administrative Law at the University of Pisa, Italy.

Kathryn Gwiazdon is Executive Director of the Center for Environmental Ethics and Law, based in Washington, DC, USA; Professor of International Law, Human Security, and Decolonialism at Northern Illinois University College of Law; Chair of the Ethics Specialist Group of the IUCN World Commission on Environmental Law; Expert Advisor to the UN Harmony with Nature program and the International Network for Epidemiology in Policy; and Steering Committee Member of the Global Ecological Integrity Group.

Antonín Hořčica is PhD candidate at the University of South Bohemia, Faculty of Economics, in České Budějovice, Czech Republic.

Bruce Jennings is Adjunct Professor of Health Policy in the Center for Biomedical Ethics and Society at Vanderbilt University and is Senior Fellow and Developmental Editor at the Center for Humans and Nature.

Shauna M. Lange is Attorney specializing in areas of elder law and public health at Lange Legal Group, PC, in Indiana, USA, and Expert Advisor to the International Network for Epidemiology in Policy.

Miloslav Lapka is Professor at the University of South Bohemia, Faculty of Economics, in České Budějovice, Czech Republic.

Kathleen Mahoney is Professor *Emerita* in Indigenous Law and Feminist Legal Theory at the University of Calgary, Faculty of Law; Fellow of the Royal Society of Canada; Senior Fellow at Lund University, Raoul Wallenberg Institute of Human Rights and Humanitarian Law; and Member of the Law Societies of Alberta and Manitoba and King's Counsel.

Massimiliano Montini is Professor of European Union Law and Sustainable Development Law at the University of Siena, Italy; Fellow at the Cambridge Centre for Environment, Energy, and Natural Resource Governance; and Vice-Chair of the Ecological Law and Governance Association.

Bradford W. Morse is Professor of Law and former Dean of Law at Thompson Rivers University, Faculty of Law, in Kamloops, British Columbia, Canada; Honorary Professor and former Dean of Law at Te Piringa, University of Waikato, Faculty of Law, in Waikato, New Zealand; and Professor *Emeritus* at University of Ottawa.

Nsikan-Abasi Odong holds an LLD from the University of Ottawa, Faculty of Law, in Ontario, Canada, and was a Rule of Law scholar and recipient of the Environment and Sustainability Scholarship and the International Doctoral Scholarship at the University of Ottawa.

Ottavio Quirico is Senior Researcher at Perugia University for Foreigners, Department of International Humanities and Social Sciences, Italy and Professor at the University of New England, School of Law and Australian National University, Centre for European Studies, in Australia.

Nikola Sagapova is Assistant Professor at the University of South Bohemia, Faculty of Economics, in České Budějovice, Czech Republic.

Maria Sarno is Professor at the University of Salerno, Department of Industrial Engineering, and Director of the Interdepartmental Centre NANO_MATES (Research Centre for Nanomaterials and Nanotechnology at the University of Salerno), in Fisciano, Italy.

Justin Sobion is called to the Bar of Trinidad and Tobago and Dominica, is currently pursuing his PhD at the University of Auckland, Faculty of Law, and is the Caribbean coordinator of the Advisory Opinion on Climate Change before the International Court of Justice.

Ngozi Finette Unuigbe is Professor in Environmental Law, Policy, and Ethics at the University of Benin in Benin City, Nigeria.

Anne Venton is a retired teacher in secondary education in Toronto, Canada, and former Head of Library at Northern Secondary School, North Toronto Collegiate, and City Adult Learning Centre, and served as an adjudicator for the Immigration and Refugee Board of Canada.

Peter Venton is President of JPV Associates, specializing in economics, peace, and public policy, and former senior policy advisor to the Ministry of Finance in Ontario, Canada.

Virginia Zambrano is Professor in Comparative Law, Intercultural Law, and Private Law at the University of Salerno, Faculty of Law, in Fisciano, Italy; Member of the UNHCR Territorial Committee for Refugees; Member of the experts' group on human rights for the National Committee on Economic Migration; and Steering Committee Member of the Global Ecological Integrity Group.

Foreword

In Solidarity

On behalf of myself and my life partner, Joan Gibb Engel (1936–2022), I would like to salute the Global Ecological Integrity Group on its 30th anniversary and express our gratitude for the opportunities it provided for us to pursue our life work, and for the lifelong friendships we formed among its members.

My first GEIG meeting was in Urbino, Italy, in 2003, held in tandem with an international gathering of the Earth Charter Initiative. Laura Westra had expressed strong support for the Earth Charter, and as a result, Brendan Mackey and I, who had served on the drafting committee for the Charter together, were invited to share our thoughts on its potential to inspire the "great transition" we were all seeking in world affairs. I titled my paper "A New Covenant for Global Democracy: The Size of the Task Before Us" and set forth my thesis as follows:

> Our present geopolitical situation removes all illusions that the vision of the Earth Charter is being carried forward by the tides of history, that historical evolution is inherently redemptive or progressive, or that the movement toward a global moral consciousness is inevitable. . . . Contemporary circumstances press us to cut to the deepest levels of human motivation. They require that we reach below the surface of our secular culture to those fundamental religious symbols and understandings of the human condition that once underpinned the disciplines of law, economics, the sciences, philosophy, and theology, yet have been obscured in the modern world.

I did not realize it at that time, but the strong endorsement I received from the members of GEIG on that occasion for the thesis that "covenant" was one of the fundamental root metaphors that needed to be reclaimed if we were to find the capacity to respond to the summons to a new axial age helped set me on the path of the primary work in which I have been engaged ever since.

Joan also received the gift of serious engagement, understanding, and support for her work from GEIG. In 2017, at our 25th anniversary meeting in Canada, she spoke on the topic "The Uses of Poetry to Effect Positive Climate-Change Policy." In the clearest prose, with illustrations from some of the poetry she most loved, she argued that "it is through poetry that we express the meaningful, humane understandings that our species has constructed from its contact with the things of Earth," and that "should we lose these things of nature," we will also "lose our humanity. . . . [O]ur love will leave with the absent rose, our 'lofty' and 'long-standing' with the absent trees." The overwhelmingly positive response she received from the members of GEIG at that meeting was a validation of her life's calling.

I share our personal experiences because I am confident that they are, by no means, unique to us but representative of the experiences of most participants in this group over the years. This has been a remarkably nurturing and sustaining community. We owe a salute to our indomitable leader, Laura Westra, who has carried its vision forward for these 30 years; to her loyal and resourceful helpmates, exemplified in the leadership of Kathryn Gwiazdon, without whom the meetings and publications would not have been possible; and finally, and by no means least of all, to each other: to the respect we have paid to one another in our meetings, in our disagreements as well as our agreements; to the blood, sweat, and tears that have gone into our presentations; to the encouragement we have given to one another; to the sacrifices many of us have made to be in attendance; and most of all, to the commitment we have shared, one that, I do believe, has truly cut to the deepest levels of human motivation, and which alone has power to save us in the end.

J. Ronald Engel

Preface

The Global Ecological Integrity Group (GEIG) is the only international, interdisciplinary professional organization that explores and advances the principle of ecological integrity in law and governance. And although we are a professional organization of academics and practitioners, we intentionally invite outstanding young scholars and late-term doctorate candidates to join and be active in our work and research. In honor of the 30th anniversary of the organization in 2022, this book brings together 33 experts from 11 nations, with diverse backgrounds in law, physics, engineering, economics, pedagogy, public health, and ethics, to explore together *Ecological Integrity and International Law: Peace, Public Health, and Global Security*.

Russia's illegal annexation and acts of aggression in Ukraine, and the most recent acts of terrorism and war in Israel and occupied Palestine, compound the harms rooted in the twin existential crises of climate change and biodiversity loss and are testing and remolding global governance institutions and the international rule of law. These are matters of life and death, justice and injustice. This book, and GEIG's continued exploration and evolution of the practical application of its scholarly research, seeks to highlight how the principles of ecological integrity, and particularly as it relates to security, peace, and public health, can help the world better respond to these interrelated and shared crises. How we treat one another – as individuals, communities, and states – is inextricably connected to how we treat our natural environment, the very foundations of life on Earth.

Ecological integrity is concerned with protecting the planet in a holistic way. Human health and environmental health are deeply interconnected, as every person on the planet should know, and as every political leader is obligated to know. What is at stake now is our ability as *Homo sapiens* to learn from experience and define our fate on the planet as linked with the fate of all living beings, and to demand this same recognition, tied to action, of powerholders and policymakers. This is what defines ecological integrity: What is good for preserving and restoring the integrity of the Earth system is good for humanity. This has been the message of GEIG for the past 30 years, consistent in principle, evolving to face current crises.

And ecological integrity looks at the foundations of life that must exist for life to not only exist but also flourish. This requires important legal and geopolitical considerations, as it is the actors, decisions, and institutions within our governance systems that determine the integrity of our lives and our living systems. As such, this also raises important questions on ethics, equity, and justice, so that we may better counter that which causes insecurity, namely, the power imbalances and hierarchies that differentiate, discriminate, and harm. On the continuum of any individual or collective existence, past to present to future, ecological integrity requires an exploration into (1) the root causes that brought a species or system to its current state; (2) how it is able to confront and adapt to harm, insecurity, and crises (i.e., its capacity and its resiliency); and (3) how to strengthen action and accountability to lessen, counter, and prevent further harm. This is equally important when considering conflict or climate change or biodiversity loss – what is our current state, how did we get here, and how can we respond in a targeted and meaningful way? A weaker integrity makes an individual or system more vulnerable to further harm and less able to heal from harm; therefore, in order to better address harm, it is important to look at what weakens and what strengthens integrity.

This book is divided into four sections which serve as permanent research points for the organization: "Ecological Integrity, Conflict, and Human Security"; "Ecological Integrity and Climate Security"; "Evolutions in Understanding Ecological Integrity"; and "Ecological Integrity for Peace." Through the lens of ecological integrity, this book seeks to offer new and compelling voices to identify and address the divisions and conflicts of the people, powers, and systems before us, so that we may have a more just, sustainable, and durable path toward peace. We welcome you to our discussions, to our methodology.

Section 1

Ecological Integrity, Conflict, and Human Security

Chapter 1

Lessons From Ukraine

The Critical Role of Ecological Integrity in Times of War, Armed Conflict, and Occupation

Kathryn Gwiazdon

Introduction

Great harm is occurring. Great violence is occurring. Great oppression is occurring. The few acquire much, and at the expense of the many – the expense of their lives, their dignity, and the foundations of those lives. This story is not new. And this harm, this violence, this oppression is an unequal harm, violence, and oppression. Not just between those in power and those without, but the disproportional impact of harm on women, children, the elderly, and people of color.

And it is not only a disproportional harm from a disproportional targeting of harm, but there is also a disproportional care. In the Middle East, there is the decades-long occupation and oppression of Palestine by Israel, seeming to culminate in recent months with the allegations of genocide by South Africa before the International Court of Justice (ICJ) (*South Africa v. Israel* 2023) and yet with the world watching as an entire people suffer acute food insecurity, famine, and erasure (IPC FRC 2024). In the United States, there is the school-to-prison pipeline of Black children (ACLU n.d.), the dehumanization of immigrants at the Mexican border (Kteily & Bruneau 2017), and for the first time in US history, once-protected human rights – the rights of life, privacy, bodily autonomy, and reproductive health for women – were taken away (*Dobbs v. Jackson Women's Health Organization* 2022). In Yemen, extreme famine (UNHCR 2024). In Ethiopia, a civil war ends, another may soon begin (Abebe & Moges 2024). There is the murder of Indigenous peoples and local community leaders in Colombia (IACHR 2023); the genocide, forced removals, and forced labor of the Uyghurs in China (OHCHR 2022; US Department of State 2024) and the Rohingya in Myanmar (UNHCR 2023); the extrajudicial killings from the "drug war" in the Philippines (Human Rights Watch n.d.); and the continued creation and predation of refugees from Syria, Haiti, and Venezuela, to name a few, and made worse by the COVID-19 pandemic (e.g., 42 USC 265 (2022); Amnesty International 2023). Where is the outrage? The statements of solidarity by world leaders? The news attention? The fundraisers? The flags flying? Where is the care?

DOI: 10.4324/9781003440871-2

But this is not – or should not be – a competition of harms and who is hurting more. Anyone is certainly capable of caring for many things at one time. Which is why the pointed identification of the difference in care and response is important. To individuals, states, and the community of nations, sometimes it seems like some harms matter and warrant attention and others do not. Of course, there are geopolitical concerns, but sometimes it seems that race, religion, gender, and ethnicity also play a part, a sort of discrimination in care.

And when these harms occur, it is also important to ask the purpose of these harms: What are the underlying causes to inflict violence, destruction, or oppression, in particular areas, upon particular people (whether united by race, religion, gender, ethnicity, age, or citizenship), and with particular methodologies? Is it for the expansion of a nation? The acquisition of resources? Greed, ego, power? Is it the same old story of imperialism, colonialism, capitalism, intersected with sexism and racism?

When Russia's military crossed Ukraine's border in 2014, and again in full force in 2022, a legal, political, and moral boundary was crossed, affecting a nation (and nations), communities, and individuals. A border was broken. A nation and its autonomy and its relations were broken. Laws and the stability and predictability they provide were broken. Communities and individuals were broken. And this "breaking" was not done with indifference but with intent – a critical consideration for criminal justice considerations. This breaking occurred because broken communities, people, and relationships equate to instability, insecurity, and an uncertain present and future. But to Russian President Vladimir Putin, this "breaking" was to help in "making" another: his own legacy and power-hold on Russia, and through harm, intimidation, and the control and rewriting and ownership of history. The acts of a dictator are never to advance a state and its people, although nationalism is often the focus of their propaganda and rhetoric, but to advance the dictator and their stranglehold on power in that state. This is largely evidenced by the citizens themselves: the vast human rights violations and the vast economic inequality.

But the destruction of one for the benefit of another is another common story in history. These harmed and broken people, places, and spaces are all made more vulnerable and less resilient through power – military or political – and through limiting the power of others. The burned villages in the Niger Delta for oil (Musa et al. 2023), the use of rape as a weapon in the Democratic Republic of the Congo for "fortress conservation" (Banerjee 2021), the destroyed olive trees in Palestine for Israeli settlements (ICJ AO Wall 2004): each destroying someone's home, life, and dignity and in order to build "more" for another – what perfect symbols of the war on peace. Yet this is the story of harm today.

And those in power own history – every student of colonialism knows this. Part of liberation is owning your own history. Or in other words, "daylighting" truth. This concept can also be seen in conservation; for example, "daylighting" is when an urban environment is built over a stream, and then that infrastructure is destroyed to let that river see light again, to be free again (e.g., Neale & Moffett 2016). This concept is also seen in the 1619 Project in the United States, or the toppling of colonialist and slave master statues (The 1619 Project 2019), the undoing of a broken past in order to liberate. These are a part of history, yes, but these slave masters and lawmakers and military leaders (often one and the same) are no longer the owners of history, or at least it is a statement or an act that what they stood for should no longer be.

And it is not just the history of a nation that the powerful own, but the history of communities and individuals. They can destroy pasts, change stories, disinform; they can purchase and publish and lobby and gaslight; and repeat, and repeat, and repeat. This is money, this is power, this is governance.

"Alternative facts" as a phrase was first used at the beginning of the Donald J. Trump presidency in 2017 (Gajanan 2017), but the concept and the methodology have been used for centuries by powerful nations, powerful people, and powerful religions. In an administration and the ongoing campaign marked by tens of thousands of lies – and many targeting the integrity of public health, governance, and the justice system – there may have been more truth to that phrase than any in the following four years: "we – the powerful – are creating alternative facts to align with the world that is ours."

And not only do the powerful own history, but they also own language and meanings. This is seen in the loss of languages – an act of cultural erasure and discrimination – but it is also seen in the definitions and determinations of success, right and wrong, what makes the rule of law, what makes a strong economy, what is "justice" or what is "human rights." Some examples: "growth is good," "capitalism is healthy, and necessary," "human rights for me, but not for thee, or even, not for 'us,' the larger human community." It is important to recognize that even the drafters of human rights documents, including the Universal Declaration of Human Rights and the US Constitution (as state constitutive documents are often the sources of human rights), had to dehumanize others in order to realize those texts (Gwiazdon 2020). And this continues today.

The powerful are not just the determinants of history and language, but of meaning, as well as the people and institutions behind enforcing that meaning. Look at the judges that defend unjust laws, or the executive officers or police that refuse to enforce just laws. The powerful have systems and structures – and a language – behind them to ensure their power-hold. And this is not new. And this is not right. But how to respond? How to liberate? What does that language look like?

And so once again, standing before humanity is a man, with a nuclear nation behind him, who seeks more, and he seeks to acquire more, through an attempted coup of history (e.g., Ukrainians are Russians, "entirely created by Russia" (Putin 2021; Chotiner 2022)), an attempted coup of truth, that is, what you see before you is not what you see before you (e.g., Russia's "special military operation" is to "denazify" Ukraine (Putin 2023, 2024; Treisman 2022)) and an attempted coup of the rule of law (e.g., "we are not the ones who started this war"; it was the West and NATO who are the aggressors (Putin 2024), and because Ukrainians are Russians, they have "no historical claim" to sovereignty (therefore rendering any claims of violated territory or crimes of aggression moot)). And these coups are attempted not only through lies and misinformation but also through direct, mass, grave, unconscionable, intentional violence to women, to children, to the elderly, and to all those they see as less than human, or "the other."

Global ecological integrity must confront issues of power and peace, of security and insecurity – food and public health, climate, and economy – of transgenerational harms and transgenerational care, and of the role of ecological integrity within it. And as we raise these terms that define this movement and define this voice for change, it is important to revisit and strengthen what is meant by *ecological integrity*. Ecological integrity is not some stable, steady state or some perfect end goal; like all things, it is an evolution in evolution, a state of nature – and humans within nature – that helps humans and nature prevent and avert harm, absorb harm, and heal from harm. The more integrity in a system, the more it is able to respond to harm, to survive harm, and to heal from harm.

Through the issues raised most recently by the acts of aggression by Russia in Ukraine, this chapter will seek to explore the critical role of ecological integrity in these power grabs – in these humans and nature "resource" grabs. Sifting through the politicization of this geopolitical storm in front of us, or even to its south in Gaza, is the same old story of imperialism, colonialism, and oppression; of ego and greed; of the grasp for a commodified nature through the destruction of a dehumanized people, and through the tools of misinformation and violence. There can never exist a critical analysis of harm – to people and nature – without a corresponding critical analysis of power, and the principles and methodologies that underlie it all.

Although this chapter will not be a list of the many violations of international law occurring since Crimea – primarily human rights law, humanitarian law, and international criminal law – it is important to note that several cases have been filed or decided that enumerate and unpack the violations (e.g., *Ukraine v. Russia (Genocide Convention)* 2024 and *Ukraine v. Russia (ICSFT and ICERD)* 2024 at the International Court of Justice; *Situation in Ukraine* (2022) at the International Criminal Court (ICC); and *Ukraine and the Netherlands v. Russia* (2022) at the European Court of Human Rights). Evidence gathering began day 1, as the world watched, and in the hopes for some future accountability. For example, environmental crimes and even ecocide are also being discussed (EUAM Ukraine 2023), and although environmental crimes are rarely prosecuted, perhaps Ukraine could set the precedent, thanks to better technology that allows for better mapping and data collection for evidence gathering.

An important development in this field came in February 2024, when Prosecutor Karim Khan of the ICC launched a program on developing policy related to environmental crimes (ICC 2024). In 2016, ICC Prosecutor Fatou Bensouda expanded the prosecutorial remit of the court to include environmental crimes (ICC 2016). However, no indictments or prosecutions followed that initial policy paper. Then, in February 2024, current ICC Prosecutor Karim Khan launched a public consultation to "advance accountability for environmental crimes" (ICC 2024). Under the Rome Statute, the ICC's jurisdiction covers individual criminal responsibility for the most serious offenses that humans can commit against other humans, including crimes against humanity, war crimes, and genocide. When Prosecutor Khan announced the initiative, he stated:

> Damage to the environment poses an existential threat to all life on the planet. For that reason, I am firmly committed to ensuring that my Office systematically addresses environmental crimes in all stages of its work, from preliminary investigations to prosecutions.
>
> (ICC 2024)

This is a crucial development in understanding that ecological integrity can threaten, and can be used to threaten (i.e., targeted in order to harm), the security and very existence of individuals, communities, and nations.

The modernization and immediacy of evidence gathering will also help when Ukraine claims reparations (see Zoï Environmental Network n.d.; Averin et al. 2024). Ukraine is also drafting a post-war recovery and development plan, including considerations for post-war "green" reconstruction (OECD 2022). This is part of the transition that is required for a "just transition," that is, recognizing the danger of commodifying nature, but also recognizing the real cost of its damage, so that it can be attempted to repair. Yes, its integrity is forever changed, even weakened, but "how to heal" is required for any discussion on justice. But likely, any environmental crimes would be included as aspects of war crimes as opposed to separate claims on the environment. But this is also part of a "just transition" that would counter the sectoral or "silo" treatment of "environmental" issues, while also raising the seriousness of their harms by considering them criminal acts. Land, fields, agriculture, water, air – critical infrastructure for life – are being intentionally or negligently destroyed, and the language and the law need to match the offense.

Many scholars and practitioners are making a substantive accounting and analysis of these legal violation, which only enriches the importance of any governance analysis on the role of ecological integrity in peace and security (and vice versa). And although Russia does not appear to respect the rule of law, as evidenced by its inarguable violation of the territorial integrity of another nation – first with the illegal annexation of Crimea, and later with the explicit act of aggression (i.e., inciting war through the unlawful use of force) against all of Ukraine – their accountability for these violations will be crucial to healing the people, nature, nation, and region, and indeed the international community of nations, and in trying to prevent future similar harms by other nations.

To assist in this larger conversation on harm and justice, it is necessary to look at the intersectionality of war, armed conflict, and occupation with ecological integrity and show how ecological integrity, as a foundation of security, is deliberately attacked to weaken a nation in order to allow for autocratic, authoritarian power grabs. And as such, the seriousness of harms to ecological integrity needs to be raised and considered as intentional acts of aggression – a crime of aggression – by states that use force against other states. In other words, an attack on ecological integrity should be considered an act of war. Attacks on territorial integrity are attacks on ecological integrity. And once the conflict begins, including the crime within existing concepts and definitions of war crimes is a logical next step.

The Intersection of Ecological Integrity and Conflict

This conflict, like all conflicts, does not stand alone. Leading to it, exacerbating it, and multiplying its harms are two other great crises of our time: COVID and climate change. Leading to it, exacerbating it, and multiplying its harms are other slow, systemic, structural violences of our age: discrimination, militarism, capitalism, and ecological destruction (see also Cusato 2021). Dr. Martin Luther King labeled the "three evils of society" to be racism, militarism, and capitalism (King 1967), and Reverend William Barber, a student of Dr. King, today adds the fourth evil of ecological destruction (Barber 2019).

It is important to revisit what is meant by "ecological integrity." An earlier understanding of *ecological integrity* largely argued for some "perfect" state of nature, absent humans, and even some policies today continue traces of this "wild" (i.e., non-human) approach. This is a dangerous path for law and governance considerations of ecological integrity, or even of broader notions of conservation or sustainability, as not only is "perfect" impossible, but it is, indeed, also the language of White supremacy and exclusion. It is the *philos* behind "find-and-fence," "fortress conservation," and even the murder and forced removal of Indigenous peoples to satisfy some Western notion of "protected area" for climate credits.

The US National Park system, often said to be the origins of the modern concept of protected areas (but of course there is evidence of protecting areas long before this; see generally Weeramantry 1992), was the result of the removal and genocide of entire peoples and cultures – and their ways of living in harmony with nature that made those beautiful areas beautiful, and sustainable, through intentional, careful management over thousands of years. Even the symbol and uniforms of the US National Park Service today still reference their early years as being a part of the US Army (Workman 1994). Nature and military were one. The protection of nature was through extreme acts of violence. This history cannot be forgotten, or repeated. This *philos* – and its realization through law and policy – is no more than an extension of colonialist, hierarchical, predatorial thinking. After all, nature does not exist separately from humans, and so the evolution of the concept of ecological integrity, to be ethically valid and scientifically accurate, must naturally embrace its humanity.

Ecological integrity is "the ability of an ecological system to support and maintain a community of organisms" – humans included – "that has species composition, diversity, and functional organization comparable to those of natural habitats within a region" (Parrish et al. 2003). This includes concepts of sustainability and resilience, and in its most basic sense, it requires physical, chemical, and biological integrity. For example, a landscape having high ecological integrity has three major characteristics: complex structure, high self-organization capacity (including adaptability and resilience), and a high level of stability. All this equates to security: a stable, secure system able to provide a flourishing, sustainable life and withstand shocks and stressors. And the concept should also be assessed using both Western science and Indigenous knowledge. When used together, these perspectives allow for a deeper understanding, or *Etuaptmumk*, a Mi'kmaw word for "Two-Eyed Seeing" (see Tremblay & Martin 2023; as related to public health, see Roher et al. 2021).

As an interrelated, interdependent systems concept, the intersection of ecological integrity and conflict is vast. It is not simply "How is the integrity of our natural environment affected by war, armed conflict, and occupation?" but also "How is the integrity of our humanity affected by war, armed conflict, and occupation?" and even "How does the integrity of our natural environment, including humanity, affect, introduce, or exacerbate conflict and oppression?" This looks at the integrity of relations, of societies; the integrity of governance structures, of democracy; and the integrity of individuals or a particular community, nation, or region, which may be geographically distant from the war, or any war, but affected by or affecting the war. It looks at the integrity of the villages in the Donbass region and in Kyiv, and in border towns and nations, such as those accepting refugees or those fearing an expansion of conflict. But it also considers those that fund the war or are entangled in legal and political security alliances, or political powerholders who learn from and mimic the methodologies of authoritarian actors or take advantage of global attention and resources focused on strife in one region so that they may cause strife or concentrate power in another region. These issues concern the integrity of all that make up the holistic, systems thinking that is ecological law, governance, ethics, and science, and it encompasses all the actors and institutions with the power to make decisions that change life and the foundations of life. Just as no human or nation is an "island" disconnected from others, no conflict in the modern era can be seen in isolation.

And it is interesting to see the issues before us: how this war began by the violation of territorial integrity, of the sovereignty of Ukraine. It was a violation of a border, and yet through the lens of ecological integrity, there are no borders in the causes and consequences of this and all wars. And just as with all conflicts, it is important to note this specific act is just a point on a longer continuum of causes and effects of what came before and after. All the elements of ecological integrity already had their own stresses before this war that this war then compounded, whether it be poverty or hunger or joblessness or homelessness or democratic deficits or geopolitical stressors nationally and internationally. There are remnants of imperialism or colonialism or past wars, resource stresses from refugees. Few things start whole before they are harmed.

And in this same vein is the role of this now conflict-ridden and conflict-focused society before the conflict began. Before Russia's act of aggression, Ukraine, known as the "breadbasket" of the world, provided wheat and humanitarian aid for other conflict-ridden and insecure areas around the world, particularly in Africa. Without this aid, not only have more lives been affected and lost, but the humanitarian crises – hand in hand with the conflicts – have also gotten worse. Ukraine has recently launched the "Grain from Ukraine" humanitarian food initiative in cooperation with the UN World Food Program, launched at the 2022 International Food Security Summit (Ukraine Ministry of Foreign Affairs 2024). The program aims to continue to combat food insecurity, but now from one war-torn country to another, such as Sudan (UN 2024). People and communities thousands of miles away are now also directly impacted by the acts of one nation unlawfully crossing one border, bringing war. But it is also good that Ukraine has found a point of solidarity – conflict – to try to still make the world a bit better of a place.

Humanitarian aid is also connected to the larger issue of budget and resource allocation. A budget once going toward education, the arts, infrastructure, and health care – strengthening the integrity of the interconnected lives and land of Ukraine – is now being diverted toward arms, military training, refugees, and crisis management. Children who once wanted to be musicians when they grew up now want to be soldiers who fight for their country. The leaders who once met in the legislature for lawmaking are now meeting in war rooms: how to counter tanks and missiles, all while protecting people that need to hide or flee or that have already been injured or captured. The "luxuries" of conservation and species protection are now de-prioritized amidst the larger struggle for survival (Hook & Marcantonio 2022). Protection for the foundations of life cannot compete when a nation is fighting for life itself – unless the nation sees that protecting the foundations of life is part of that fight.

And on this temporal scale of past, present, and future is also the proximity scale of immediate, acute threats and long-term harms, both physical and mental, and including the fear of future instability and harms. Think of the very threat of war as a stress to daily lives: what to protect, what to risk, what could be destroyed. And these harms are intimate: How could my life be changed, my home be lost, my future be changed, my future be lost? Will my body or my city or my children be taken by Russia (regarding the unlawful mass deportation of children, see *Situation in Ukraine* 2022)?

When identifying the issues raised, it is not only the present and future harm to the integrity of the environment – humans and nature – caused by the war itself, including the particular risks of sensitive sites such as Chernobyl or pathogen laboratories, but also the use, or threat of use, of chemical and biological agents, including the threat to use nuclear weapons. Putin has threatened to use nuclear weapons on several occasions since the full-scale Russian invasion into Ukraine, most recently in March 2024, often tying it to involvement by "the West" (Putin 2024; see also Cimbala & Korb 2024). The use, or threat of use, of nuclear weapons can never be seen in isolation. If they are used, and even if the weapons are targeted to a specific area, there is not only the immediate and long-term impact and fallout to consider, but it would also forever change the face of war on this planet, making it easier for other nations to see the plausibility of using nuclear weapons. Even the rhetoric leads to fear, and the normalization of the rhetoric increases the risk of replication – in rhetoric or use – of other nuclear-wielding nations. For example, US Senator Lindsey Graham recently stated that Israel should do "whatever" it needed to do to end the war, comparing their "existential" threat to why it was "OK" to drop nuclear bombs on Hiroshima and Nagasaki (Marquez 2024). Nuclear annihilation is ecological annihilation. Nuclear annihilation is life annihilation. And it is not a threat to Ukraine but to the world, to all life and to all the elements and relations that make up our interconnected, interdependent world.

The destruction of ecological integrity during war is the destruction of humans and nature, mentally and physically. It is the stress of danger and death and loss, but it is also an invasion of their homes through bombs, their bodies through rape, and their minds through fear. It is a denial of their identity and an attempted erasure of their selves, their stories, and their histories. And with forced removals and concentration camps already in place (e.g., *Situation in Ukraine* 2022), genocide is just a few small steps away, with the institutions, process, and principles already in place. Genocide of the Ukrainian people under Russian control is an act that could occur with a single order.

Harms also include conflict-induced air, water, and soil pollution, as seen with vast infrastructure damage, oil and gas fires, cluster bombs, and thermobaric rockets. And again, this is a transgenerational harm. These concerns were noted early in the conflict, with the Bulletin of the Atomic Scientists reporting that

> [t]he widespread destruction caused by the Russian invasion will inevitably lead to the contamination of land, water, and inhabited environments, where they can persist with a lifespan likely much longer than this conflict. While those risks may pale in comparison to the immediate horrors of war, their risk to human health is real, enduring, and will require focused international attention long after acute violence ceases.
>
> (Hook & Marcantonio 2022)

Due to the 2023 bombing of the Nova Kakhovka Dam, the country experienced "the worst [environmental catastrophe] in Ukraine since Chornobyl," with "unprecedented environmental consequences," and is preparing additional claims for violations of the Ukraine Criminal Code as well as the Rome Statute of the ICC (Watts 2023; Tignino et al. 2023).

It is clear that the environment was at the front of their concerns since the beginning of the full-scale invasion. Ukraine's Ministry of Environmental Protection and Natural Resources was immediately concerned of water and energy sites, including nuclear facilities, as well as destroyed military equipment and ammunition, missiles and mines, unexploded ordnances, and air, soil, and groundwater contamination, including with heavy metals (Hook & Marcantonio 2022). And the Geneva-based Zoï Environmental Network was already beginning to map incidents of war-related damage or disruption (Zoï Environmental Network n.d.; Marsi 2022), with their first 12-month report released in 2024 (Averin et al.). With advancements in the rule of law, and particularly war crimes, Ukraine started early to note harms and gather evidence in hopes of some future accountability.

The Russian aggressors and oppressors, like those before them in history, are intentionally attacking the integrity of the life, culture, and place of Ukraine because integrity is required for human security. Every element of human security – economic security, food security, health security, environmental security, personal security, community security, political security – is affected by ecological integrity (UNDP 1994). Harming ecological integrity creates an insecure people and nation, ripe for predation and oppression. It is by no coincidence that infertile soil provides fertile soil for authoritarian power grabs. Look at the deliberate attacks on agriculture (Welsh & Glauber 2024), the sustenance of a people; or on schools (Human Rights Watch 2023), the future people and the future of a people; or on critical infrastructure (note: also considered a war crime), the relations and infrastructure of a people that gives their lives stability and security and a future (UNSC 2024). Look at the mockery of the rule of law, of the UN Security Council, of human rights, of democracy – the bodies and systems and rights that give those who oppose this war the arguments to counter these unjust, violent actions

(Lederer 2023). Look at the types of weapons used and how they impact, even torture, civilians and combatants, and how they embed in the water, soil, and air (Zoï Environmental Network n.d.). Look at the cause of the conflict: acquisition of the land, the people, the agriculture, the oil, the ports.

In 2014:

> Putin launched an offensive into Ukraine against their primary coal and steel producing area – the Donbass [region] – firing artillery into the fertile stretch of land and paralyzing many aspects of environmental protection. This included the management of coal mines, where pumping, necessary to prevent toxic water from filling the mine shafts and polluting ground and drinking water, was halted.
>
> (Marsi 2022; see also Britannica 2024; IEA 2021)

In 2022, "he refocused Russia's nationwide invasion on this region, home to about 4,500 mining, metallurgical, and chemical enterprises" (Marsi 2022). It is not for history that Putin seeks to "reunite" Ukraine with Russia, but for resources, and all the profit and power that come with it. It is also important to note Putin's close history with the oil and metals industries while in local government roles in St. Petersburg, and the personal and political power – through organized crime and corruption – that they create (see Belton 2020: 211–240).

This is why experts know to look past the rhetoric and disinformation – disinformation, a tactic perfected by the Soviet Committee of State Security (KGB), of which Putin was notoriously a member (Belton 2020; Pomerantsev 2019) – and instead look at what he is targeting for destabilization and control. Whether an oil and gas region or an entire people or nation, the process is the same: target, destabilize, control. And the purpose is the same: power, through expansion and control of human and natural resources. These are the same roots of imperialism and colonialism.

The Courage to Care: The Responsibility of Scholars, States, and Society

Everything that builds stability and resilience – and good – in human and ecological systems is threatened or destroyed by war, armed conflict, and occupation. Scholars, academics, and practitioners must make these ties; they must care. The nexus of ecological integrity and international humanitarian law, including human rights law (as it applies during peace and wartime), must be made more evident. Therefore, the rest of this chapter will give insight into two efforts in global governance that seek to bring these worlds together. The first is a declaration from the International Network on Epidemiology in Policy (INEP) titled "Declaration on human rights and health in war, armed conflict, forced displacement, occupation, and violence: statement of solidarity with the peoples of Ukraine and around the globe" (Declaration) (INEP 2022). The other is a draft publication from the 73rd session of the International Law Commission, the UN body mandated with codifying international law, titled "Protection of the environment in relation to armed conflicts: Draft principles on protection of the environment in relation to armed conflicts" (ILC 2022).

To set the context of ecological integrity, it is important to understand the basic concepts of international humanitarian law: *jus ad bellum*, *jus in bello*, and the newly evolving consideration of *jus post bellum*. Respectively, these are the just conditions for war or the use of armed force, the laws during war, and the laws after war. In modern times, concepts of a "just war" are largely considered archaic, where attacking first is often seen as a crime of aggression (i.e., unlawfully starting a war); in its place, however, is the vast, legally gray area of "self-defense," which includes peremptory and anticipatory self-defense. Nations are often given wide latitude when they use force for self-defense, but there is also cause for concern as "self-defense" is often used as a pretext for what would otherwise be seen as an act of aggression. For example, one of Putin's many justifications for his large-scale assault into Ukraine was protecting "Russians" in Ukraine from the Ukrainian government, distorting both the principles of self-defense and the responsibility to protect (Ashby 2022). It is important to note that prior to the incursion, Russia handed out Russian passports to Ukrainian citizens, just as they did in Georgia in 2002 (Kapanadze 2022). This was to lay future claim to the people and land, as well as to provide justification that because "their" citizens were being harmed (in both Ukraine and Georgia, Russia claimed genocide was occurring), military action to defend them was necessary.

For *jus in bello*, three principles are meant to constrain action (including action in self-defense), rooted in both The Hague and Geneva Conventions, and now part of customary international law: necessity, humanity, and proportionality. For *necessity*, it is doing only what is necessary to end the conflict and to not cause unnecessary suffering. This ties into the methods of warfare and is also linked to *humanity* – the roots of humanitarian law – which understands that war brings out the worst in people and nations, but some level of dignity and protection is to be afforded to civilians and those under your control. *Proportionality* requires that acts be targeted and proportionate in response. An important part of the law during wartime is also justice, or accountability, which brings us to the evolving field of *jus post bellum*. This would not just be accountability for actors who committed war crimes but would also identify responsibility and remedies, including rebuilding and restoring. This leaves crucial considerations for ecological integrity, when both the healing of humans and nature is required.

INEP's Declaration on Human Rights and Health in War, Armed Conflict, Forced Displacement, Occupation, and Violence: Statement of Solidarity With the Peoples of Ukraine and Around the Globe

The International Network for Epidemiology in Policy is an international organization of epidemiological and public health societies that promotes the use of epidemiology to develop evidence-based policies to improve equity, health, and well-being for the global population. They focus on policy briefs and declarations covering a wide range of issues, including but not limited to gender equity, climate change, gun violence, and armed conflict. A cofounder of the Global Ecological Integrity Group (GEIG), Dr. Colin Soskolne, cofounded INEP as an organization rooted in public health, incorporating its wider implications for ecological integrity, in order to more strongly advance robust, just, sustainable policy.

Reviewing excerpts from the "Declaration on human rights and health in war, armed conflict, forced displacement, occupation, and violence: statement of solidarity with the peoples of Ukraine and around the globe," a public health–oriented document, it is important to note the types of integrity addressed and how it can be strengthened to advance ecological integrity in law and policy and for human security. After the preambulatory language of solidarity and the recognition that other violent crises are occurring – a recognition that led to the 2024 "A Call for Courage: declaration on human rights and health: statement of solidarity with the civilians of Israel and Palestine and with all those affected by war, armed conflict, forced displacement, occupation, and violence globally" (INEP 2024) – the document opens with a statement on discrimination. Understanding the intersectionality of harms is key to a way forward (INEP 2022, emphasis added):

> Noting that just as all harms are not inflicted equally upon human populations, there is discrimination regarding the attention and compassion shown towards people of different race, sex, class, age, and ethnicity, from different cultural and religious communities, non-Indigenous and Indigenous communities, and from high-income countries, middle-income, and low-income countries.

And:

> Noting the duty of all in the field of public health to advocate for the highest attainable standard of health for all, without discriminating factors; and, that *the health of all peoples is fundamental to the attainment of peace and security* and is dependent on the fullest co-operation of individuals and states.

It continues with a focus on key tenets of international humanitarian law, as well as its relation to human rights, "[e]xpressing grave concern on the evidence of the targeted and indiscriminate harm to the civilian population and infrastructure in Ukraine by Russian forces, *upon which human rights and public health rely*," and turns to the destruction of infrastructures of life and care, "[n]oting particularly the evidence of the targeted and indiscriminate harm to humanitarian and transportation corridors, hospitals, schools, and shelters," as well as the caretakers and cornerstones of our communities, "[n]oting particularly the evidence of rape and sexual abuse of women, children, and the elderly," and also "the reports of discrimination occurring at the borders towards refugees of African and Asian descent" and "the existence of pathogen laboratories in Ukraine, as well as nuclear energy facilities, including Chernobyl, and the particular harm that could occur if targeted or damaged" (INEP 2022, emphasis added). Even as we see parallels among aggressors and oppressors, it is crucial to note the specificity of place. Ukraine runs 15 nuclear reactors, over 1,600 chemical, petrochemical, and pharmaceutical enterprises, and 150 coal mines (World Nuclear Agency 2024; Ukraine Invest n.d.; IEA 2021). To a predatory and imperialistic nation, this translates to power.

The Declaration then turns to the action items, or the recommendations for states and policymakers (INEP 2022):

(1) The ongoing, systematic documentation of the magnitude of impacts, and implementation and evaluation of effective interventions that account for and address:
(2) The physical and mental impact of the harms and threat of harms on all people for present and future generations;
(3) The effects on all vulnerable populations, including but not limited to children, institutionalized adults, and low-income populations;
(4) The intersectional and intergenerational effects on families and family separation, destruction of lives, homes, ways of life and future stability and viability of impacted sub-populations;
(5) The effects on healthcare institutions, workers, and patients;
(6) The effects on food security and agriculture and capacity to sustain human and other life;
(7) The effects on air, water, and soil integrity and capacity to sustain human and other life;
(8) The effect on sanitation;
(9) The mass migrations of people through secure humanitarian corridors;
(10) The responsibility of nations to protect and receive refugees and allow for asylum, without discrimination;
(11) The spread of COVID-19 and other infectious diseases amid war, conflict, or occupation;
(12) Environmental exposure to known and suspected carcinogens due to destruction of infrastructure leading to increased exposures in air, water and contaminated food;

(13) The protection of pathogen laboratories;
(14) The protection of nuclear facilities; and
(15) The use of torture, imprisonment, starvation, rape, murder, and other acts of aggression by belligerent troops.

Again, elements of integrity are throughout this document, and explicitly so. Scholars, academics, and practitioners must be deliberate not just in how ecological integrity is used but in its connection to public health, security, and conflict. This document, although just the product of a single non-governmental organization, helps advance a more progressive understanding of the role of ecological integrity in conflict and occupation, rooted in science and human rights, but speaking the language of policy.

ILC's Draft Principles on Protection of the Environment in Relation to Armed Conflicts

Next will be a brief review of the text and context of the International Law Commission (ILC) document on the protection of the environment in relation to armed conflicts (a field sometimes referred to as PERAC) and its role in these evolving notions of ecological integrity and security. The ILC was established by the UN General Assembly in 1947 to advance and codify international law. The "Draft principles on protection of the environment in relation to armed conflicts" (ILC Draft Principles) opens with standard language on future generations and the evolution of environmental law and then highlights that environmental harm must be taken into consideration with the laws of armed conflict, including domestic and international conflicts and occupation (2022). This is important to note because "state sovereignty" historically has often been used to prevent the application of international humanitarian law in "internal" matters, even if grave harms were occurring. For example, Israel tried to argue before the International Court of Justice (ICJ) in the *Advisory Opinion on the Legal Consequences of the Construction of a Wall in the Occupied Palestinian Territories* that both international humanitarian law and human rights law did not apply to the occupied territories of Palestine (2004). The ICJ soundly rejected both contentions. In human rights law, and as noted earlier, there is also the "responsibility to protect," which calls on states and the international community to respond to grave violations of human rights of states to their own citizens, including through the use of force, but it has rarely been invoked (e.g., Libya; see UNSC 2011). Instead, states often look to diplomacy, political and trade isolation, or economic sanctions to address repressive and oppressive states, or they may request an investigation to the UN Human Rights Council or the ICC or file proceedings before the ICJ.

After the preambulatory language, the ILC Draft Principles begins to list its principles, with the first encompassing the temporal aspects of international humanitarian law noted earlier: "The present draft principles apply to the protection of the environment *before, during, or after* an armed conflict, including in situations of occupation" (2022, emphasis added). It then focuses on the creation of protected zones and areas of cultural importance, with language and purpose paralleling the already-protected UNESCO sites in times of war. It also gives particular consideration to harms on Indigenous land and state responsibility to work with Indigenous peoples to remedy such harms. It asks that provisions on environmental care be included in military agreements and peace operations, that is, include a recommendation that military operations and troop presence do not negatively affect the environment.

The ILC Draft Principles also highlights environmental protection in areas where people have been displaced by conflict, but it fails to note how environmental harms from conflict are a cause for displacement. This has already been noted extensively by states and scholars concerning climate change (e.g., Brosig et al. 2019; Palazzo 2022), and it should have been explicitly noted here. Instead, the language of this principle seems to focus on environmental harms or resource depletion resulting from formal or informal refugee camps, and it also spreads responsibility to "states, international organizations and other relevant actors" to "prevent, mitigate and remediate harm" (ILC 2022). Just as tanks and bombs lead to mass displacement, so does the harm from environmental damage. Refugees are an important consideration in human, national, and global security and are often the result of a destroyed or damaged ecology, unable to sustain life. This is a glaring omission in the ILC Draft Principles.

The next couple of principles builds on a growing movement that argues that private, corporate actors have human rights obligations, and particularly related to environmental harms (e.g., *Milieudefensie v. Shell* 2021). Principles 10 and 11 concern "due diligence by business enterprises" and "liability of business enterprises," respectively, in relation to the environment and human health and corporate liability, including subsidiaries (ILC 2022). The role of private actors in wartime and conflict, from contractors to arms industries, or in the case of occupation, private consumer businesses, such as those in the illegal Israeli settlements in occupied Palestine (see OHCHR 2023), is an important consideration, as these actors often try to shield themselves from accountability due to their private, corporate nature. Yet "private" actors are playing an increasing role in conflicts, whether through funding or normalizing the harms or by being directly involved in the harms. As the war in Afghanistan began to wind down, for example, there were more private contractors in Afghanistan than there were soldiers (Watson Institute 2021). Another example is Blackwater Security Consulting, a US-based private military company hired by the US government whose employees were convicted for the Nisour massacre of civilians in Iraq in 2014 (note: they were later pardoned by President Trump) (Wamsley 2020). This shows how individuals in private companies may commit grave atrocities, and with important geopolitical repercussions, even though they are not an official arm of the state.

The document then turns to another concerning principle: "No part of the environment may be attacked, unless it has become a military objective" (ILC 2022). As shown earlier, attacking the foundations of life, the integrity of our human and natural systems, is often a military objective in war. It is an intentional act against the security of a nation. And so with Principle 13, not many "attacks" would be limited by the ILC Draft Principles. Here, as with too many international documents, state sovereignty – and what they consider to be just or unjust acts, or "military objectives" – seems to reign supreme, and at the expense of people and the planet. Principle 13 effectively neuters the potential impact of this document, but at the very least, the principles show a point in evolution of the connections between environment and security.

The document continues that acts against the environment as reprisals are prohibited, as well as the pillage of natural resources, which follows the customary international law prohibition concerning the "spoils of war." But again, what if "pillage" is the aim of the use of force? Turning back to Ukraine and its land, people, energy, industries, agriculture, trade routes, etc. – what if an entire nation is the "spoil" of war? Under international law, you cannot benefit from a violation of international law (e.g., from a crime of aggression or the violation of territorial integrity), but Putin's intention is imperialism: It is the expansion of the Russian state through the violation of international law. If he sees any value in draft principles related to PERAC, it would be to protect the land and resources that he seeks to acquire through his use of force.

The ILC Draft Principles do recommend that environmental restoration be included as part of the peace processes, but this should be a consideration before conflict, that is, how ecological integrity can be used as a source of foreign policy to prevent war, whether through the protection of transboundary forests, rivers, etc., or even embracing the science of ecosystems and watersheds – that species, soils, and air are affected by everything around them, and so even what appears to be "internal" acts have "external" effects. Natural areas do not recognize man-made geopolitical boundaries, but man-made legal systems force the idea of ownership and property upon natural areas. A reconciliation of these understandings would be bilateral and multilateral agreements to protect a shared space for a shared future, or peacemaking and peacekeeping through safeguarding ecological integrity. This would be in support of such concepts as Leonardo Boff's "liberation ecology"; after all, with a system of oppression, we need a system of liberation (Boff 1995). This could also be an expansion or strengthening of existing concepts in law, such as the extraterritoriality of harms and state's interests to acknowledge them and be held accountable for them. Whether with climate change or conflict, and whether through a scientific or geopolitical lens, no harm is created or felt in isolation.

Part Four, Principles 19, 20, and 21, addresses the responsibilities of an occupying power to protect the environment, prevent harm that may affect the health and well-being of the occupied peoples, and prevent transboundary harm *from*, not into, the occupied territory (ILC 2022). Although there are territorial disputes throughout the world, the two most notable occupations in recent history are Russia in Crimea and Israel in Palestine. Both are considered *de facto* annexations in violation of international law by international bodies, states, and human rights experts, and both were created and are maintained through the use of force. Since the annexation of Crimea in 2014, damage to the landscape has been extensive, with "chemical contamination of the soil and groundwater," particularly in places where they conduct military drills (Hryhorczuk et al. 2024). Crimea has become a critical location for Russia's full offensive into Ukraine, and the land and people have suffered. Land once used for agriculture is now used for military exercises and ordnance testing.

In 1979, the USSR invaded and occupied portions of Afghanistan. By the time they left in 1989, the land had been sowed with millions of land mines. Today, Afghanistan is considered one of the most heavily afflicted nations by land mines "planted indiscriminately over most of the country. Agricultural farms, irrigation canals, residential areas, grazing areas, roads and footpaths, both in urban and rural areas are contaminated" (Landmine & Cluster Munition Monitor n.d.). A nation has no chance to stabilize after a conflict if its very foundations have been short- and long-term destabilized. This is evidenced by the invasion of Afghanistan in 2001 by the US and coalition forces, and their withdrawal, and the Taliban takeover, in 2022. This is also the perfect example of the need, and disregard, of *jus post bellum*: What is the responsibility of Russia today to remedy the harms of their Soviet past? What is the responsibility of the United States after 20 years of war?

The adoption of the ILC Draft Principles is an important point in the evolution to better understand the interrelated harms to humans and nature that occur during war, conflict, and occupation. But it has concerning limits, unfortunately common to many global governance documents, with unequal burdens or assumptions placed on harmed peoples and, ultimately, a supplication to state sovereignty. It is the product of unequal powerholders at an unequal bargaining table in an inaccessible-to-all bargaining room. It represents business-as-usual, and so major changes in behavior by powerholders should not be expected. Evidence to this could be the complete destruction of the people and land in Gaza by Israel since the 7 October 2023 terrorist attack by Hamas. Principles disregarded, humanitarian and human rights law disregarded, but with cases before both the ICJ and indictments from the ICC, perhaps there will be some future accountability that takes into account ecological integrity and its existential role in human security.

Conclusion

In conclusion, the relationship between ecological integrity and war, armed conflict, and occupation is highlighted because of recent acts of aggression and war in Ukraine, and even more recently Gaza, but it is applicable to all acts of violence and oppression. Behind these great harms are often the same types of actors, ruled by the same principles, with the same end goals – even nature herself. From history to present day, acts of war have aligned themselves with acts against the environment, where the intentional targeting of ecological integrity – to destroy or to use – is a methodology for destruction, expansionism, and imperialism. Target the integrity of an ecological system, its law and governance included, and you weaken a people and a nation. Which is why this intentional targeting must be highlighted as such and named as such. The gravity of the harm must honor the breadth of all that ecological integrity gives to life on Earth.

Reference List

42 U.S.C. 265 (2022) *Suspension of Entries and Imports from Designated Places to Prevent Spread of Communicable Diseases*. Available at: www.govinfo.gov/app/details/USCODE-2022-title42/USCODE-2022-title42-chap6A-subchapII-partG-sec265.

The 1619 Project (2019) *The New York Times Magazine*. Available at: www.nytimes.com/interactive/2019/08/14/magazine/1619-america-slavery.html.

Abebe, A.K. & Moges, Z. "Ethiopia's Amhara Conflict Could Spark Civil War," *Foreign Policy*. Available at: https://foreignpolicy.com/2024/03/06/ethiopia-amhara-conflict-civil-war/.

Advisory Opinion on the Legal Consequences of the Construction of a Wall in the Occupied Palestinian Territory [ICJ AO Wall] (2004) International Court of Justice. Available at: www.icj-cij.org/case/131.

American Civil Liberties Union [ACLU] (n.d.) *School-to-Prison Pipeline*. Available at: www.aclu.org/issues/juvenile-justice/juvenile-justice-school-prison-pipeline.

Amnesty International (2023) *USA: At Border, Humanitarian Delegation Witnesses an End of Title 42 Without Chaos and Cruelty of New Asylum Ban*. Available at: www.amnesty.org/en/latest/news/2023/05/usa-humanitarian-delegation-end-title-42/.

Ashby, H. (2022) "How the Kremlin Distorts the 'Responsibility to Protect' Principle," *United States Institute of Peace*. Available at: www.usip.org/publications/2022/04/how-kremlin-distorts-responsibility-protect-principle.

Averin, D., et al. (2024) "The Environmental Consequences of the War Against Ukraine: Preliminary 12-Month Assessment (February 2022–February 2023)," *Conflict and Environment Observatory & Zoï Environmental Network*. Available at: https://zoinet.org/product/osce-ukraine-one-year/.

Banerjee, S. (2021) "Murder, Rape, and Torture: Fortress Conservation on Trial," *CounterPunch*. Available at: www.counterpunch.org/2021/11/05/murder-rape-and-torture-fortress-conservation-on-trial/.

Barber, W. (2019) *Forward Together, Not One Step Back*, Keynote Delivered at Berkeley Talks. Available at: https://news.berkeley.edu/2019/04/14/berkeley-talks-transcript-rev-dr-william-j-barber-ii.

Belton, C. (2020) *Putin's People: How the KGB Took Back Russia and Then Took On the West*, New York: Picador.

Boff, L. (1995) *Ecology and Liberation: A New Paradigm*, New York: Orbis Books.

Britannica (2024) *Donbas*. Available at: www.britannica.com/place/Donbas.
Brosig, M., et al. (2019) *Implications of Climate Change for the U.S. Army*, US Army War College. Available at: https://climateandsecurity.files.wordpress.com/2019/07/implications-of-climate-change-for-us-army_army-war-college_2019.pdf.
Chotiner, I. (2022) "Vladimir Putin's Revisionist History of Russia and Ukraine: The historian Serhii Plokhy Discusses the Russian President's 'Very Imperial Idea' of His Country, and the Potential for Ukrainian Resistance," *New Yorker*. Available at: www.newyorker.com/news/q-and-a/vladimir-putins-revisionist-history-of-russia-and-ukraine.
Cimbala, S. & Korb, L. (2024) "Putin's Nuclear Warnings: Heightened Risk or Revolving Door?" *Bulletin of the Atomic Scientists*. Available at: https://thebulletin.org/2024/03/putins-nuclear-warnings-heightened-risk-or-revolving-door/#post-heading.
Cusato, E. (2021) *The Ecology of War and Peace: Marginalizing Slow and Structural Violence in International Law*, Cambridge: Cambridge University Press.
Dobbs, State Health Officer of the Mississippi Dept. of Health et al v. Jackson Women's Health Organization et al [Dobbs v. Jackson Women's Health Organization] (2022) 597 U.S. 215. Available at: www.supremecourt.gov/opinions/21pdf/19-1392_6j37.pdf.
European Union Advisory Mission for Civilian Security Sector Reform Ukraine [EUAM Ukraine] (2023) *Ecocide in Ukraine Won't Go Unpunished. United for Justice.* United for Nature. Available at: www.euam-ukraine.eu/news/ecocide-in-ukraine-won-t-go-unpunished-united-for-justice-united-for-nature/.
Gajanan, M. (2017) "Kellyanne Conway Defends White House's Falsehoods as 'Alternative Facts'," *Time Magazine*. Available at: https://time.com/4642689/kellyanne-conway-sean-spicer-donald-trump-alternative-facts/.
Gwiazdon, K. (2020) "State global responsibility for environmental crises: The ethical and legal implications of a state's failure to protect human rights," in: L. Westra, K. Bosselmann, & M. Fermeglia (eds) *Ecological Integrity in Science and Law*, Cham: Springer. Available at: https://doi.org/10.1007/978-3-030-46259-8_10.
Hook, K. & Marcantonio, R.D. (2022) "Grappling with Environmental Risks in the Fog of War," *Bulletin of the Atomic Scientists*. Available at: https://thebulletin.org/2022/03/grappling-with-environmental-risks-in-the-fog-of-war/.
Hryhorczuk, D., et al. (2024) "The environmental health impacts of Russia's war on Ukraine," *Journal of Occupational Medicine and Toxicology*, 19(1). Available at: https://occup-med.biomedcentral.com/articles/10.1186/s12995-023-00398-y.
Human Rights Watch (2023) *Tanks on the Playground: Attacks on Schools and Military Use of Schools in Ukraine*. Available at: www.hrw.org/sites/default/files/media_2023/11/ukraine1123web_0.pdf.
Human Rights Watch (n.d.) *Philippines' 'War on Drugs'*. Available at: www.hrw.org/tag/philippines-war-drugs.
Integrated Food Security Phase Classification [IPC] Famine Review Committee [FRC] (2024) *Famine Review Committee: Gaza Strip, March 2024, Conclusions and Recommendations*. Available at: www.ipcinfo.org/fileadmin/user_upload/ipcinfo/docs/IPC_Famine_Committee_Review_Report_Gaza_Strip_Acute_Food_Insecurity_Feb_July2024_Special_Brief.pdf.
Inter-American Commission on Human Rights [IACHR] (2023) *IACHR Condemns Murders of Indigenous Adolescents of the Murui-Muina People by an Illegal Armed Group in Colombia*. Available at: www.oas.org/en/IACHR/jsForm/?File=/en/iachr/media_center/PReleases/2023/120.asp.
International Criminal Court [ICC] (2016) *Policy Paper on Case Selection and Prioritization*. Available at: www.icc-cpi.int/news/policy-paper-case-selection-and-prioritisation.

International Criminal Court [ICC] (2024) *The Office of the Prosecutor Launches Public Consultation on a New Policy Initiative to Advance Accountability for Environmental Crimes Under the Rome Statute*. Available at: www.icc-cpi.int/news/office-prosecutor-launches-public-consultation-new-policy-initiative-advance-accountability-0.

International Energy Agency (2021) *Ukraine Energy Profile: Energy Security*. Available at: www.iea.org/reports/ukraine-energy-profile/energy-security.

International Network on Epidemiology in Policy [INEP] (2022) *Declaration on Human Rights and Health in War, Armed Conflict, Forced Displacement, Occupation, and Violence: Statement of Solidarity with the Peoples of Ukraine and Around the Globe*. Available at: www.epidemiologyinpolicy.org/ukraine.

International Network on Epidemiology in Policy [INEP] (2024) *A Call for Courage: Declaration on Human Rights and Health: Statement of Solidarity with the Civilians of Israel and Palestine and with All Those Affected by War, Armed Conflict, Forced Displacement, Occupation, and Violence Globally*. Available at: www.epidemiologyinpolicy.org/palestine.

Kapanadze, S. (2022) "Putin's Invasion Playbook All Too Familiar to Georgia," *Center for European Policy Analysis*. Available at: https://cepa.org/article/putins-invasion-playbook-all-too-familiar-to-georgia/.

King, M.L. (1967) *The Three Evils of Society*, Address Delivered at the National Conference on New Politics. Available at: www.nwesd.org/ed-talks/equity/the-three-evils-of-society-address-martin-luther-king-jr/#.

Kteily, N. & Bruneau, E. (2017) "Backlash: The politics and real-world consequences of minority group dehumanization," *Personality and Social Psychology Bulletin*, 43(1), pp. 87–104. Available at: https://doi.org/10.1177/0146167216675334.

Landmine & Cluster Munition Monitor (n.d.) *Afghanistan*. Available at: http://archives.the-monitor.org/index.php/publications/display?url=lm/1999/afghanistan.

Lederer, E. (2023) "Russia Rejects U.S. Claims Its UN Security Council Presidency Is an 'April Fool's Joke'," *PBS*. Available at: www.pbs.org/newshour/politics/russia-rejects-u-s-claims-its-un-security-council-presidency-is-an-april-fools-joke.

Marquez, A. (2024) "Sen. Lindsay Graham Says Israel Should Do 'Whatever' It Has to While Comparing the War to Hiroshima and Nagasaki," *NBC News*. Available at: www.nbcnews.com/politics/politics-news/sen-lindsey-graham-says-israel-whatever-comparing-war-gaza-hiroshima-n-rcna151828.

Marsi, F. (2022) "Can Ukraine Hold Russia Accountable for Environmental Crimes?" *Al Jazeera*. Available at: www.aljazeera.com/features/2022/5/16/can-ukraine-hold-russia-accountable-for-environmental-crimes.

Milieudefensie et al. v. Royal Dutch Shell plc. (2021) Hague District Court. Available at: https://climatecasechart.com/non-us-case/milieudefensie-et-al-v-royal-dutch-shell-plc/.

Musa, S.S., et al. (2023) "The public health implications of the oil discovery in Northern Nigeria," *Clinical Epidemiology and Global Health*, 21, p. 101276. Available at: https://doi.org/10.1016/j.cegh.2023.101276.

Neale, M. & Moffett, E. (2016) "Re-engineering buried urban streams: Daylighting results in rapid changes in stream invertebrate communities," *Ecological Engineering*, 87, pp. 175–184. Available at: https://doi.org/10.1016/j.ecoleng.2015.11.043.

Organization for Economic Cooperation and Development [OECD] (2022) *Environmental Impacts of the War in Ukraine and Prospects for a Green Reconstruction*. Available at: www.oecd.org/ukraine-hub/policy-responses/environmental-impacts-of-the-war-in-ukraine-and-prospects-for-a-green-reconstruction-9e86d691/.

Palazzo, A. (2022) *Climate Change and National Security: Implications for the Military*, Kansas: Army University Press. Available at: www.armyupress.army.mil/

Portals/7/combat-studies-institute/csi-books/Command/ClimateChg-Palazzo%20 interactive%20w%20cover%2010Aug2022.pdf.
Parrish, J.D., et al. (2003) "Are we conserving what we say we are? Measuring ecological integrity within protected areas," *BioScience*, 53, pp. 851–860. Available at: https://academic.oup.com/bioscience/article/53/9/851/311604.
Pomerantsev, P. (2019) *This Is Not Propaganda: Adventures in the War Against Reality*, London: Faber.
Putin, V. (2021) "Article by Vladimir Putin 'on the historical unity of Russians and Ukrainians'," *The Kremlin, President of Russia*. Available at: http://en.kremlin.ru/events/president/news/66181.
Putin, V. (2023) "Presidential address to federal assembly," *The Kremlin, President of Russia*. Available at: http://en.kremlin.ru/events/president/news/70565.
Putin, V. (2024) "Presidential address to the federal assembly," *The Kremlin, President of Russia*. Available at: www.en.kremlin.ru/events/president/transcripts/messages/73585.
Roher, S., et al. (2021) "How is etuaptmumk/two-eyed seeing characterized in indigenous health research? A scoping review," *PLoS ONE*, 16(7). Available at: www.ncbi.nlm.nih.gov/pmc/articles/PMC8291645/.
Situation in Ukraine (2022) International Criminal Court. Available at: www.icc-cpi.int/situations/ukraine.
South Africa v. Israel (2023) International Court of Justice. Available at: www.icj-cij.org/case/192.
Tignino, M., et al. (2023) "International law and accountability for the Nova Kakhovka Dam disaster," *Lieber Institute West Point*. Available at: https://lieber.westpoint.edu/international-law-accountability-nova-kakhovka-dam-disaster/.
Treisman, R. (2022) "Putin's Claim of Fighting Against Ukraine 'Neo-Nazis' Distorts History, Scholars Say," *National Public Radio*. Available at: www.npr.org/2022/03/01/1083677765/putin-denazify-ukraine-russia-history.
Tremblay, M.C. & Martin, D.H. (2023) "Etuaptmumk/two-eyed seeing: A guiding principle to respectfully embrace indigenous and western systems of knowledge," in: D. Jourdan & L. Potvin (eds) *Global Handbook of Health Promotion Research*, Vol. 3, Cham: Springer. Available at: https://doi.org/10.1007/978-3-031-20401-2_10.
Ukraine and the Netherlands v. Russia (2022) European Court of Human Rights. Available at: https://hudoc.echr.coe.int/eng#{%22itemid%22:[%22001-222889%22]}.
Ukraine Invest (n.d.) *Chemicals*. Available at: https://ukraineinvest.gov.ua/en/industries/chemicals/.
Ukraine Ministry of Foreign Affairs (2024) *Grain for Ukraine*. Available at: https://mfa.gov.ua/en/grain-ukraine.
Ukraine v. Russian Federation (Allegations of Genocide under the Convention on the Prevention and Punishment of the Crime of Genocide [Genocide Convention]) (2024) International Court of Justice. Available at: www.icj-cij.org/case/182.
Ukraine v. Russian Federation (Application of the International Convention for the Suppression of the Financing of Terrorism [ICSFT] and of the International Convention on the Elimination of All Forms of Racial Discrimination [ICERD]) (2024) International Court of Justice. Available at: www.icj-cij.org/case/166.
UN (2024) *Wheat from War-Torn Ukraine to Feed Families Affected by Conflict in Sudan*. Available at: https://news.un.org/en/story/2024/02/1147012.
UN Development Programme [UNDP] (1994) *UN Development Report*. Available at: https://hdr.undp.org/content/human-development-report-1994.
UN International Law Commission [ILC] (2022) *Protection of the Environment in Relation to Armed Conflicts: Draft Principles on Protection of the Environment in Relation to Armed Conflicts*. Available at: https://documents.un.org/

doc/undoc/ltd/g22/348/04/pdf/g2234804.pdf?token=CvsqTgY4GxoC6PmXuk&fe=true.
UN Office of the High Commissioner on Human Rights [OHCHR] (2022) *OHCHR Assessment of Human Rights Concerns in the Xinjiang Uyghur Autonomous Region, People's Republic of China*. Available at: www.ohchr.org/en/documents/country-reports/ohchr-assessment-human-rights-concerns-xinjiang-uyghur-autonomous-region.
UN Office of the High Commissioner on Human Rights [OHCHR] (2023) *Database Pursuant to Human Rights Council Resolution 31/36: OHCHR Update of Database of All Business Enterprises Involved in the Activities Detailed in Paragraph 96 of the Report of the Independent International Fact-Finding Mission to Investigate the Implications of the Israeli Settlements on the Civil, Political, Economic, Social, and Cultural Rights of the Palestinian People Throughout the Occupied Palestinian Territory, Including East Jerusalem*. Available at: www.ohchr.org/en/hr-bodies/hrc/regular-sessions/session31/database-hrc3136.
UN Refugee Agency [UNHCR] (2023) *Rohingya Emergency*. Available at: www.unhcr.org/us/emergencies/rohingya-emergency.
UN Refugee Agency [UNHCR] (2024) *Yemen Crisis Explained*. Available at: www.unrefugees.org/news/yemen-crisis-explained/.
UN Security Council [UNSC] (2011) *Resolution 1973, Imposes Additional Measures in Connection with the Situation in the Libyan Arab Jamahiriya; Establishes a Panel of Experts to Assist in Monitoring the Relevant Measures*. Available at: www.un.org/securitycouncil/s/res/1973-%282011%29.
UN Security Council [UNSC] (2024) *Escalating Attacks on Ukraine's Civilian, Energy Infrastructure Making Humanitarian Aid Delivery Even More Dangerous, Relief Chief Tells Security Council*. Available at: https://press.un.org/en/2024/sc15695.doc.htm.
U.S. Department of State (2024) *2023 Country Reports on Human Rights Practices*. Available at: www.state.gov/reports/2023-country-reports-on-human-rights-practices/.
Wamsley, L. (2020) "Shock and Dismay After Trump Pardons Blackwater Guards Who Killed 14 Iraqi Civilians," *National Public Radio*. Available at: www.npr.org/2020/12/23/949679837/shock-and-dismay-after-trump-pardons-blackwater-guards-who-killed-14-iraqi-civil.
Watson Institute of International and Public Affairs, Brown University [Watson Institute] (2021) *Corporate Power, Profiteering, and the "Camo Economy"*. Available at: https://watson.brown.edu/costsofwar/costs/social/corporate.
Watts, J. (2023) "Dam Breach Could Be Ukraine's 'Worst Ecological Disaster Since Chornobyl'," *The Guardian*. Available at: www.theguardian.com/world/2023/jun/06/dam-breach-could-be-ukraines-worst-ecological-disaster-since-chornobyl.
Weeramantry, C. (1992) *Gabčíkovo-Nagymaros Project (Hungary v. Slovakia): Separate Opinion of Vice President Weeramantry*. Available at: www.icj-cij.org/case/92/judgments.
Welsh, C. & Glauber, J. (2024) "Food as the 'Silent Weapon': Russia's Gains and Ukraine's Losses," *Center for Strategic & International Studies*. Available at: www.csis.org/analysis/food-silent-weapon-russias-gains-and-ukraines-losses.
Workman, R.B. (1994) "National Park Service Uniforms: In search of an identity 1872–1920," *National Park Uniform Series*, Harpers Ferry: US National Park Service History Collection. Available at: www.nps.gov/subjects/hfc/upload/No-2_In-Search-of-an-Identity.pdf.
World Nuclear Agency (2024) *Country Profile: Nuclear Power in Ukraine*. Available at: https://world-nuclear.org/information-library/country-profiles/countries-t-z/ukraine.
Zoï Environmental Network (n.d.) *War on Ukraine*. Available at: https://zoinet.org/topics/war-on-ukraine/.

Chapter 2

Unexpected Challenges for the European Green Deal

COVID-19, War in Ukraine, and the Energy Crisis

Eva Cudlínová, Roman Buchtele, Nikola Sagapova, Miloslav Lapka, and Antonín Hořčica

Introduction

Europe is often accused of not having big goals and ideas, of not being innovative enough, and of being an open-air museum. One of the most vocal critics is Edmund Phelps: "It's incredible how incompetent Europe is when it comes to innovation" (Kovanda 2009).

Is the European Green Deal (EGD) also an outdated idea, with no ambition to change the world economy and culture? Is the goal of Europe to be the first climate-neutral continent in the world by 2050 feasible without innovation? These questions go far beyond the economic debate and affect nearly everyone in Europe. Let us try to look at the current framework of conditions and assumptions within which the EGD is operating. It must be said that recent years' challenges mentioned in the title represent conditions that were completely unexpected when the EGD was adopted.

The EGD contains three key principles for the whole clean energy transition: (1) ensuring a secure and affordable EU energy supply; (2) developing a fully integrated, interconnected, and digitalized EU energy market; and (3) prioritizing energy efficiency, improving the energy performance of our buildings, and developing a power sector based largely on renewable sources (European Commission 2019).

The EGD presents itself as a bold challenge to the world, as an example of how to tackle the carbon footprint and as a global solution to environmental problems. In this respect, the 2019 EGD seems to be a new, innovative, and even revolutionary perspective of changes that cross over the European Union (EU) boundaries.

Such an ambitious plan like the EGD requires hidden assumptions for its successful fulfilment. First, that relative stability of economic and social life in the EU ensures stable energy prices, stable institutions, and social services. Second, that external countries have security under the three pillars of the sustainable development system, which means that the economy, social life, and environment in Europe will not be under threat from unexpected great changes. Third, that external sources of economic growth will still be

DOI: 10.4324/9781003440871-3

available and can be converted into a market mechanism, with assumptions that environmental changes are reversible. And lastly, that all new evolutionary expected challenges will be solved by a proven market model for growth and prosperity; in other words, traditional economic tools are applicable for the new, changing reality. New challenges that appeared during 2021 tested all these assumptions.

New Challenges Related to COVID-19, the War in Ukraine, and the Energy Crisis

COVID-19 has affected a number of variables that will ultimately affect the economy of the regions: the balance between work and family, the diminishing role of the public sphere, the extraordinary executive powers of government (Benhabib 2021). It caused a major economic shock, which was particularly evident in the GDP indicator (Hove 2020). COVID-19 also has some positive outcomes – it significantly strengthened integration within the EU. Pressure was exerted on EU institutions and member states, leading to tough intergovernmental bargaining over the distribution of recovery benefits and crisis-related costs. The second positive impact was a significant strengthening of the EGD priority (Crnčec 2020).

Figure 2.1 shows that the highest trend in new cases was between October 2021 and approximately April 2022. The chart selects the largest countries in Europe but also includes an EU-wide figure.

The development of COVID-19 is also strongly linked to changes in macroeconomic indicators. Figure 2.2 shows the change in economic growth through GDP in 2019, 2020, and 2021. It is already evident at a glance that in 2020, this indicator showed negative values in most EU countries. The most significant declines were in Spain, Italy, and Greece. The drop in GDP for most countries is mainly in 2020, which was at the onset and main wave of COVID-19, and by 2021, EU countries had already started to recover economically.

Chiaramonti and Maniatis (2020), who examine the links between the pandemic and energy security, analyze the impacts of the COVID-19 crisis on different sectors. In particular, short-term plans to respond to the immediate impact of COVID-19 on the fuel industry were an adequate solution. It also protected jobs and economic activity. Rowan and Galanakis (2020) suggest opportunities for sustainable agri-food production and for accelerating eco-innovation through ambitious techno-economic restoration plans.

Even before the war in Ukraine, the relationship between the EU and the Russian Federation in the context of the EGD was already under consideration. The EU's relations with Russia also played an important role, especially regarding trade in energy resources, which are dominated by fossil fuels. The EGD poses a serious challenge to the traditional EU–Russia energy trade pattern but also offers new opportunities for cooperation and a more sustainable

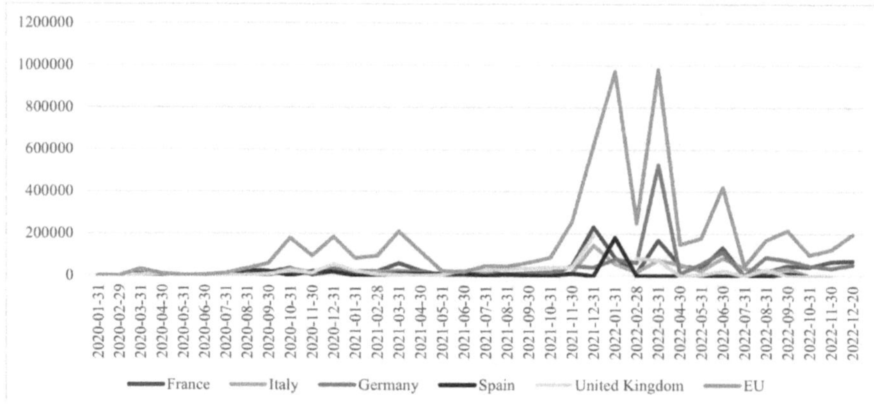

Figure 2.1 Daily new confirmed COVID-19 cases.

Source: Authors, with data obtained from Johns Hopkins University CSSE COVID-19 Data 2022.

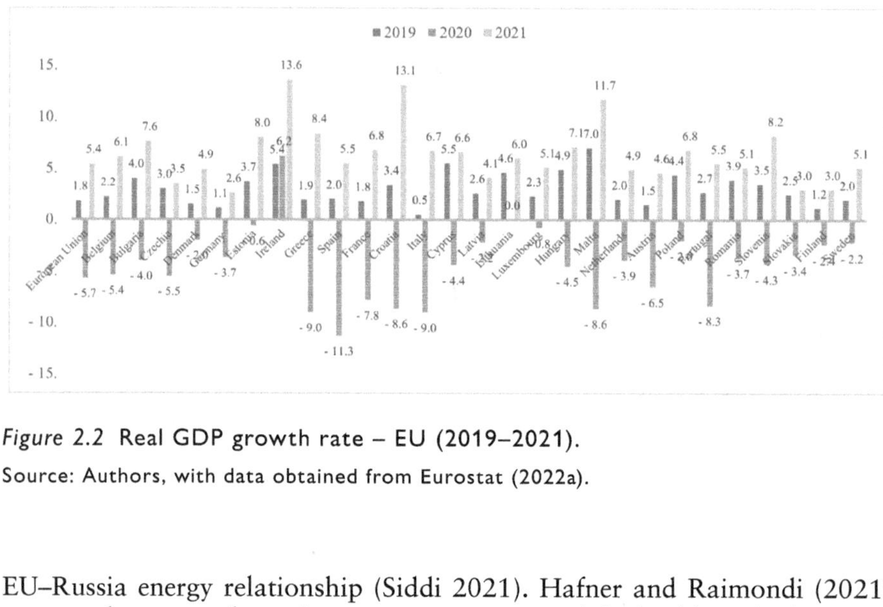

Figure 2.2 Real GDP growth rate – EU (2019–2021).

Source: Authors, with data obtained from Eurostat (2022a).

EU–Russia energy relationship (Siddi 2021). Hafner and Raimondi (2021) assume that natural gas is seen as a transitional fuel, although there is a discourse about the long-term future of natural gas and the need to decarbonize it. In response to the war in Ukraine, a number of countries have announced economic sanctions against Russia. The introduction of these sanctions would have a negative impact on most OECD countries (Chepeliev et al. 2022).

The energy crisis that started with rising prices during winter 2021 was escalated by the war in Ukraine. As the EU is the largest consumer of energy from Russia, real household incomes could fall by 0.7% to 1.7%, and energy prices could rise by up to 11% (Chepeliev et al. 2022). The absence of a common external energy security policy within the EU is also often discussed. This was most evident in the winter between 2021 and 2022, when there was a shortage of natural gas. This absence also made it difficult for the EU as a whole to take a clear position on the war in Ukraine. In addition to increasing the pace of decarbonization, there is also a need to develop a common external energy security policy for the transition period until the EU has sufficient domestic low-carbon energy capacity (Mišík 2022). The evolution of energy prices is presented in Figure 2.3. The war in Ukraine supported this trend, and prices started to rise even more steeply.

Based on the figures and subconclusions presented, it can be summarized that the COVID-19 pandemic has significantly affected macroeconomic indicators and energy price developments. The war in Ukraine and related developments have further exacerbated this trend. As the conflict is still ongoing, the exact impacts are not known yet, but the trend of the indicators under review can be inferred based on developments in 2022.

Contemporary EGD Situation: Two Views on Its Future

The EU has been advancing its commitments and policies to combat global climate change for the past decades (Dupont et al. 2020). Even before the current disruptions posed by COVID-19, the war in Ukraine, and the energy crisis, there were some clear challenges for the EGD and its implementation.

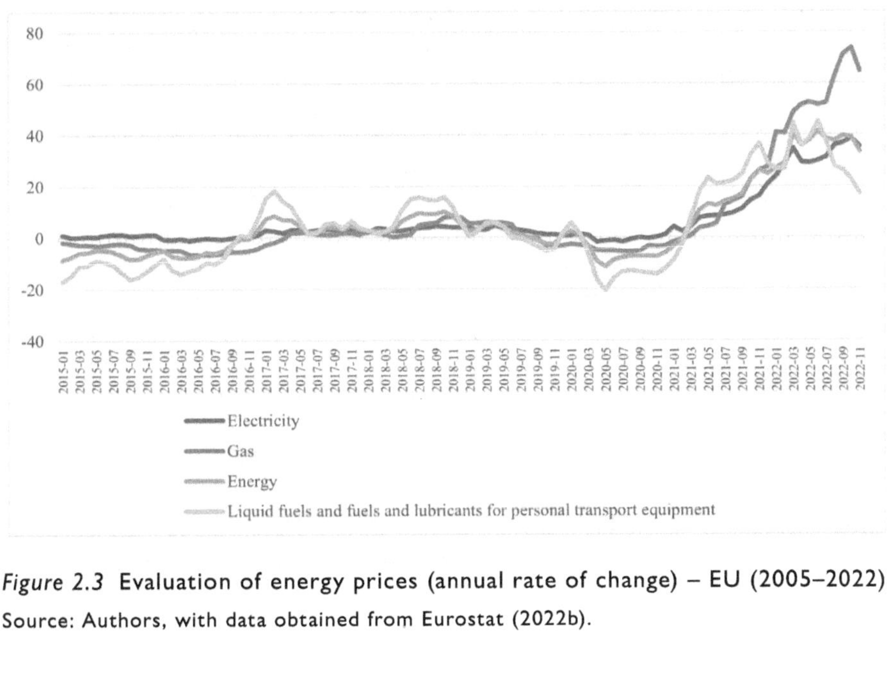

Figure 2.3 Evaluation of energy prices (annual rate of change) – EU (2005–2022).
Source: Authors, with data obtained from Eurostat (2022b).

Challenges such as the need for more coherent green diplomacy, better coordination of the EU and its member states, as well as the innovation and development of low-carbon industrial technologies (Engström 2022). Given the crises the EU has had to face recently, it is quite understandable that the status and realization of the EGD were questioned by many states, politicians, as well as enterprises. The voices calling for its scrapping, abandoning, postponing, or at least delaying of some of the climate measures and initiatives, or the discontinuing of EU Emissions Trading System (EU ETS), etc., were raised (appearing) from the time of the EGD's inception (Elkerbout et al. 2020; Pîrvu 2020). Moreover, due to the hardship caused by energy security and the large increase in energy prices, some countries have returned to fossil fuels, such as coal, tending to increase its share in the energy mix, which is a completely opposite action than to decarbonizing the economies (Borowski 2022). There are also some other questions discussed in relation to the transition toward a greener future and renewables. The main arguments are related to a new energy security risk caused by the minerals and metals necessary for the manufacturing of renewable energy power plants, batteries, or electric vehicles, as there is lack of their substitutes (Mark et al. 2021), or the risk derived from the Jevons paradox (Trincado et al. 2021). On the other hand, there are still many supporters of the EGD, who highlight its potential role in the overall economic recovery, economic as well as ecological modernization, and green growth (Elkerbout et al. 2020; Haas et al. 2022). Supporters of the EGD apply for behavioral changes (Cassetti et al. 2022), believe in strengthening the EU's global leadership role (Oberthür & Dupont 2021), and evaluate the additional pillar to the EU economic model. The EGD could be understood as an enhancement of sustainability and integration and strengthens political, institutional, and social dynamics (Bongardt & Torres 2022), as well as the significant acceleration of emissions reduction while strengthening European cohesion (Wolf et al. 2021). Another positive outcome of the EGD could bring a lowering of Europe's external dependency (Milani & Chaves 2022) or avoiding the costs and risks of a delay in reducing emissions (Dolge & Blumberga 2021). Table 2.1 shows a brief summary of the main views presented by selected research experts in the case of the future development of the EGD. As their opinions are not unanimous, the interpretation shows two marginal positions, optimistic and pessimistic.

Although the challenges bring new threats and opportunities, many scholars claim that the result will depend on the overall implementation of the EGD and performed actions in these turbulent times. The recent uncertainty and high energy prices were determined by the concurrence of events and crises (Benato et al. 2022). To lower the impact on society, there is a need for a just transition (Somosi & Megyeri 2022) and combating energy poverty (Siksnelyte-Butkiene 2022) that needs to be reflected.

Table 2.1 Fulfillment of the EGD goals under the burden of three challenges (COVID-19, War in Ukraine, energy crisis)

	Arguments of the optimists	*Arguments of the pessimists*
COVID-19	Positive impact on EU climate policy; the previous political trends were reinforced and developed (Dupont et al. 2020), strengthening European cohesion (Wolf et al. 2021), integration (Crnčec 2020), and coordination (Engström 2022), accelerating eco-innovation, sustainable agri-food production, creation of new disruptive technologies (Rowan & Galanakis 2020).	Need for engaging countries and regions of the world (Smol 2022), including the Global South, in a joint effort (Engström 2022); energy poverty (Siksnelyte-Butkiene 2022).
War in Ukraine	Impetus to accelerate green transition and to develop wider Eurasian cooperation (Astrov et al. 2022); new opportunities for cooperation and a more sustainable EU–Russia energy relationship (Siddi 2021); new competitors and ways of energy supplying will appear (Somosi & Megyeri 2022); innovation and development of low-carbon technologies (Engström 2022), lowering Europe's external dependency (Milani & Chaves 2022).	Absence of a common external energy security policy within the EU (most evident in the winter between 2021 and 2022, when there was a shortage of natural gas), which made it difficult for the EU as a whole to take a clear position on the war in Ukraine (Mišík 2022); energy poverty (Siksnelyte-Butkiene 2022).
Energy crisis	More integration within the EU, strengthening the EGD priority; policy changes leading to a more decisive and ambitious EU energy policy and climate action (Crnčec 2020); a just transition (Somosi & Megyeri 2022), bolstering the EU's systemic resilience to energy market volatility (Popkostova 2022); developing renewable energies capacity and investments in RES (Dragomir et al. 2022).	A serious challenge to the traditional EU–Russia energy trade pattern (Siddi 2021); return to fossil fuels (Borowski 2022); calls for delays; abandoning of EGD and its measures (Elkerbout et al. 2020; Pîrvu 2020); energy poverty (Siksnelyte-Butkiene 2022); new security risk (Mark et al. 2021); hardship to find the balance between security, affordability, and sustainability of energy (Popkostova 2022).

Considering the Situation From an SD Perspective: New Challenges vs. Old Solutions

Sustainable development principles have been a part of European policy since the Lisbon Treaty was signed. The EU made a commitment to the Sustainable Development Goals Agenda 2030 in 2015 (United Nations 2015).

Globally, the impact of recent challenges can be assessed using the world average of the Sustainable Development Index (SDG Index), as reported in the latest Sustainable Development Report of 2022 (Sachs et al. 2022). Based on the data provided by Sachs et al. (2022), it can be said that there is a significant upward trend between 2010 and 2012 in the world average value of the SDG Index.

In the EU itself, there has been no decline in the SDG Index based on data from Lafortune et al. (2022) at the EU-27 level. The trend is the same as at the world level. As the 2022 report does not yet include the impact of the energy crisis also associated with the Russian aggression in Ukraine, a return to the original trajectory cannot be expected in the following years.

Europe has traditionally been used to solving problems with money distributed through special funds. The EU is taking the same approach to tackling new crises and challenges. There were created several special funds focused directly on COVID-19, the war in Ukraine, and the energy crisis. The main funds are (1) a recovery plan called the Recovery and Resilience Facility (RRF), which aims not only to repair post-COVID social and economic damage but also to support efforts to make the economy more sustainable and to prepare society for the green and digital transformation (European Commission 2021); (2) the Fit for 55 package (European Council 2021), with which the EU is building on its legislative commitment to achieve a 55% reduction in emissions by 2030, as well as another similar instrument, REPower EU (European Commission 2022a), by which the EU aims to achieve independence from Russian gas supplies by 2030 while pursuing the goal of fossil fuel substitution to accelerate the energy transition; and (3) efforts to take joint operational action, such as emergency measures to reduce energy prices (European Council 2022), support for the development and approval of vaccines in response to COVID-19 (European Council 2020), and the implementation of conditions for their purchase by EU member states (European Commission 2022b).

The question is whether the standard measures that the EU has developed and is able to adapt to new challenges in an operational manner are sufficient at a time when crises are beginning to accumulate and build on each other. Let us look at the threats to the environment by global climate change, the financial crisis of 2007–2008, the energy crisis, the food crisis, and the conflict triggered by Russian aggression, which is only deepening the impact of previous crises.

Discussion and Conclusions

A discussion about the EGD goals is a typical example of the mainstream economy's shortcomings – a conflict between the support for nature and of the economy. On one side of the Pareto optimal curve are COVID-19, war in Ukraine, and the energy crisis; on the other side are environmental and global climate change EU goals.

For a relatively long period, the EU has seemed to be an exception in the world – an island of economic prosperity, social welfare, and environmental protection. The hidden question is, At what price? The EU must pay for its privileged position – to be rich and prosperous and to take care of the environment – with a high carbon footprint and consumption at the expense of the rest of the world, as Figure 2.4 presents.

The EU is in a difficult situation; thus, it is essential to tackle all three main crises while maintaining the set path toward building a sustainable Europe, which must necessarily promote all three dimensions of sustainability. It must avoid increasing the national and regional disparities and strengthening of centripetal tendencies leading to the risk of European disintegration.

It is also worth considering whether or not to start moving toward institutional changes that could lead to the elimination of the economy based on GDP parameters and how to finance different solutions. The scenarios as regards to the EU and the EGD could be the following: (1) implementation of the EGD, (2) postponing the EGD for possible better future conditions, and (3) resignation and abandonment of the idea of the EGD.

The answer to which scenario will be the most probable depends on answering the following two questions: (1) What type of economy will prevail

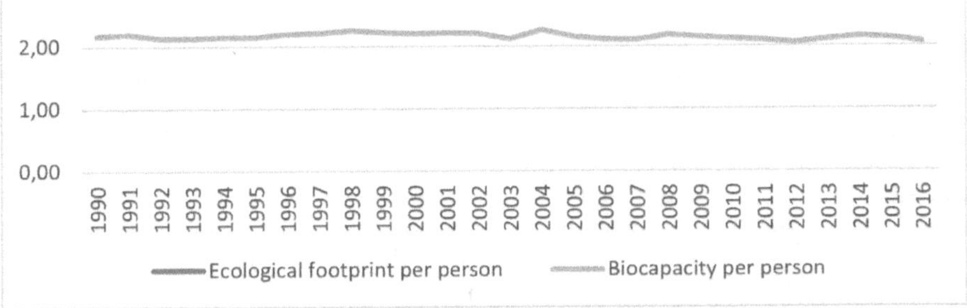

Figure 2.4 EU-27 and UK – ecological footprint, biocapacity, and ecological deficit per person in Europe, 1990–2016.

Source: Authors, with data obtained from the European Environment Agency (n.d.).

in the future – green, circular, bioeconomy, economy as usual? (2) What type of development will be supported by sustainable development – green growth, degrowth, economic growth measured by GDP? Some promising ways of changes can be found in the activity of the Beyond GDP initiative working within the framework of the EU and the OECD (Stiglitz et al. 2018) or a degrowth economy direction (European Environment Agency 2021).

Acknowledgments

This chapter was supported by the project CEE2ACT – Empowering the Central and Eastern European Countries to Develop Circular Bioeconomy Strategies and Action Plans (HORIZON-CL6-2021-GOVERNANCE-01-1 0, HORIZON project ID: 101060280). Authors would like to thank Justin Calvin Schaefer for proofreading.

Reference List

Astrov, V., Ghodsi, M., Grieveson, R., Holzner, M., Landesmann, M., Kochnev, A., Pindyuk, O., Stehrer, R., & Tverdostup, M. (2022) *Russia's Invasion of Ukraine: Assessment of the Humanitarian, Economic and Financial Impact in the Short and Medium Term (publication): Vol. Policy Notes and Reports No. 59*. The Vienna Institute for International Economic Studies. Available at: https://wiiw.ac.at/p-6132.html.

Benato, A., D'Alpaos, C., & Macor, A. (2022) "Possible ways of extending the biogas plants lifespan after the feed-in tariff expiration," *Energies*, 15(21), p. 8113. Available at: https://doi.org/10.3390/en15218113.

Benhabib, S. (2021) "Democracy, science and the state: Reflections on the disaster(s) of our times," *Philosophy and Social Criticism*, 47(4), pp. 477–485. Available at: https://doi.org/10.1177/01914537211006771.

Bongardt, A. & Torres, F. (2022) "The European Green Deal: More than an exit strategy to the pandemic crisis, a building block of a sustainable European economic model," *Journal of Common Market Studies*, 60(1), pp. 170–185. Available at: https://doi.org/10.1111/jcms.13264.

Borowski, P.F. (2022) "Mitigating climate change and the development of green energy versus a return to fossil fuels due to the energy crisis in 2022," *Energies*, 15, p. 9289. Available at: https://doi.org/10.3390/en15249289.

Cassetti, G., Boitier, B., Elia, A., Le Mouël, P., Gargiulo, M., Zagamé, P., Nikas, A., Koasidis, K., Doukas, H., & Chiodi, A. (2022) "The interplay among COVID-19 economic recovery, behavioural changes, and the European Green Deal: An energy-economic modelling perspective," *Energy*, 263, pp. 125798. Available at: https://doi.org/10.1016/j.energy.2022.125798.

Chepeliev, M., Hertel, T., & van der Mensbrugghe, D. (2022) "Cutting Russia's fossil fuel exports: Short-term economic pain for long-term environmental gain," *World Economy*, 45(11). Available at: https://doi.org/10.1111/twec.13301.

Chiaramonti, D. & Maniatis, K. (2020) "Security of supply, strategic storage and Covid19: Which lessons learnt for renewable and recycled carbon fuels, and their future role in decarbonizing transport?" *Applied Energy*, 271, p. 115216. Available at: https://doi.org/10.1016/j.apenergy.2020.115216.

Crnčec, D. (2020) "COVID-19 crisis: More EU integration and a step forward for EU energy policy and climate action?" *Teorija in Praksa*, 57(4), pp. 1105–1123.

Dolge, K. & Blumberga, D. (2021) "Economic growth in contrast to GHG emission reduction measures in Green Deal context," *Ecological Indicators*, 130, pp. 108153. Available at: https://doi.org/10.1016/j.ecolind.2021.108153.

Dragomir, V.D., Gorgan, C., Calu, D.-A., & Dumitru, M. (2022) "The relevance and comparability of corporate financial reporting regarding renewable energy production in Europe," *Renewable Energy Focus*, 41, pp. 206–215. Available at: https://doi.org/10.1016/j.ref.2022.03.002.

Dupont, C., Oberthür, S., & von Homeyer, I. (2020) "The COVID-19 crisis: A critical juncture for EU climate policy development?" *Journal of European Integration*, 42(8), pp. 1095–1110. Available at: https://doi.org/10.1080/07036337.2020.1853117.

Elkerbout, M., Egenhofer, C., Núñez Ferrer, J., Cătuți, M., Kustova, I., & Rizos, V. (2020) *The European Green Deal after Corona: Implications for EU Climate Policy*, Brussels: CEPS Policy Insights, no. 2020–06.

Engström, M. (2022) *A Green Deal, Open to the World. Swedish Institute of European Policy Studies*, Stockholm: Swedish Institute for European Policy Studies. Available at: www.sieps.se/en/publications/2022/a-green-deal-open-to-the-world/.

European Commission (2019) *Energy and the Green Deal.* Available at: https://ec.europa.eu/info/strategy/priorities-2019-2024/european-green-deal/energy-and-green-deal_en (Accessed 2 January 2023).

European Commission (2021) *Recovery and Resilience Facility.* Available at: https://ec.europa.eu/info/business-economy-euro/recovery-coronavirus/recovery-and-resilience-facility_en (Accessed 2 January 2023).

European Commission (2022a) "REPowerEU: A Plan to Rapidly Reduce Dependence on Russian Fossil Fuels and Fast Forward the Green Transition*," *European Commission.* Available at: https://ec.europa.eu/commission/presscorner/detail/en/ip_22_3131.

European Commission (2022b) *EU Vaccines Strategy.* Available at: https://commission.europa.eu/strategy-and-policy/coronavirus-response/public-health/eu-vaccines-strategy_en (Accessed 2 January 2023).

European Council (2020) *The EU's Response to the COVID-19 Pandemic.* Available at: www.consilium.europa.eu/en/policies/coronavirus/ (Accessed 2 January 2023).

European Council (2021) *Fit for 55.* Available at: www.consilium.europa.eu/en/policies/green-deal/fit-for-55-the-eu-plan-for-a-green-transition/ (Accessed 2 January 2023).

European Council (2022) *Council Formally Adopts Emergency Measures to Reduce Energy Prices.* Available at: www.consilium.europa.eu/en/press/press-releases/2022/10/06/council-formally-adopts-emergency-measures-to-reduce-energy-prices/ (Accessed 2 January 2023).

European Environment Agency (2021) *Growth Without Economic Growth.* Publications Office of the European Union. Available at: https://data.europa.eu/doi/10.2800/492717.

European Environment Agency (n.d.) *Ecological Footprint of European Countries.* Available at: www.eea.europa.eu/data-and-maps/indicators/ecological-footprint-of-european-countries-2/assessment (Accessed 2 January 2023).

Eurostat (2022a) *Real GDP Growth Rate – Volume.* Available at: https://ec.europa.eu/eurostat/databrowser/view/tec00115/default/table?lang=en (Accessed 22 December 2022).

Eurostat (2022b) *HICP – Monthly Data (Annual Rate of Change).* Available at: https://ec.europa.eu/eurostat/databrowser/view/PRC_HICP_MANR__custom_2160811/bookmark/table?lang=en&bookmarkId=2597def3-51fc-4f00-a59e-77753ab94b90 (Accessed 22 December 2022).

Haas, T., Syrovatka, F., & Jürgens, I. (2022) "The European Green Deal and the limits of ecological modernization," *CPE*, 7(2), pp. 247-261. Available at: https://doi.org/10.5771/2566-7742-2022-2-247.

Hafner, M. & Raimondi, P.P. (2021) "Priorities and challenges of the EU energy transition: From the European Green Package to the new Green Deal," *Russian Journal of Economics*, 6(4), pp. 374–389. Available at: https://doi.org/10.32609/J.RUJE.6.55375.

Hove, J.V. (2020) *Impact of State Aid on Competition and Competitiveness During the COVID-19 Pandemic: An Early Assessment*, Study for the Committee on Economic and Monetary Affairs, Policy Department for Economic, Scientific and Quality of Life Policies, Luxembourg: European Parliament. Available at: www.europarl.europa.eu/thinktank/en/document/IPOL_STU(2020)658214.

Johns Hopkins University CSSE COVID-19 Data (2022) "Coronavirus (COVID-19) Cases," *Our World in Data*. Available at: https://ourworldindata.org/covid-cases (Accessed 22 December 2022).

Kovanda, L. (2009) *Interviews with Nobelists*, Week No. 39/2009. Available at: www.lukaskovanda.cz/rozhovory-s-nobelisty/edmund-phelps/ (Accessed 6 January 2023).

Lafortune, G., Fuller, G., Bermont-Diaz, L., Kloke-Lesch, A., Koundouri, P., & Riccaboni, A. (2022) *Achieving the SDGs: Europe's Compass in a Multipolar World. Europe Sustainable Development Report 2022*, Paris: SDSN and SDSN Europe. Available at: https://eu-dashboards.sdgindex.org/.

Mark, L., Pisani-Ferry, H., Shapiro, J., Tagliapietra, S., & Wolff, F.G.B. (2021) *The Geopolitics of the European Green Deal, Bruegel Policy Contribution*, No. 04/2021, Brussels: Bruegel.

Milani, C.R.S. & Chaves, L.N. (2022) "How and why European and Chinese pro-climate leadership may be challenged by their strategic economic interests in Brazil," *Asia Europe Journal*, 20, pp. 403–422. Available at: https://doi.org/10.1007/s10308-021-00645-z.

Mišík, M. (2022) "The EU needs to improve its external energy security," *Energy Policy*, 165, p. 112930. Available at: https://doi.org/10.1016/j.enpol.2022.112930.

Oberthür, S. & Dupont, C. (2021) "The European Union's international climate leadership: towards a grand climate strategy?" *Journal of European Public Policy*, 28(7), pp. 1095–1114. Available at: https://doi.org/10.1080/13501763.2021.1918218.

Pîrvu, E.L. (2020) "The European Green Deal – a feasible ambitious initiative?" *Journal of Danubian Studies and Research*, 10(1), pp. 276–291. Available at: https://dj.univ-danubius.ro/index.php/JDSR/article/view/483/649.

Popkostova, Y. (2022) "EUROPE'S energy crisis conundrum: Origins, impacts and way forward," *European Union Institute for Security Studies*, 2. Available at: www.iss.europa.eu/sites/default/files/EUISSFiles/Brief_2_Energy%20Crisis.pdf.

Rowan, N.J. & Galanakis, C.M. (2020) "Unlocking challenges and opportunities presented by COVID-19 pandemic for cross-cutting disruption in agri-food and green deal innovations: Quo Vadis?" *Science of the Total Environment*, 748. Available at: https://doi.org/10.1016/j.scitotenv.2020.141362.

Sachs, J., Kroll, C., Lafortune, G., Fuller, G., & Woelm, F. (2022) *Sustainable Development Report 2022* (1st ed.), Cambridge: Cambridge University Press. Available at: https://doi.org/10.1017/9781009210058.

Siddi, M. (2021) "A green revolution? A tentative assessment of the European Green Deal," *International Organizations Research Journal*, 16(3), pp. 1–33. Available at: https://doi.org/10.17323/1996-7845-2021-03-04.

Siksnelyte-Butkiene, I. (2022) "Combating energy poverty in the face of the COVID-19 pandemic and the global economic uncertainty," *Energies*, 15(10), p. 3649. Available at: https://doi.org/10.3390/en15103649.

Smol, M. (2022) "Is the green deal a global strategy? Revision of the green deal definitions, strategies and importance in post-COVID recovery plans in various

regions of the world," *Energy Policy*, 169, p. 113152. Available at: https://doi.org/10.1016/j.enpol.2022.113152.
Somosi, S. & Megyeri, E. (2022) "A moving target: changing priorities in the energy policy of the European Union," *International Journal of Energy Economics and Policy*, 12(4), pp. 542–552. Available at: https://doi.org/10.32479/ijeep.13052.
Stiglitz, J.E., Fitoussi, J.-P., & Durand, M. (2018) "Beyond GDP: Measuring What Counts for Economic and Social Performance," *OECD*. Available at: https://doi.org/10.1787/9789264307292-en.
Trincado, E., Sánchez-Bayón, A., & Vindel, J.M. (2021) "The European Union green deal: clean energy wellbeing opportunities and the risk of the jevons paradox," *Energies*, 14(14), p. 4148. Available at: https://doi.org/10.3390/en14144148.
United Nations. (2015) *Transforming our World: The 2030 Agenda for Sustainable Development*, New York: United Nations. Available at: https://sustainabledevelopment.un.org/post2015/transformingourworld/publication.
Wolf, S., Teitge, J., Mielke, J., Schütze, F., & Jaeger, C. (2021) "The European Green Deal – more than climate neutrality," *Intereconomics*, 56, pp. 99–107. Available at: https://doi.org/10.1007/s10272-021-0963-z.

Chapter 3

Shattered Rights, Emerging Economies and Laws, and New Awareness

What Ethical Visions in a Stormy Reality? The Trajectory of Emergency Powers

Marco Ettore Grasso

Introduction

The critical times we are living in shed light on a necessary awareness that primarily concerns the desire to contribute to the creation of a more just world. This desire can come true if economic, political, and financial powers – and in particular, "emergency powers" – do not prevail over the lives of citizens and if the rights of the weakest are increasingly protected. In these recent years, several emergency situations, such as war, pandemics, or climatic disasters, have contributed to the birth of new economies, which represent specific utilitarian interests, as well as laws and rights, which are linked to the causes and effects of the emergency. What philosophical visions should be considered within an emergency vision? What scenarios, ethical perspectives, or visions should be considered within this framework, which is characterized by political and social confusion, loss of individual and collective identity, inequalities, and utilitarianism? In this context, a serious review is also needed in order to safeguard the freedom of the individual.

Emergency Law

When we talk about emergency law, we refer to the set of national and supranational legislations that govern an emergency situation. The state of exception we are experiencing seems to have already been announced by the literature of Aldous Huxley (1932) or by George Orwell (1949). According to Carl Schmitt (1985), the concept of "emergency" concerns a case of "extreme peril, a danger to the existence of the state, or the like," whereas the International Law Association (1985) defines it as an

> exceptional situation of crisis or public danger, actual or imminent, which affects the whole population or the one of the area to which the declaration applies and constitutes a threat to the organized life of the community of which the state is composed.

DOI: 10.4324/9781003440871-4

The concept of emergency is often characterized by vagueness and can be interpreted in many ways, with some scholars against a universal definition. According to Gross and Ní Aoláin (2006), for instance, any definition of *state of emergency* would be amorphous and therefore unable to build useful criteria to distinguish between times of emergency and times of peace.

This chapter argues that *state of emergency*, as a legal and social concept, is useful to reach a specific answer when facing exceptional challenges, applied to actual threats and imminent dangers. It is inversely proportional to *sovereignty*, affirming itself when national sovereignty is endangered. This particular condition can be dangerous, however, particularly related to human rights and freedoms, as it could be an excuse for abuse.

In Thomas Hobbes's opinion (Hobbes 1651), all law in its entirety has an exceptional character. Emergency law, however, is largely considered an extraordinary law which gives extraordinary power, effective only as long as the emergency stands. For this reason, some countries which are more acquainted with political abuses by those in power have armed themselves with more detailed Constitutions in order to protect their citizens. For instance, the Italian Constitution mentioned *emergency discipline* in Article 78, concerning war emergency. It specifies that during a war, the government can adopt decrees which can exceptionally derogate freedoms and rights.

Emergency presupposes a derogation. Hans Kelsen (1979) explored the notion of derogation as a deviation from ordinary behavior, justified under some particular circumstances. The Latin verb *derogare* (to derogate) is opposed to *abrogare* (to abuse), and it is different from *obrogare* (to delete). Cicero (1943), in this regard, writes: "Huic legi nec obrogari fas est, neque derogari aliquid ex hac licet, neque tota abrogari potest." Therefore, it is not possible to substitute this law with a new one; neither is it possible to partially derogate from it, nor to abrogate it in its entireness. Carl Schmitt (1931) distinguished, for example, between *Verfassung* (the Constitution) and *Verfassungsgesetze* (constitutional norms). The latter can be suspended in order to stabilize the first one.

As Ferejohn and Pasquino (2004) state, the norm, the derogation, and the justification of the derogation are all connected with the emergency, which can justify the derogation or the suspension and, moreover, the higher principle that one may want to shield, protect, or safeguard under extraordinary circumstances. From an ontological point of view, it might be possible to think that norm and exception are two different realities of the world, complementary to each other.

The concept of state of emergency is presupposed on the dichotomy between norm and exception. In this regard, two different philosophies are to be found in this field: (1) the rule of law school, which favors the establishment of clear rules for state of emergencies, and (2) the sovereignty school, which presupposes the idea that a state of emergency is an extrajuridical condition. The concept of "state of emergency" is indirectly introduced in Articles 42 and 48 of the French Constitution, while in the German Constitution, *emergency* is used to mention an imminent danger to the free democratic order or to the existence of the Federation, or the *Länder*. The Hungarian Constitution also includes state of danger, state of national crisis, preventive defense emergency, as well as state of war.

Two distinct normative systems exist in the emergency field: the ordinary system of rights and the normative system operating in the state of emergency. Coherent, emergency legislation should never question democratic stability but should only limit itself to answering social needs. If this does happen, any judge should intervene to invalidate the act issued for the emergency. Infringements on rights is a great limit of an emergency state. Some cases of these infringements concern the rights to freedom of expression, privacy, assembly, and liberty.[1] In accordance with the Italian Constitution, for instance, the right to health – referred to as a collective right – infers the possibility to limit other constitutionally guaranteed freedoms, such as the freedom of movement (Art. 16), the freedom of assembly (Art. 17), as well as the freedom to exercise one's faith (Art. 19), or the right to education (Art. 34), and the right to entrepreneurship (Art. 41).

Therefore, "exceptional law" or "emergency law" can be defined as that group of rules that have to be adopted in order to face an extraordinary situation in the ordinary evolution of the legal order. Indeed, the historical origin of emergency law can be traced to ancient Roman law.[2] If an emergency condition arose, the Senate could give two consuls the power to appoint a dictator for up to six months. The dictator was authorized to suspend any right, and he could issue temporary decrees. This administrative power has been praised by thinkers such as Niccolò Machiavelli, Jean-Jacques Rousseau, and James Harrington.[3]

Who Benefits From the Emergency?

It is generally understood that during emergency periods, the rule of law can be preserved if politicians use good sense. Emergency powers are often

1 Typically in countries like Kenya, Sri Lanka, Ethiopia, Nepal, and Central African Republic, where freedom is most compromised.

2 The neo-Roman tradition annoverates emergency powers within the Constitution (e.g., Art. 48 of the Weimar Constitution and Art. 16 of the French Constitution).

3 Concerning emergency law literature, see, for example, Sascha Mueller, "Turning Emergency Powers Inside Out: Are Extraordinary Powers Creeping Into Ordinary Legislation?" in *Flinders Law Journal*, 2016, 18; Emmanuel Gross, "How to Justify an Emergency Regime and Preserve Civil Liberties in Times of Terrorism," in *South Carolina Journal of International Law and Business*, vol. 5 (1), 2008; Anna Jonsson Cornell, Janne Salminen, "Emergency Laws in Comparative Constitutional Law: The Case of Sweden and Finland," in *German Law Journal*, vol. 19 (2), 2018; and Ugo Mattei, Liu Guanghua, Emanuele Ariano, "The Chinese Advantage in Emergency Law," in *Global Jurist*, 21 (1), 2021.

passively welcomed by citizens, which unknowingly contribute to forge a new normality strongly desired by powerful interests. This begs the question: Who benefits from an emergency? In a public health crisis, the largest pharmaceutical companies seem to benefit the most, while in times of armed conflict, it is the war industry, or the industry that deals with the production of weapons, equipment, and military technologies. Today, the most powerful lobbies are directly or indirectly connected to these industries, with utilitarian interests always looming. It is as if these very interests are leading us toward emergency conditions. It is a troubling presupposition to consider that powerful interests may create crises in order to benefit from the same crises.

And it is no surprise that this blend of utilitarianism, capitalism, and state of emergency often leads to mismanagement, against the common good and well-being of the citizens. For example, during the height of the COVID-19 pandemic, health was largely measured only by whether or not you contracted the virus. To better illustrate, Aristotle distinguished life as *bios* from life as *zoé*. *Zoé* is merely life, whereas *bios* is the quality of life itself (e.g., Agamben 2002). Accepting the lockdown for a few months meant giving up the *bios* in the name of *zoé*.

Emergency and Security

In times of emergency, every citizen has an equal interest in giving priority to security over certain fundamental rights, with the temporary transfer of certain human rights to the state for the preservation or protection of security. Unfortunately, however, the derogation of rights does not fall equally on the citizenry. According to the "privileged indifference" thesis, emergency measures could bring inegalitarian burdens: elites already know that it is not they who will be bearing the brunt of harms during times of crisis. In this perspective, emergency security measures therefore then become instruments of discrimination instead of instruments of democracy.[4] This is why emergency security measures, in language and effect, should always be balanced with antidiscrimination laws. The relationship between security and equality concerns not just the rule of law, aimed at protecting freedom, but the role of law to maintain a liberal-democratic society of citizens recognized as equals (Milstein 2020).

Democratic equality is often identified with principles of non-domination or non-subjection. It is the main tool of inclusion, by which citizens come

4 According to Milstein (2020: 853), "if the purpose of emergency measures is only to curtail liberal-democratic guarantees in order to preserve them thereafter, then preservation and reassurance are goods whose value must extend to the life of the community beyond the emergency itself. This recommends a wariness of measures that purport to save a community while shearing the bonds that hold it together."

together to constitute a *demos*, namely, a democratic "we" (Milstein 2020: 839). In Habermas's opinion, liberal and political rights are "co-original": Human rights and democracy are not only both primary but also even equiprimordial and related. The co-originality of these rights would guarantee the *demos* as a community of equals (Habermas 1996).

The Emergency Creates New Rights and New Economies

The emergency creates new rights that emerge from the exceptional conditions. These rights have to do with the rights of those who are oppressed by emergency power and that see their own fundamental rights violated, in particular their right to dignity. When speaking of new rights, this also includes the rights inherent to emergency power. For example, in some countries, the emergency requires the use of a digital identity, which could be defined as the subject of a new right. In fact, this particular identity was designed in order to protect the identity of the person so that his/her identity cannot be illegally exploited by others. However, a dimension in which a citizen cannot access essential public services if he does not have a digital identity (which in Italy is called *Spid*) would fall into innumerable acts of discrimination and violate his own life choice, which moreover would revive Foucault's nightmare concerning the concepts of biopolitics and social control.[5] In this dimension of new rights, it is also important to mention the rights of those who have reacted – perhaps with little success – through the right to civil disobedience.

The new rights are flanked by new economies which emerge from this emergency ethical perspective. In this sense, it is sufficient to remember all the economic and financial interests that are at stake in an emergency condition. The powerful lobbies, in fact, invest in vaccines, masks, disinfectants, and green technologies (which are considered essential for "smart working," or working from home using technological tools). To better understand these new rights and economies, however, it is necessary to trace a path that examines emergency law and, above all, its limits.

The Climate Emergency

Remaining in the field of the exploitation of power, even with regards to the climatic emergency,[6] the logic of utilitarianism is absolutely evident. For instance, Vandana Shiva (2022), physicist and social and ecological justice

5 According to Michel Foucault, medicine is a biopolitical strategy, that is to say, an instrument of social control that can invade the field of politics.

6 Darrel Moellendorf (2022) provides a climate justice approach with a focus on global poverty, highlighting the need to develop climate "hopeful politics."

scholar, argues that the solutions to climate change as offered by Bill Gates, founder of Microsoft and one of the world's wealthiest people, would only intensify the problem. For example, large amounts of greenhouse gas (GHG) emissions (responsible for warming the climate) are caused by fertilizers used in the chemical industry; nonetheless, Gates has pointed to these very fertilizers as part of the solution. He has also proposed industrial farming, an industry that creates food insecurity but something that he claims will counter climate change. It is important to note that Gates owns hundreds of thousands of acres of agricultural land in the United States, a common investment for the ultrawealthy (The Land Report 2022; Shiva 2022).

Yet despite these misalignments of private profit and public solution, at the 2015 Paris Conference of the Parties (COP) for the UN Framework Convention on Climate Change (UNFCCC), and at all COPs thereafter, it was made very clear how Gates and other powerful actors and industries hold more negotiating power than heads of states. They are using their power, often gained from the actions and industries that caused the crisis, to advance their own monied interests as a solution to the crisis. This is also directly counter to the responsibility of the state, whose role it is to advance the interests of the common good in the face of a crisis. This makes private interests in public response even more concerning, and particularly during times of great harm.

Shiva (2022) describes three kinds of pandemics: The first is the virus, the second is hunger, and the third is means of subsistence. These three pandemics generate a bigger pandemic: inequity. The globalized economy is directed from the engines of greed. Corporate interests grow, and their shares peak. In India, where Shiva was born and now lives, the so-called "green revolution" has destroyed lands, water, and lives. In 1984, Indian farmers led strikes in the streets as industrialized agriculture exploited lands and farmers. As agriculture was industrialized, many people were dying out of hunger, yet the less people were working lands, the more agriculture was considered to be productive. In this scenario, according to Shiva, no difference would be made if we had more or less food (Shiva 2022).

To powerful interests, we must destroy the climate to save the climate, as shown by people like Gates who now arguably control the COPs, or as shown by companies like Crop Life, a manufacturer of herbicides, that become official partners to them. But multinational corporations (MNCs) not only want to control the COPs but also want power over the future of seeds in the world. This means that crops will be substituted by genetically modified seeds, which end up in animal feed and therefore our plates, giving birth to a series of chronic illnesses. Synthetic food, sometimes referred to as "fake food" because of its nutritional bankruptcy, is responsible for as much as 75% of illnesses on this planet (Shiva 2022). To these powerholders, what matters is not a real alimentary system, or even a more secure state, but the creation and control over the exchange of goods and capital. And with avenues to all decision-making in government, they have effectively created a food totalitarianism, declaring to the public what is safe or not safe to eat, dependent upon their own financial interests.

War Emergency

New rights and new economies include the right to wage war in emergency scenarios, as well as war economies. In this regard, the war industry is the most profitable industry. A natural law approach could still be of help. Hugo Grotius (1625) affirmed that natural law must prevail during wartimes as well. However, natural law, which includes unwritten rights linked to individuals – as rights, interwoven with the human being – is the first to be suppressed during wartimes.

At a time when war undermined law's validity, Grotius (1625) theorized a solution to the problem of the validity of the principles governing relations between states in the event of war. He argued that natural law, as universal, can never fail in the event of war, and it is the only authority to which one can defer in relations between states. *Pacta servanda sunt* – it is mandatory to keep faith with the pacts because this obligation derives from the natural law upon which civil coexistence between peoples must be based. According to Grotius, war is not contrary to human nature but rather belongs to it; his work is therefore aimed not at war abolishment but to its regulation.

Immanuel Kant, for his part, is averse to just war principles and is a critic against his predecessors, who attempted to develop the doctrines of just war, while St. Thomas Aquinas states that a just war must satisfy some conditions: War must be prosecuted by sovereign political authority, and it must have a just cause. John Rawls, however, understands the doctrines of "just war" theory thanks to an extension of the original position, where parties represent "different nations who must choose together the fundamental principles to adjudicate conflicting claims among states" (Rawls 1999: 331). According to Rawls, it would be unrealistic to hope for an international society comprised entirely of liberal people (Moellendorf 2014).

In Roman law, the declaration of war took place according to the rite of the *Fetiales*. In accordance with this custom, people reached other people's lands' boundaries with a bloody shaft, and after having recited the formula with which they declared war, the shaft was then launched in enemy territory. Nowadays, declarations of war are more hidden; sending weapons, for example, could actually constitute a declaration of war.[7]

7 The law relating to war could be exemplified with The Hague conventions of 1899 and 1907 and the international law of armed conflicts with the Geneva Conventions of 1949. Nowadays, we speak of international humanitarian law, which is the set of rules of international law concerning the protection of the victims during armed conflicts. It includes the rules which protect people who do not take or no longer take part in hostilities and places limits on the use of weapons, means, and methods of warfare.

Humanitarian law can be divided into two big categories: *jus ad bellum* and *jus in bello*. The first refers to the law to declare war,[8] and the second to the law that controls the belligerent behaviors during an armed conflict.[9] *Jus ad bellum* is an exceptional law, as Article 51 of the United Nations Charter legitimates war only on condition of self-defense, which should be proportioned as much as possible to the offense. The very failure of the UN regarding the veto powers held by the five permanent countries of the UN Security Council helps to better comprehend the weakness of international law, founded on faculties and powers that do not mirror the current need of global justice.[10]

Article 11 of the Italian Constitution states that "Italy repels war . . . as an offensive mean." However, even though Italy is not in a legitimate defense condition, the country is advancing war by financing and sending weapons to Ukraine, forgetting to repel war, as stated in Article 11. It is not possible, in fact, to send weapons and call for peace at the same time: It is a contradiction we must not accept.

The Role of Science in the Emergency Vision

Emergency science is implicated in every field in which there might be an emergency. As such, we have a belligerent science, a sanitary-epidemical science, and an ecological/climatic-energetic science. Science demands a scientific approach, to keep in consideration the thesis and confute previous ones, offering new ones.

During the pandemic scenario, a positivistic science ruled that did not allow contradictions, even those rooted in science. Whether it be the imposition of a vaccine or the dispatch of weapons, such as what happened in countries such as Italy, this leads to critical thought because politicians often do not act according to peoples' needs, not realizing the principle of popular sovereignty.

Paul Karl Feyerabend (1975) described *science* as an anarchic system obsessed by its own myth, which believes itself to be true, no matter what. This philosopher was outraged about the compliant behavior kept by many scientists toward alternative traditions. According to him, given a norm – no matter how relevant or necessary to science – there always are circumstances where it is opportune to not only ignore the norm but also adopt its opposite.

Another philosopher important to mention is Imre Lakatos, who distances himself from Karl Popper's thoughts and intervenes in a debate provoked by

8 It concerns the conditions under which a war may be initiated.

9 It concerns the means that may be used in prosecuting a war. See, for example, Darrel Moellendorf (2015).

10 Law is made up of a good amount of utilitarianism, as it protects many interests.

Thomas Kuhn. He substituted the concept of paradigm with one of research programs, which resemble scientific theories, coherent with one another, that obey rules and methodologies shared by a community of scientists. These theories constitute a heuristic frame, which is considered not possible of counterfeiting. In Imre Lakatos's philosophy, science proceeds in two ways: One side keeps unaltered the heuristic nucleus of research program, substituting instead the other new theories, capable of resolving new problems (Lakatos 1978); the other side substitutes the entire heuristic nucleus, and as a matter of fact, it can change the entire research program if the original system is not able to resolve the problems and give thorough answers. Moreover, he states that the history of science is not a simple struggle between opposite theories, between the void theory and the science theory, one of which must be falsified, as Popper said (Popper 2013). *Science* is thus an eternal confrontation between different research programs, each of which is kept alive as long as it serves a heuristic function, as long as it is possible for new theories to be created.

Science is often invoked as the remedy for every evil; it is the new totem of postmodernism, where everything seems to be allowed in its name. But which science are we talking about? Real science has always confronted both thesis and antithesis. In modern days, it seems like there is always one thesis and every scientific contribution against that very thesis lives to be hidden when it could help resolve the scientific problem.

The concept of science could very well be manipulated. If a given situation is to be qualified as scientific, there is a risk of promoting a sort of social totalitarianism. According to Abel Rey, part of the fault is to be given to who accepts the system without objections and becomes tyrant of his/her own self. This consideration aligns with Hannah Arendt's "banality of evil," highlighting the responsibility of the ones who remain silent, contributing to the reinforcement of totalitarianism (Arendt 2006). In Rey's opinion, philosophy cannot add anything new to science; it can only describe its methods and outcomes (Rey 1903).

If science were to replace politics, it could rise to a biopolitical tool of control and surveillance which uses technology to control the lives of citizens (Foucault 2008). It is true that scientific paradigms do not reflect objective truth (Kuhn 1962), in the same way that scientific formulations are provisional and falsifiable (Popper 1934). In the name of science, we have witnessed the creation of new categories which tend to extend certain social gaps, such as "anti-vaxxer," for example. If science becomes a reason for social division, what credibility should it have?

In order for science to be ethically sound, it should not harm human dignity, nor should it be seen as God, but more a mere wise counselor. It should therefore foster criticism and freedom of thought. If this does not happen, it would mean that science would no longer be science and, therefore, the source of humble knowledge which does not demand but advises. Rather, it would become the means by which to pursue evident utilitarian interests, to which power aspires. Only if science is able to guide – and not to condition – human action can it then be judged sensible. Otherwise, it would become an anarchic system, as Feyerabend theorized, which would not guarantee the minimum requirement that Lakatos spoke of, represented by the legitimate antinomy between conflicting theses.

The Limit of the Emergency Law: The Right to Dignity

Emergency law should never harm human dignity, which therefore should be considered as a limit to emergency power. If it would seem legitimate for such power to compress the content of human rights by virtue of an emergency, it would not at all seem legitimate to allow this power to harm human dignity.

During the pandemic, unfortunately, we witnessed the crying of children who could not leave the house or see their grandparents for months. This would be enough to argue that the right to dignity has been sufficiently harmed. To better understand what is meant by *right to dignity*, it is important to understand the concept of dignity itself. On one hand, the term "dignity" ontologically indicates the special position of man in the cosmos; on the other hand, it indicates a value, namely, the position held by man in social life.

To the extent that man contributes to the material or spiritual progress of society, he increases the value of his own dignity. The value of every man is what others recognize socially. In Hobbes's *Leviathan*, it is clear that the value of a man is equivalent to his price that corresponds to what would be given for the use of his power. It is not an absolute thing but depends on the need and judgment of others.

In Samuel von Pufendorf's view, however, *dignity* pertains to the moral freedom of man. Man is the only one able to independently set limits to his actions, to submit to laws that he himself has given himself (Pufendorf 1672). In his opinion, the dignity of human nature and its primacy over other living beings require that human actions be undertaken in accordance with a certain norm and without order, civilization, and beauty would not be possible. The concept of dignity can also be about contemplation. Pascal (1941), in this regard, believes that all dignity and all merit lie in man's thought.

In Kant's view, *dignity* means that man is a being capable of acting morally, as he has an inestimable value. According to him, what is superior to any price does not admit anything equivalent, since it has a dignity, an intrinsic value which is different from the mere "price" (Kant 1785). Georg Forster (1794) also points to the need to make happiness fall from its pedestal, in order to put dignity in its place, which is a true signpost of life.[11]

11 "Echte wegwei ser des lebens," that is to say, "real guideposts of life."

Although the concept of dignity has, by now, penetrated into German culture and literature, it struggles to emerge in the legal sphere. With Georg Wilhelm Friedrich Hegel's philosophy, dignity can be interpreted as the duty to respect men, and in terms of a legal imperative (Hegel 1991). However, only in the modern era – after the Second World War – did the principle of dignity find full legal legitimacy. In line with Arendt (2006), for example, *dignity* is the right to have rights. It needs a new guarantee, which can only be found in a new "law on land," destined to apply to all humanity.

In the German Constitution, the recognition of human dignity becomes a kind of *grundnorm* that is placed at the top of the entire legal system. It is an objective legal norm which, unlike fundamental rights, is not subject to limitations. The first paragraph of Article 1, in fact, emphasizes that the dignity of man is intangible. Respecting it (*zu achten*) and protecting it (*zu schützen*) are the duty of all state power. Dignity, therefore, is intangible (*unantastbar*) and is linked to respect for fundamental rights.

If dignity is the limit not to be exceeded, can emergency law compress fundamental rights? Yes, but to the extent that this compression of rights does not harm human dignity. If power violates this moral freedom and intrinsic value, which is the object of a supreme right aimed at demanding respect for certain fundamental rights, then this power loses any semblance of legitimacy. Therefore, it becomes unlawful and should be collectively rejected so that the conscience of the people – represented by a popular feeling that does not renounce human dignity – resurrects and takes over. How could we measure the degree of dignity violation[12] in order to deem emergency law illegitimate or unjust?

The content of dignity could be glimpsed in the exercise of fundamental freedoms, according to the thesis of John Stuart Mill (1859), whereby the exercise of one's freedom should not cause harm to another's freedom. A second way of examining the issue could be through the principle of non-discrimination, where one cannot cause a disadvantage to a social group if this disadvantage is not objectively well justified. The emergency, before even concerning a crisis situation, then concerns the conflict between two antithetical ways of thinking. A conflict, however, that leads to the annihilation of one of these visions, or the one that contradicts positivistic science.

Ultimately, emergency law should be judged unfair and illegitimate if it is a discrimination of some (it does not matter if a few or many; the fact that they are beings alone is enough), to the point of violating the fundamental core of human dignity. The emergency could represent the attempt made by the elites to reshape power in such a way as to present themselves as the new monarchs.

12 Concerning "human dignity," see, for example, Roberto Mancini (2014), Paolo Becchi (2009).

The reason a state of emergency is extended can be found in the necessity to dilate the permanence of the very government of the emergency.[13] Totalitarian governments need fearful and submissive men in order to maintain their power, and men need governments to justify their fear and submission. Governments then no longer govern but limit themselves to taking advantage of continuous emergency states to legitimize themselves or, if they do not exist, to be invented in order to maintain their power. When democracy dies, it becomes only a *fassadendemokratie* (façade democracy). Is it possible to build an authentic *gemeinwesen*, a human community based on new habits, such as those of social distancing?

Often, power uses words to actually refer to their opposite. If, for example, the word "sustainability" is used, then it could mean that it is likely to adopt unsustainable policies; if the word "security" is used, it is likely that it is helping create insecurity through totalitarian control; if it says "peace," then perhaps it is planning a war; while if it uses the word "equity," it may be trying to create conditions of inequality, through social distancing, for example, or the establishment of new categories. In this regard, Thucydides (1998) affirmed that in the face of a dramatic event, words lose their meaning. The suspicion, however, is that the ruling class uses different words in order to confuse the citizens and better implement its ultimate utilitarian goal. Behind an apparent end, in fact, there is another end. This is the plot of the emergency narrative in these times. In this sense, the emergency hides primary and secondary ends. The primary ends are the apparent ones: (1) the protection of identity, (2) the protection of health, and (3) the protection of peace and international security. Then there may be the more secondary purposes, such as (1) the exercise of technological control and surveillance,[14] (2) the realization of a "transhuman"[15] project, and (3) the creation of conflict and war.

Some Conclusions

In this chapter, the concept of emergency was often linked to sustainability, invoking a sort of ethics of balance that governs, or should govern, an emergency. This new theoretical framework, emergency sustainability, advances a sustainability vision with an emergency dimension, dealing with the study of humanitarian, ethical, environmental, health, and social emergencies which are particularly connected to every crisis situation. In this sense, it argues for a new branch of sustainability that aims to overcome the classic bipartition

13 A state of emergency without a real emergency is Foucault's nightmare.

14 In addition to Foucault, it could be linked to silent capitalist surveillance, which recalls Bentham's "Panopticon" (Bentham 1791).

15 Transhumanists favor genetic engineering and believe that artificial intelligence will surpass human intelligence.

between weak and strong sustainability, while also contemplating a series of concepts, such as those of epistemic health and responsible catastrophism, which give a certain ethical direction to this particular sustainable vision. Within this vision, the concept of resilience emerges overwhelmingly. In fact, any emergency presupposes a form of psychological, social, ethical, ecological, anthropological, political, or economic resilience.[16]

However, an ethical vision of the emergency does not presuppose a justification for the emergency, so much so that it leaves room without any limitation to the emergency law. An emergency dimension presupposes precisely an ethical approach, and therefore respect for the human being, their relationships, and for human dignity, also taking into account a minimum-security dimension. The role of science in the vision of the emergency must be reassessed, since the emergency situation must be analyzed with scientific eyes, but this does not mean excluding ethical science or other sciences that make up the puzzle of the human sphere. And as already pointed out, a real emergency sustainability presupposes respect for dignity, which must be examined according to a heuristic approach. The emergency equilibrium requires a resilience capable, above all, of looking at the human dimension and its wounds (Grasso 2021).

Policymakers too often appear to take advantage of an emergency to be able to exercise power in an arbitrary way, ignoring the various social, personal, cultural, and environmental differences. This will to power must be limited through forms of civil disobedience, since wherever this power finds valid complacency, then the banality of evil will be increasingly accentuated, so much so as to contribute to an irreversible oblivion.[17] In the name of health, sometimes new medicines are tested on human beings; in the name of ecological-climatic emergency, sometimes unsustainable solutions are made up; and in the name of peace, sometimes the war industry is financed. It is also true – and this is my hope – that at least a certain degree of emergency utilitarianism could be useful for the real well-being of humanity.[18]

Acknowledgments

This research was carried out at the Goethe-Universität Frankfurt am Main (Normative Ordnungen, Frankfurt University) while I was a visiting scholar during the summer semester of 2022. Special thanks to the university for its kind hospitality and to Professor Darrel Moellendorf for his helpful suggestions. A synthesis of this research was also successfully presented during

16 In this regard, see Marco Ettore Grasso (2021).

17 Numbers on weapons sale are unwithstandable; in the last year, Italy has spent over 30 billion euros on weapons.

18 In this sense and about utilitarianism, see Georg Wilhelm Friedrich Hegel (1848).

the GEIG International Conference at the University of Salerno (28 June–02 July 2022) on “Ecological Integrity, Peace, and Public Health.”

Reference List

Agamben, G. (2002) *Homo Sacer. Die Souveräne Macht und Das Nackte Leben*, Frankfurt am Main: Suhrkamp Verlag.

Arendt, H. (2006) *Eichmann in Jerusalem: A Report on the Banality of Evil*, London: Penguin Classics.

Becchi, P. (2009) *Il principio dignità umana*, Brescia: Morcelliana.

Bentham, J. (1791) *Panopticon or the Inspection-House*, London: T. Payne.

Cicero, M.T. (1943) *De Republica, De Legibus, III, 22*, Cambridge, MA: Harvard University Press.

Ferejohn, J. & Pasquino, P. (2004) “The law of the exception: A typology of emergency powers,” *International Journal of Constitutional Law*, 2(2), pp. 222, 223.

Feyerabend, P. (1975) *Against Method: Outline of an Anarchistic Theory of Knowledge*, London: New Left Books.

Forster, G. (1794) *Über die Beziehung der staatskunst auf das glück der menschheit*, Frankfurt am Main: Insel Verlag.

Foucault, M. (2008) *The Birth of Biopolitics. Lectures at the Collège de France 1978–1979*, Palgrave Macmillan.

Grasso, M.E. (2021) *Resilience and Sustainability in Law: Theoretical and Critical Approaches*, Verlag: Cambridge Scholars Publishing.

Gross, O. & Ní Aoláin, F. (2006) *Law in Times of Crisis, Emergency Powers in Theory and Practice*, Cambridge: Cambridge University Press.

Grotius, H. (Huig de Groot) (1625) *De iure belli ac pacis*, Paris.

Habermas, J. (1996) *Between Facts and Norms: Contributions to a Discourse Theory of Law and Democracy*, Cambridge, MA: MIT Press.

Hegel, G.W.F. (1848) *Vorlesungen über die Philosophie der Geschichte*, Berlin: Verlag von Duncker und Humblot.

Hegel, G.W.F. (1991) *Elements of the Philosophy of Right*, ed. by A.W. Wood, trans. by H.B. Nisbett, Cambridge: Cambridge University Press.

Hobbes, T. (1651) *Leviathan: Or the Matter, Forme and Power of a Commonwealth, Ecclesiasticall and Civil*, London.

Huxley, A. (1932) *Brave New World*, New York: Harper Brothers.

International Law Association. (1985) “The Paris minimum standards of human rights norms in a state of emergency,” *Journal of International Law*, 75(AM), pp. 1072–1075.

Kant, I. (1785) *Aus der Grundlegung zur Metaphysik der Sitten: Übergang von der gemeinen sittlichen Bernunfterkenntniß zur philosophischen*. Kants Populäre Schriften, ed. by Paul Menzer, 1911, Berlin: De Gruyter.

Kelsen, H. (1979) *Allgemeine Theorie der Normen*, Wien: Manz.

Kuhn, T.S. (1962) *The Structure of Scientific Revolution*, Chicago: University of Chicago Press.

Lakatos, I. (1978) “The Methodology of Scientific Research Programmes,” in: *Philosophical Papers*, Vol. 1, Cambridge: Cambridge University Press.

The Land Report (2022) *The Land Report 100*. Available at: https://editions.mydigitalpublication.com/publication/?i=778214.

Mancini, R. (2014) *Trasformare l'economia, Fonti culturali, modelli alternativi, prospettive politiche*, Milano: Franco Angeli.

Mill, J.S. (1859) *On Liberty*, Oxford: Oxford University Press.

Milstein, B. (2020) "Security and democratic equality," *Contemporary Political Theory*, 20(4), p. 853.
Moellendorf, D. (2014) "Just War," in: Jon Mandle and David Reidy (eds) *Companion to Rawls*, Malden: Wiley-Blackwell.
Moellendorf, D. (2015) "Two Doctrines of Jus ex Bello," *Ethics*, 125(3), Chicago: The University of Chicago Press.
Moellendorf, D. (2022) *Mobilizing Hope. Climate Change and Global Poverty in the 21st. Century*, Oxford: Oxford University Press.
Orwell, G. (1949) *Nineteen Eighty-Four*, London: Secker & Warburg.
Pascal, B. (1941) *Pensees and the Provincial Letters*, New York: Modern Library.
Popper, K. (1934) *Logik der Forschung*, Wien: Julius Springer.
Popper, K. (2013) *Logik der Forschung*, Berlin/Boston: Akademie Verlag.
Pufendorf, S. (1672) *De Jure Naturae et Gentium, I, I, 5*, Lund: Adam Junghans.
Rawls, J. (1999) *A Theory of Justice*, Cambridge, MA: Harvard University Press.
Rey, A. (1903) *Éléments de philosophie scientifique et de morale*, Paris: Cornély.
Schmitt, C. (1931) *Der Hüter der Verfassung*, Berlin: Dunker & Humblot.
Schmitt, C. (1985) *Political Theology. Four Chapters on the Theory of Sovereignty*, trans. by George Schwab, Cambridge, MA: MIT Press.
Shiva, V. (2022) *Dall'avidità alla cura. La rivoluzione necessaria per un'economia sostenibile*, Verona: Emi.
Thucydides (1998) *The Peloponnesian War*, Indianapolis: Hackett.

Chapter 4

Global Food Systems

Challenges for an Ecological Perspective

Cristiane Derani

Introduction

Food production is essential to human existence, and how this production is organized directly reflects on life on planet Earth. Regulating global food systems, and restructuring their national and international governance, is essential to ensure the production of healthy food that respects ecosystems and is sufficiently fairly distributed to all human beings (SAPEA 2020). These policies are essential so that the entire productive network that surrounds it – from planting to post-consumer waste – is adjusted. The idea is to adjust the production and consumption process into the planetary boundaries, adapting to the planet's absorption capacity and promoting the restoration of ecosystems and low-entropy energy flows (Georgescu-Roegen 1971). This is the process that builds food security while respecting food sovereignty and justice.

Food systems are political, historical, and cultural creations in complex networks of relationships, whose objective is the production and consumption of food. They are transversal to the whole society and comprise a flow of nature, energy, work, and finance. Furthermore, food systems are understood as the

> entire range of actors and their interlinked value-adding activities involved in the production, aggregation, processing, distribution, consumption and disposal of food products that originate from agriculture, forestry or fisheries, and parts of the broader economic, societal and natural environments in which they are embedded.
>
> .(FAO 2018)[1]

Their diversity depends on the different ways of organizing nature, culture, and material conditions of production. Although the process of cultural

1 The food system is composed of subsystems (e.g., farming system, waste management system, input supply system, etc.) and interacts with other key systems (e.g., energy system, trade system, health system, etc.) (FAO 2018).

DOI: 10.4324/9781003440871-5

homogenization and the liberal capitalist mode of production engendered a global, one might say hegemonic, system of food production, this system is at a breaking point. Food production is responsible for roughly a quarter of global greenhouse gas emissions and is also the top contributor to water pollution and biodiversity collapse (Loring 2022).[2]

As sustained by Rose (2021):

> [T]oday's food systems are not working for human and planetary health. Whether we examine the factory farming of livestock, the proliferation of ultra-processed and unhealthy foods and sugary beverages, or the links between food system financialization, large-scale land acquisitions and highly polluting agri-business, the underlying theme is clear: the contemporary global food system has generated a pandemic of non-communicable diseases and produced environmental devastation on a barely comprehensible scale.

This gloomy picture is supported by political, regulatory, and institutional mechanisms through which the system imposes itself and expands in time and space. Due to the essential nature of what they produce, food systems are the productive structure that sustains society. Therefore, the legal framework that affects them must be structured in detail with a view to ecological, social, and economic justice. Food availability is a human right. Also, it is the prior condition to justice. Therefore, it cannot be considered a mere marketable good.

The hypothesis is that the indispensable change in food systems depends on a normative and political framework that recognizes the specificities of these systems and leads to fair and ecologically sustainable production. Using seven sustainability axes (Sachs 2003), this chapter demonstrates how building sustainable food systems[3] results in ecological gains and social justice. Concluding, it establishes some legal concepts able to adjust policies and regulations for an ecological legal order, in reaction to the financialization of global food systems, targeting the production and distribution of food consistently with planetary boundaries and its essentiality to human societies.

2 Approximately 75% of the planet's food crop diversity was lost in the 20th century as farmers ceased to cultivate an assortment of local crops in favor of the genetically uniform, high-yielding varieties of wheat, rice, maize, and potato introduced by the Green Revolution. This loss of genetic diversity increased the vulnerability of the global food system to pests, drought, floods, and other external shocks, including those associated with climate change. Agriculture currently generates more anthropogenic greenhouse gas emissions than any other sector of the economy (Gonzalez 2015).

3 "A sustainable food system is a food system that delivers food security and nutrition for all in such a way that the economic, social and environmental bases to generate food security and nutrition for future generations are not compromised" (EC 2021; FAO 2018).

A Network for Building Relationships

Food is the result of ecological, economic, social, and cultural relationships, requiring different legal, political, and governance arrangements. The most essential issues of the human species are connected to them, such as survival, sociability, and health (physical, mental, and spiritual).

Food systems are diverse and may or may not be integrated into a global system. The advance of globalization has taken over most existing systems, transforming them and subjecting them to a unified logic of production, circulation, and distribution (Harvey 2001, 2007). One of the consequences of this standardization and generalization of food production across different countries, territories, cultures, and economies is the abstraction of food for its commercialization (Rabbi et al. 2021).

The globalization of food systems was based on the privatization of the elements that constitute them and the financialization of their value. All the values that concretely populate these systems are silenced and despised, suffocated by global value chains woven by the financial flow that links relationships. Globalized food systems are organized into value chains (Vroegindewey & Hodbod 2018), and their flow is measured by their substantially defined monetary representation in activities parallel to the concrete production of this commercialized object. The financialization of the food value chain creates a "parallel reality," in which the monetary flow is detached from the real productive forces, a spectrum that ultimately determines concrete action itself.

Food systems must be considered beyond the global value chain because they bring with them a range of essential values for human life and planetary integrity. They must be understood in a holistic way of environmental, social, political, and economic interaction. For they bring

> to light reinforcing and balancing feedback loops, tensions between the different components and flows of food systems, and interactions that are cyclical, multilayered and multi-scale. It is a way of thinking about the world that seeks to identify the linear and non-linear relationships between the different components of the system.
>
> (IPES-Food 2015)

With the construction of food systems committed to nature and the society in which they are inserted, they can leave their role as the main extractor of nature and energy, to become an important movement for the recovery of the environment, integration, and environmental and social justice. For this, it is necessary to understand and regulate the use of the soil, including the forms of rural property, the traditional cultures, and the technologies used. Going through what is produced, how it is produced, and what seeds and inputs are used. Transport, storage, packaging, and waste must also be revised and transformed, especially regarding the necessary banning of plastic packaging that contaminates the entire environment and harms human health (Kan & Miller 2022).

The goal of the sustainability of food systems is to make this production network a responsible organization toward nature, fair toward human societies, and a promoter of human and planetary health. Sustainable global food systems must be configured as a system guided by ecological values, based on a right that includes human organization founded on the development of a harmonious life with nature. A "social-ecological system" is required to achieve an accurate estimate of sustainability. This is a system that emerges through the interaction of society and the natural environment (Rabbi et al. 2021). As there is no life without food, the productive organization of food systems must be, at the same time, a network for building relationships with nature and among human beings in a harmonious way, as sustained by the first objective of the Rio Declaration.[4]

The building of responsible global food systems is based on seven axes:

1. *Geographic.* Production is built in territories of life. Its occupation and mode of use will mark the entire production system.
2. *Political.* Food systems are presented in extracts and are interconnected in a global web of relationships with different subjects, organized by law, politics, and national, regional, and international governance.
3. *Economic.* Production is organized in a web of financial commitments and mercantile relations of work and goods, producing private goods, for which a price is paid for their acquisition, while generating public diseconomies, that is, negative externalities, freely delivered to the society.
4. *Social.* Social relations are shaped according to a society's access to food production and consumption. Places of production and distribution, forms of conservation, and production of waste are the result of practices that expose the degree of economic efficiency and food justice within societies.
5. *Cultural.* Food production is the first catalyst of human societies, constitutive of their cultural identity. Different food systems are linked to cultural diversity. The variety of agricultural and pastoral cultures, and food customs, is an expression of a society's culture.
6. *Ecological.* The basis of food production is nature. The connection between water, energy, biodiversity, and climate is life. The primordial condition for the existence of these systems. It is nature, before all human practice, that offers the possibilities of production and consumption. Likewise are the food systems that constitute, restore, and destroy nature.
7. *Ethical.* Food is the source of life, and its availability has to be fair and equitably shared. It is important to honor the food and its source, valuing living and nonliving nature responsible for bringing food to the plate.

4 Principle 1: "Human beings are at the centre of concerns for sustainable development. They are entitled to a healthy and productive life in harmony with nature" (UN 1992).

The legal treatment of global food systems must take ownership of all seven axes, aiming at building healthy food production systems accessible through a process that responds to the interests of the society that produces and consumes it. As well as respecting the carrying capacity of natural conditions. The integration of food into nature's reproduction cycle is the essential condition for building a sustainable global food system and, of course, for the very continuity of human life on Earth.

Global Food Systems and Public Good

The formation of complex legal objects needs constant political and regulatory reinforcement to guide their treatment by jurisdictional bodies, as well as by public policies and change agents. This is precisely what happens with global food systems that require attention and treatment that lead them to the projected direction of sustainability, in line with their characteristic of being a public good, indispensable to the maintenance of life, human, and planetary well-being. Sustainable and fair food systems are public goods, as they are capable of generating direct benefits (supplies of sufficient and healthy food) and indirect benefits (income generation, recovery of ecosystems, reaffirmation of social and cultural values), which are not mutually exclusive and generate continuous benefits. This public good is based on the appropriation of nature, whose elements are considered common or common pool resources (CPR) (Ostrom 1990; Ostrom & Ostrom 1977) to all societies, to produce food, which is necessary for the existence of all human beings, configuring, from then on, its position as a human right.

The recognition of global food systems as public goods leads to the necessary reorganization of these systems toward social and ecological efficiency, in contrast to what has occurred with global food systems, swayed by efficient monetary gains in global value chains, organized by the commodification of food. The commodification process and the disregard for the ecological and ethical aspects of this public good made global food systems the biggest producers of negative externalities (environmental and social) (Pigou 1920), becoming the opposite of what is expected of a public good.

As they are public goods, global food systems demand a political and legal practice aimed at dealing with the *res publica*, highlighting the public values that circulate and are produced in these relationships. It is in this context that the six pillars of sustainability must be present, orienting relationships from the appropriation of nature, through the management of territories, appropriation of cultures and traditional knowledge, to the discussion of international commercial and political agreements, to ensure an effective public (collective) good. As a public good, global food systems need regulation that moves away from the commoditization of their values and reinserts the plurality of social, cultural, environmental, and political values not only to eliminate the generation of negative externalities but also to produce positive externalities.

The Blindness of the Economists' Calculations

The basis of food systems is the appropriation of the commons, here understood as nature, culture, and territory. Hence, the socialization of costs (negative externalities) and the exclusive privatization of gains (commoditization) are unacceptable. As a result of the incentives to externalize costs to produce "more for less," the environmental costs of agriculture exceed the market value of production (FAO 2017). Nevertheless, it is necessary to emphasize that a public good must be built in the public interest (Blomquist & Ostrom 1985).

Despite their essentiality, nature and other commons are invisible to economists' calculations, market liberalization policies, and the structure of prices in value chains. There is a gratuity and indifference in the appropriation of the commons that will not be solved through arbitrary pricing. It is up to global food systems to transform their relationships with nature by generating positive externalities from these commonwealth values that are distributed "communally" (*commonizing*), guaranteeing equitable access to the food produced. Global food systems seek pockets of nature to consume and extinguish them while filling private coffers with monetary means. There is a perennial transfer of commonwealth to consumable private monetary wealth. This is the real *tragedy of the commons*.[5]

The greater the appropriation of the commons, the more profitable the production, completely distorting the result, transferring collective wealth (commons) to private accounts, and worse, not providing fair and healthy food. The more globalized the food system, the more vulnerable the sustainability axes[6] become. Chemical additives from the soil to the plate permeate the entire chain. Plastic and food wastes, in addition to intense energy expenditure, are polluting and highly entropic. Although the actors who dominate this system justify it by increasing productivity aimed at eradicating hunger, often large plantations and extensive areas of livestock create belts of hunger and poverty.

From the composition of public goods by the private appropriation of CPR (Ostrom & Ostrom 1977), the so-called negative externalities are generated. The exclusive, privatized use of common resources within global food systems, that produce socialized negative outcomes, as a counterpart

5 Paraphrasing Hardin's expression but bringing it another content. The destruction of the commons is due to their exclusionary privatization in a production process that must be aimed at the entire community. The use of the commons in the private food production system does not have a priceable value, and most of the time this is not even attempted. Hence the great tragedy. Its disappearance through its privatization in a process that hides its condition deteriorates or extinguishes its existence (see Hardin 1968).

6 See earlier, the seven axes of sustainability.

to privatized positive outcomes, gains greater relevance and results in even greater injustice and inequity.[7]

It is not the matter of producing private goods through the freedom of economic initiative that, during its process, generates public diseconomies while protecting individual economies. When it comes to food systems, the generation of negative externalities as production residue is enhanced by the exhaustion of the commons in the same productive process.

One cannot equate, for example, atmospheric pollution from a car factory with soil and water exhaustion brought on by food production. Both are negative externalities, but unlike the classic perception of externalities arising from a factory's emissions, the production of unhealthy foods and the depletion of common resources are diseconomies caused during the making of a public good, essential and intended for everyone, without which humanity ceases to exist. It means that the correction of this kind of externalities is not solved by a single process of internalization by the private investor, as Pigou stated. It is about going to the roots of the process and readdressing the way the public good is produced to obtain a compatible public gain.

Food as a Question of Values

Literature is filled with scientific and economic studies demonstrating the need to move away from a food system founded on commodity production (Jackson et al. 2021; Coghlan et al. 2022). A report from the Group of Chief Scientific Advisors to the European Commission recently concluded that the path to a more sustainable food system requires "moving from food as a commodity to food as [more of] a common good" (Jackson et al. 2021).

Food is not a simple commodity available to anyone willing to pay or interested in acquiring it. It is a vital good, indispensable to life. In this world of capitalism, where everything is levelled to a monetary equivalent, it is necessary to separate the wheat from the chaff, or the commodity from a human right. Designating food as a human right is a necessity to differentiate it from an essentially private commodity.

A *good* is defined as a commodity by the way it is produced. An orange will always be an orange, but it can be a commodity, a basic consumer item, an ecological recovery product, or a result of a social economy. The different economic, normative, social, and ecological meanings of the orange do not

7 Externalized environmental costs include degradation of soils (Amundsonet et al. 2015), biodiversity loss (Newboldet al. 2016), unsustainable use and pollution of water (Liuet et al. 2012; Dalinet et al. 2017), and driving climate change, where agri-food emits as much as 30% of global greenhouse gases (Bajželjet et al. 2013; Eshelet et al. 2014; Poore and Nemecek 2018). Quoted from (Benton & Bailey 2019).

depend on its physical characteristics but on the social productive arrangement responsible for its existence and value in society.

Thus, the treatment of food as a commodity is due to the global production of this food, based on intensive agro-industrial manufacturing destined for a global market, whose price is negotiated on future exchanges. Following this reasoning, the decommodification of food is built with profound changes in food systems. At the macro level, this decommodification also requires a relativization of the GNP as a reliable measure of a nation's wealth. The GNP distorts the amount of global wealth and does not reflect the degree of economic (in)justice in a society, since it is unable to measure the exhaustion of nature, the appropriation of these commons, and the production of negative externalities.

As Philip McMichael (2005) observes:

> The development project incorporated post-colonial states into a universal system of national accounting methods, standardizing the measurement of material wellbeing (GNP), and the "externalization" of a variety of environmental degradations and social catastrophes. Only monetized transactions were counted as productive, devaluing subsistence, cooperative labour, indigenous culture, seed saving, and managing the commons as unproductive, marginalized, and undeveloped activity. As a consequence, the world's rural population decreased by some 25 percent in the second half of the twentieth century, with the steady displacement of peasant cultures.

The valuation of sustainable and fair food systems requires a metric compatible with the values involved that are non-monetary, such as the value of the commons, the nature that is incorporated into production, the value of producing a public good essential to the existence of society, and the value of an existential human right that represents the material version of the right to life. The commonization of food systems is a normative and political institutional activity of economic reorientation and the real construction of democracy, going beyond the hegemonic neoliberal ideology that accepts the end of the world but rejects the end of capitalism.[8]

Sustainable Food Systems and Public Goods

Food, beyond the commodity, has dimensions that refer to human rights, such as the rights to health and life. Thus, it needs to find its normative space consistently with this concept that goes beyond mere monetary value

8 As Fisher (2009) said, "it is easier to imagine the end of the world than it is to imagine the end of capitalism."

for private gain. To do so requires different governance structures and institutions that are socially inclusive, fair, and environmentally sustainable. This certainly goes much further than neoliberal preconceptions that market forces are an efficient means of allocating resources (Hayek 1945) to food production. The regulation of food as a human right requires, for its effectiveness, the transformation of global food systems to place them on sustainable bases. In other words, the efficacy of this human right must be anchored on the seven axes of sustainability: ecological, cultural, social, political, geographical, economic, ethical.

Accordingly, sustainable food systems are public goods based on the appropriation of the commons, aimed at food production, which is a right of every human being. The right to food is as obvious as the right to life, the former being a concrete expression of the latter. Ensuring the availability of healthy food for all is a social obligation, a legal duty, a political task, and an ethical injunction of justice. At the core of the expressions *food security*, *food sovereignty*, and *food availability* is the human right to ensure that everyone has sufficient and quality food on a sustainable ecological basis.

The right to food as a human right is present in international documents, from the Universal Declaration of Human Rights (UDHR) to the International Covenant on Economic, Social, and Cultural Rights (ICESCR). Article 25 of the UDHR states that everyone has the right to food (UN 1948), and Article 11 of the ICESCR goes further by providing for the right to *adequate* food (ICESCR 1967). This right is also protected through Article 6(1) of the International Covenant on Civil and Political Rights (ICCPR), which guarantees the right to life and has been interpreted to require the implementation of affirmative measures to eliminate chronic undernourishment (UN 2019). Additionally, Article 1 of both the ICESCR and the ICCPR prohibits states from interfering with a population's means of subsistence.

To provide authoritative guidance on the right to food, the UN Committee on Economic Social and Cultural Rights published General Comment 12, which clarifies the obligations of states to respect, protect, and fulfil the right to food (UN 1999). States have both positive and negative obligations. They must ensure the availability of food as well as must refrain from actions that interfere with the ability of communities and individuals to feed themselves. For example, states must respect the subsistence rights of small farmers by protecting them from foreign dumping of subsidized food products (Warner 2015).

The realization of this human right necessarily involves reorienting the commercial norms that guide food systems. In the case of global food systems, any structural modification aimed at greening these systems, which includes the implementation of the human right to food, depends on local restructuring with national norms. It also depends on international arrangements, such as bilateral or multilateral trade agreements and international treaties that govern the production and distribution of food fairly and sustainably.

The human right to the environment, food justice, and finally, greening of economic systems calls for clear economic actions in relations between countries and between peoples within states. Appropriate, in this sense, is the expression of John Knox, United Nations Independent Expert on Human Rights and the Environment, who defends a "diagonal human right."[9]

In international law, human rights are norms of *jus cogens*.[10] Therefore, the violation of the right to healthy and sustainable food is an unacceptable infraction. While law defines access to food as a peremptory norm, reality disdains it, making food insecurity and unavailability an example of injustice, against which we must speak out.

The Complexity of a Sustainable Food System

To conclude, building sustainable food systems is also a guarantee of food security, food availability, and food sovereignty, from which ecological and food justice are derived.[11] In this construction, the nexus of food and water, energy, biodiversity, climate, land distribution, conservation of traditional knowledge, satisfaction, and physical and psychological health is respected. Food systems are both the reflection and fabric of our civilization. A healthy civilization produces, and is a consequence of, healthy and sustainable food systems.

The construction of food/ecological security and justice needs to move away from the Western reductionist view that monetizes everything and that imposes the rationality of scale gain, fast consumption, and lower cost on food production. Security and justice cannot be built by exhausting nature and submitting food systems to a monetary logic.

9 *Diagonal human rights* are understood as "rights held by individuals against foreign governments for the extraterritorial consequences of their aid, trade, finance, and investment policies, including the power they wield in international financial institutions" (Warner 2015).

10 As stated by former ICJ Judge Tanaka: "If we can introduce in the international field a category of law, namely *jus cogens*, recently examined by the International Law Commission, a kind of imperative law which constitutes the contrast to *jus dispositivum*, capable of being changed by way of agreement between states, surely the law concerning the protection of human rights may be considered to belong to the *jus cogens*" (ICJ 1966).

11 For Gonzalez (2015), "food justice is the right of communities to grow, sell, and consume healthy, nutritious, affordable, and culturally appropriate food produced through ecologically sustainable methods, and their right to democratically determine their own food and agriculture policies. In other words, food justice is grounded in the human right to food, and is based on the principles of intergenerational equity, intragenerational equity, public participation in decision-making, and economic self-determination. Like the food sovereignty movement's call for local and national control of food-producing resources, this definition "does not negate trade, but rather it promotes the formulation of trade policies and practices that serve the rights of peoples to safe, healthy and ecologically sustainable production."

Food security through the lens of justice requires food availability and food sovereignty. In other words, that food not only exists but is also equally distributed and respects the health, quality, cultures, production methods, and political and ecological relationships involved. *Food security* and *food justice* refer to sustainable modes of food production built on ecological justice. It involves the democratization of land availability and the reinforcement of autonomy in terms of obtaining seeds and using fertilizers and insecticides. This is presented through production, distribution, storage, and consumption networks with low emissions, low generation of waste, without misuse, and resulting in healthy food.

Food systems are public goods that appropriate the commons (nature). From this stems the duty of all members of this system to make it ecologically sustainable and fair. New practices are built by new normative and political structures designing governance that adapts food production to planetary limits, the reconstruction of ecological systems, and distributive justice within the relations of these systems.

Principles of international law point to the construction of this concept of food justice that associates food security and sovereignty. The principle of self-determination, as well as that of permanent sovereignty over natural resources, converges so that states act responsibly for the management of the commons and the organization of the public goods that are sustainable food systems, to guarantee the collective benefit of the food in an ecologically balanced environment.[12]

Conclusions

Global food systems are the result of a global mode of production, based on monetary circulation, that subjects public goods, nature, culture, and rights to the same mercantile logic. This chapter argues that global food systems, which have developed over time and intensified their ability to deplete nature in recent decades, need to be rebuilt with the help of norms and policies that respect the importance and essentiality of food production. It is necessary to improve small-scaling food systems. Regional and local alternatives need to be created in order to support more sustainable food systems. Reducing the food chain and multiplying different local and regional food systems can also mitigate greenhouse gas emissions by minimizing the use of fossil fuel; can regenerate ecosystems, changing from a production based on pesticides and chemical

12 "Instead of the traditional focus on the rights of states over their countries' natural resources, this approach would emphasize the duties of states to discharge their fiduciary obligations to their citizens in good faith, the democratization of control over productive resources, and the obligations of states and citizens to promote ecologically sustainable use of these resources" (Gonzalez 2015).

fertilizers; and can evolve based on a variety of crops and seeds, promoting adaptation and resilience to drought, floods, and pests (Gonzalez 2015).

These norms must be in line with the public good character of food systems and strengthen the cultural and democratic bonds of the relationships that constitute them. That they are ecological norms that organize the use of the commons in the private production of food fairly and sustainably. And that they restore food justice, taking into account the fundamental right to healthy food produced through social and ecological justice.

For this, it is important to reformulate the legal system, especially in the field of environmental law and international economic and commercial law, toward organizing food systems so that they produce healthy food and, at the same time, restore ecosystems. These changes cannot occur without social conviction about the need for new directions for civilization. Food systems are the backbones of a society. Consequently, by bending the structures of the global food systems to an ecological organization, the whole society will reposition itself in change. So also, the argument of Ali et al. (2021) highlights the power of transforming food systems:

> [T]here is a strong affinity between food systems theory and the eco-centric cosmologies of Indigenous peoples, who "feel responsible for living in harmony with nature and creating a culture that reciprocally protects and nourishes life in all its forms." Such ontologies and forms of knowing "differ fundamentally from the prevailing totalitarian, positivistic, rationalist, mechanistic and reductionist Western worldviews."

A law that organizes itself to reorient human activities and remove them from the environmental collapse needs to appropriate the multiple values that surround them, decommodifying them and highlighting the maximum present social values. It is a law attentive to the more harmonious relationships with nature present in other cultures that managed to protect themselves from the Western-monetized reductionist view. Only with this commitment is it possible, while conserving and rebuilding the natural bases of production, to provide food security and justice.

Reference List

Ali, T., Buergelt, P., Maypilama, E., Paton, D., Smith, J., & Jehan, N. (2021) "Synergy of systems theory and symbolic interactionism: A passageway for non-indigenous researchers that facilitates better understanding indigenous worldviews and knowledges," *International Journal of Social Research Methodology*, pp. 1–16. Available at: https://doi.org/10.1080/13645579.2021.1876300.

Benton, T.G. & Bailey, R. (2019) "The paradox of productivity: Agricultural productivity promotes food system inefficiency," *Global Sustainability*, 2(e6), pp. 1–8. Available at: https://doi.org/10.1017/sus.2019.3.

Blomquist, W. & Ostrom, E. (1985) "Institutional capacity and the resolution of a commons dilemma," *Policy Studies Review*, *5*(2). Available at https://scholarworks.iupui.edu/bitstream/handle/1805/25445/Blomquist1985Institutional.pdf?sequence=1.

Coghlan, C., Proulx, P., & Salazar, K. (2022) "A food-circular economy-women nexus: lessons from Guelph-Wellington," *Sustainability*, 14, p. 192. Available at: https://doi.org/10.3390/su14010192.

European Commission [EC] (2021) *Food Systems – Definition, Concept and Application for the UN Food Systems Summit: A Paper from the Scientific Group of the UN Food Systems Summit.* Available at: https://knowledge4policy.ec.europa.eu/publication/food-systems-definition-concept-application-un-food-systems-summit-paper-scientific_en.

Fisher, M. (2009) *Capitalism Realism: Is There No Alternative?* Winchester: Zero Books.

Food and Agriculture Organization [FAO] (2017) *The Future of Food and Agriculture – Trends and Challenges*, Rome. Available at www.fao.org/3/i6583e/i6583e.pdf.

Food and Agriculture Organization [FAO] (2018) *Sustainable Food Systems: Concept and Framework*. Available at www.fao.org/3/ca2079en/CA2079EN.pdf.

Georgescu-Roegen, N. (1971) *The Entropy Law and the Economic Process*, Cambridge, MA: Harvard University Press.

Gonzalez, C. (2015) "Food Justice: An Environmental Justice Critique of the Global Food System," in: S. Alam, S. Atapattu, C. Gonzalez, & J. Razzaque (eds) *International Environmental Law and the Global South*, pp. 401–434, Cambridge: Cambridge University Press. Available at: https://doi:10.1017/CBO9781107295414.020.

Hardin, G. (1968) "The Tragedy of the Commons," *Science, New Series*, 162(3859), pp. 1243–1248.

Harvey, D. (2001) *Spaces of Capital: Towards a Critical Geography*, Edinburgh: Edinburgh University Press.

Harvey, D. (2007) *A Brief History of Neoliberalism*, Oxford: Oxford University Press.

Hayek, F. (1945) "The use of knowledge in society," *The American Economic Review*, *35*(4), pp. 519–530. Available at: www.jstor.org/stable/1809376.

International Court of Justice [ICJ] (1966) *Judge Tanaka's Dissenting Opinion in the South West Africa Case (Ethiopia v. South Africa; Liberia v. South Africa), Second Phase, Judgment ICJ Rep 298*. Available at: www.icj-cij.org/public/files/case-related/47/047-19660718-JUD-01-06-EN.pdf.

International Panel of Experts on Sustainable Food Systems [IPES-Food] (2015) *The New Science of Sustainable Food Systems: Overcoming Barriers to Food Systems Reform: First Report of the International Panel of Experts on Sustainable Food Systems*. Available at: www.ipes-food.org/_img/upload/files/NewScienceofSusFood.pdf.

Jackson, P., Rivera Ferre, M.G., Candel, J., et al. (2021) "Food as a commodity, human right or common good," *Nat Food*, 2, pp. 132–134. Available at: https://doi.org/10.1038/s43016-021-00245-5.

Kan, M. & Miller, S. (2022) "Environmental impacts of plastic packaging of food products," *Resources, Conservation and Recycling*, 180, pp. 106–156. Available at: https://doi.org/10.1016/j.resconrec.2022.106156.

Loring, P. (2022) "Long-Standing Systems for Sustainable Farming Could Feed People and the Planet – If Industry Is Willing to Step Back," *The Conversation*. Available at: https://theconversation.com/long-standing-systems-for-sustainable-farming-could-feed-people-and-the-planet-if-industry-is-willing-to-step-back-183506.

McMichael, P. (2005) "Global development and the corporate food regime," in: F.H. & P. McMichael (eds) *New Directions in the Sociology of Global Development (Research in Rural Sociology and Development, Vol. 11)*, pp. 265–299, Bingley: Emerald Group Publishing Limited. Available at: https://doi.org/10.1016/S1057-1922(05)11010-5.

Ostrom, E. (1990) *Governing the Commons: The Evolution of Institutions far Collective Action*, Cambridge: Cambridge University Press.

Ostrom, V. & Ostrom, E. (1977) "Public goods and public choices," in: E.S. Savas (ed) *Alternatives for Delivering Public Services*, pp. 7–49, Boulder: Westview Press.

Pigou, A.C. (1920) *The Economics of Welfare*, London: Macmillan and Co.

Rabbi, M.F., Hasan, M., & Kovács, S. (2021) "Food security and transition towards sustainability," *Sustainability*, 13, p. 12433. Available at: https://doi.org/10.3390/su132212433.

Rose, N. (2021) "From the cancer stage of capitalism to the political principle of the common: The social immune response of 'food as commons,'" *International Journal of Health Policy and Management*. Available at: https://doi.org/10.34172/ijhpm.2021.20.

Sachs, I. (2003) *Caminhos para o Desenvolvimento Sustentável*, Rio de Janeiro: Garamond.

Science Advice for Policy by European Academies [SAPEA] (2020) *A Sustainable Food System for the European Union*, Berlin: SAPEA. Available at: https://doi.org/10.26356/sustainablefood.

UN Committee on Economic Social and Cultural Rights [CESCR] (1999) *General Comment No. 12: The Right to Adequate Food*, 12 May 1999, UN Doc. E/C.12/1999/5.

United Nations [UN] (1992) *Report of the United Nations Conference on Environment and Development*, A/CONF.151/26 (Vol. I). Available at: www.un.org/en/development/desa/population/migration/generalassembly/docs/globalcompact/A_CONF.151_26_Vol.I_Declaration.pdf.

United Nations [UN] (2019) *General Comment No. 36* on the International Covenant on Civil and Political Rights, CCPR/C/GC/36.

United Nations International Covenant on Economic, Social and Cultural Rights [ICESCR] (1967) New York, 16 December 1966, in force 3 January 1976 GA res. 2200A (XXI), 21 UN GAOR Supp. (No. 16) at 49, UN Doc. A/6316 (1966); 993 UNTS 3; 6 ILM 368 (1967).

United Nations Universal Declaration of Human Rights [UDHR] (1948) Paris, 10 December 1948, GA res. 217A (III), UN Doc A/810 at 71 (1948).

Vroegindewey, R. & Hodbod, J. (2018) "Resilience of Agricultural Value Chains in Developing Country Contexts: A Framework and Assessment Approach," *Sustainability*. Available at: https://doi:10.3390/su10040916.

Warner, E. (2015) "South of South: Examining the international climate regime from an indigenous perspective," in: S. Alam, S. Atapattu, C. Gonzalez, & J. Razzaque (eds) *International Environmental Law and the Global South*, pp. 451–468, Cambridge: Cambridge University Press. Available at: https://doi:10.1017/CBO9781107295414.022.

Chapter 5

Food Security and Ecological Integrity

European Responses to Growing Global Instability

Alicia Epstein

Introduction

Ecological decline and rising geopolitical tensions have increasingly highlighted the precarious reliance of many nations on global agricultural supply chains and the growing threats to food security, especially following the disruptions of the 2020 COVID-19 pandemic (Swinnen & McDermott 2020). In early 2022, the pressures on the international food system were further compounded by Russia's invasion of Ukraine and the de facto grinding to a halt of agricultural exports from the latter (Fink 2022). Given the role of Ukraine as one of the largest exporters of grains and fertilizer components, this conflict has had global, as well as local, implications for food security in both the short and medium terms (FAO 2022a).

In the short term, the war had the immediate effect of drastically pushing up food prices for a number of key commodities, including dairy, cereals, oils, and sugars, with the latter reaching its highest month-to-month rise, in February 2022, since tracking of these products began in 1961 (FAO 2022b). This has particularly affected countries that rely heavily on food imports to meet domestic demand and, in so doing, undermined the ability of the poorest members of these societies to access adequate food to meet their nutritional needs, due to lack of affordability (Lin et al. 2023). In the medium term, the war in Ukraine has the potential to pose productivity challenges in countries, including a number of European Union (EU) member states (MS), that rely on the import of Ukrainian agricultural products, such as fertilizer components, seeds, and feed, to attain projected levels of output (European Parliament 2022a, 2022b).

The long-term and ecological consequences of the current global food crisis will largely depend on the way governments respond to these pressing challenges and the types of actions pursued to this end. In the case of the EU, the response has largely focused on addressing market and price instability for agricultural commodities (Santeramo & Kang 2022), while far less attention has been paid to considering the environmental implications of these measures. Indeed, the global food crisis has not only been used to legitimize

DOI: 10.4324/9781003440871-6

a renewed focus on increasing EU agricultural productivity but has also provided a basis for suspending environmental standards and rules applicable to the agricultural sector. For instance, in March 2022, the European Commission introduced legislation to allow for the suspension of environmental rules that are otherwise applicable to recipients of farm subsidies under the EU's Common Agricultural Policy (Commission Implementing Decision (EU) 2022/484). While the derogations were intended to be temporary and limited to the 2022 payment year, these have since been extended, raising serious questions about the EU's ability to meet its own environmental and climate objectives in the years to come.

Against this background, this chapter considers the ecological implications of the EU's response to the global food crisis and some of the potential consequences for food security. The chapter is divided into two parts, with the first of these looking at the context of the 2022 global food crisis and the impacts on food security in the short to medium term. The second part of the chapter focuses on the impact of the food crisis on EU agricultural sectors and the measures taken in response. In particular, it looks at the environmental impacts of these measures and argues that the EU's emphasis on increasing productivity is likely to produce only marginal gains, while exerting a disproportionally large environmental cost.

The 2022 Food Crisis: A Perfect Storm

Multiple crises came to a head in 2022. Indeed, protracted market instability, along with extended droughts and climate-induced productivity challenges in many parts of the world, were already putting notable upward pressure on food commodity prices when the COVID-19 outbreak became widespread at the start of 2020 (Benton et al. 2022; Swinnen et al. 2022; Béné et al. 2021). The pandemic-related trade disruptions that followed (Laborde et al. 2020), as well as soaring energy and fertilizer costs, served to further drive up food prices, which rose at historically high rates during the subsequent two years (European Commission 2022a, 2022b). According to the Food and Agricultural Organization (FAO), food commodity prices grew by over 23% in 2021 from the year before, which at that time was the fastest pace in a decade (FAO 2022a).

The impacts of the war in Ukraine were instant and significant, given the importance of both Russia and Ukraine as key exporters of staple agricultural commodities, as well as fertilizer and fertilizer components used to grow crops in importing countries (European Commission 2022b). Combined, the two countries accounted for about half of all the maize and sunflower oil supplied to the global market in 2021, with Ukraine alone having supplied for a 44% share of global exports of the latter (Jagtap et al. 2022). They are both also major exporters of wheat and barley, much of which is imported by countries with high levels of poverty and food insecurity (Lin et al. 2023). Importantly, moreover, Ukraine supplied almost half of the wheat distributed in 2021 by the UN World Food Program to highly food-insecure populations in Yemen, Somalia, and Sri Lanka, among others (Behnassi & El Haiba 2022; e.g., World Food Programme 2022).

While Russia faced a number of significant logistical challenges, the export of agricultural products was still possible following its invasion in February of 2022. On the other hand, Ukrainian production was severely restricted, especially in the eastern parts of the country, where fighting was most intense (Fink 2022). Moreover, exports of Ukrainian agricultural products were essentially cut off by Russia's blockade of the Black Sea, which is the main transport route for getting these commodities to world markets, leading to supply shortages for countries with high shares of imports from Ukraine (Fink 2022).

Scarcity of important agricultural commodities was exacerbated in March 2022, when Russia implemented a temporary ban on the export of grains and fertilizers, leading to further market volatility caused by medium-term concerns over productivity in regions heavily reliant on the import of these products (Benton et al. 2022). This move was particularly impactful for the EU, given that Russia was its main supplier of potash, which was covered by the 2022 ban and is a crucial component for fertilizer production (Joint Research Centre 2022).

By May of 2022, average global food prices had risen by over 30% from the year before (FAO 2022b), with the high rate of food price inflation unevenly impacting countries around the world. For instance, food prices rose by about 8.1 and 8.6% in the EU and United States (US), respectively, during the same time (European Parliament 2022a; Bureau of Labor Statistics 2022). Whereas these levels of inflation are significant even for developed countries, they may be contrasted with impacts in countries that were already experiencing serious pandemic-related supply problems, as well as conflict and societal unrest prior to the war in Ukraine (Sirengo et al. 2022).

Implications for Food Security

The most immediate effects of the production and supply chain disruptions exacerbated by the war were undoubtedly felt by the Ukrainian population, which became increasingly dependent on food aid following the invasion and heavy bombardment by the Russian military in various parts of the country (European Parliament 2022c). Likewise, countries that relied heavily on imports from Ukraine lost a primary supplier and needed to secure alternative sources of import on an increasingly volatile international market to meet domestic food demands (Pereira et al. 2022; Jagtap et al. 2022; Glauben et al. 2022). While global food stocks ensured the availability of sufficient supplies to meet immediate food demands in 2022, the extremely high rates of food price inflation led to these supplies becoming increasingly unaffordable (United Nations 2022).

In this light, the effects of rising food prices have primarily been felt in countries where a significant share of the disposable income is spent on food (Okou et al. 2022). Sub-Saharan Africa is by far the most affected, with the average household spending around 40% of its disposable income on food, and one in four people was already regarded as food-insecure prior to 2022 (FAO et al. 2021; Okou et al. 2022). According to the FAO, the price hike experiences in 2022 have radically increased the number of people at serious risk of hunger and famine, especially in regions already experiencing violent conflict, such as Ethiopia and Yemen (FAO 2022b). Undoubtedly, this has served to hamper progress on attaining the UN Sustainability Goals (Pereira et al. 2022).

Access to food was also impacted in developed countries, including a number of EU MSs, due to increased unaffordability, albeit on an entirely different scale to that experienced in the Global South (FAO 2022a; Santeramo & Kang 2022). However, while EU food supplies were not immediately at risk, the EU Commission expressed its concern that high production costs had the potential to affect the certainty of food supplies in the long term if farmers "are not compensated by higher prices" (EU Commission communication March 2022: 5; Ben Hassen & El Bilali 2022).

The war in Ukraine has also highlighted the growing risks of armed conflicts for global and regional food security. Using its strategic advantage in the early stages of the war, Russia purposefully and tactically targeted Ukrainian food exports, agricultural infrastructure and production, as well as food storage facilities, with a combination of bombings and blockades, in order to inflict maximum damage and destabilization (Fink 2022). According to the Organization for Security and Cooperation in Europe, many of these actions were deliberately aimed at the civilian population and are likely to amount to possible war crimes (OSCE 2022). Undoubtedly, Russian acts of aggression made it significantly more difficult to carry out the basic activity of farming, leading to lower levels of output in 2022 than initially expected (Lin et al. 2023).

Although the targeting of food and agricultural sectors gained heightened attention in the wake of the 2022 invasion, it is not the first time that Russia has weaponized food to exert political pressure and reprisal. Indeed, in 2014, Russia retaliated against EU sanctions, imposed following its annexation of Crimea, by specifically targeting European agricultural sectors of importance for EU trade with Russia (Rankin 2014). Without warning, Russia unilaterally banned the import of all fruits and vegetables from the EU in August of that year, along with import bans on a range of food products from the United States and other exporting countries (EURACTIVE 2014). This had immediate and dramatic consequences for European producers, given Russia's position as one of the top importers of EU goods in these sectors, and was surely intended to cause as much economic harm as possible (European Parliament 2014).

EU Responses to the 2022 Food Crisis

Against this background, the EU took a number of actions to address the most immediate consequences of food price inflation at the beginning of 2022. Amongst the first of these was the introduction of measures aimed at ensuring the proper functioning of trade, trade infrastructure, and transportation networks needed to supply food for the Ukrainian civilian population (EU Commission 2022c, 2022d; European Council 2022). The EU also acknowledged the risks of growing food price inflation, both for its own MSs and those mostly developing countries that relied heavily on imports of staple foods (European Parliament 2022a). To this effect, it introduced measures to address supply chain bottlenecks and support the flow of agricultural products to global markets in order to meet the demand of import-reliant countries and counteract rising food prices (European Commission 2023c). In particular, the EU and other international institutions repeatedly emphasized the need to avoid imposing unilateral "measures to protect domestic food supply by preventing exports," as such actions risked sparking a domino effect of reciprocal trade restrictions by other governments and, thereby, worsening the crisis (EU Commission 2022a: 6).

The EU also adopted a number of sanction packages, targeting key sectors, entities, and individuals with financial ties to the Russian war offensive, in a concerted effort to weaken the country economically (Lonardo 2022). It is, however, important to emphasize that these sanctions did not target agricultural sectors, and according to the Council of the EU and the Commission, care was taken to ensure these measures would not affect Russian food exports and that, where relevant, sectoral prohibitions include specific exceptions for those products (Council of the EU 2022a). Undoubtedly, this approach was crucial for countering the Russian weaponization of food exports and for ensuring that Russia could continue to supply world markets with grain and fertilizer components, as the alternative would have significantly worsened the food crisis and amplified the risks of further food price volatility (Benton et al. 2022).

That said, the majority of the EU's efforts and resources were directed toward supporting European farmers impacted by the shortages of grain imports and rising fuel and production costs following the Russian invasion (European Commission 2022a; European Parliament 2022a). Importantly, the EU adopted a 500-million-euro package in March 2022 that allowed MSs to provide support for those agricultural sectors most affected by the crisis (European Commission 2022a). While these funds were intended to be prioritized for "farmers engaged in sustainable practices," the EU Commission also instructed that they should "target the sectors and farmers who are the hardest hit by the crisis" (EU Commission 2022a: 7). Thus, rather than requiring MSs to take into account specific environmental considerations when granting emergency financial aid to agricultural producers, the measures allow for considerable discretion to direct funding based on criteria that are not linked to environmental impacts.

The lack of environmental conditions for granting emergency financial support to farmers is especially noteworthy, given that several of the sectors most affected by the crisis account for significant ecological pressures in the EU (European Parliament 2022c). Indeed, the resulting feed shortages had a major impact on pig sectors in MSs that relied heavily on imports from Ukraine, including Spain, Poland, and Italy, whose governments lobbied for the EU to address these challenges and provide assistance to affected producers (Webber et al. 2022). In particular, calls were made for the EU to suspend environmental standards relating to pesticide residue levels in feed in order to allow imports from producers, such as the United States and a number of Latin American states, that would otherwise face technical barriers to accessing the EU market (Pinedo 2022). Spain even unilaterally decided to make use of emergency powers to substitute the shortfall of Ukrainian exports with feed from Argentina before the EU officially announced the suspension of regulatory standards on feed imports to address the supply challenges facing the pig- and pork-producing sectors (Gualtieri & Soto 2022).

There were also growing calls from the EP's largest parliamentary group, the center-right Europe People's Party, for the EU to revise aspects of its Green Deal with relevance to agriculture and farming (Fortuna 2022). This included weakening or suspending MS's obligations to contribute toward the EU's 2030 climate and environmental targets, specified for agricultural sectors in the Farm to Fork and Biodiversity Strategies (European Commission 2020a, 2020b). Likewise, agricultural ministries in several MSs argued for changes to be made to the CAP framework that would allow them to revise environmental commitments and measures specified in their national strategic plans for the 2023–2027 programming period (Fortuna et al. 2022). Not only were these intended to be made as downward revisions, but they also had the potential to be extended far beyond the crisis, considering the running time of CAP programming.

Derogations From EU Environmental Standards

While the EU did not go so far as to revise the objectives of the Green Deal, it did adopt measures to allow farmers to derogate from environmental obligations that they would otherwise have to observe in order to receive agricultural subsidies under the CAP. In its Decision of 23 March 2022, the EU Commission authorized MSs to allow for derogations from the so-called "greening measures," which require farmers to observe certain land management practices intended to provide benefits for the climate and environment as a precondition for receiving farm subsidies (Commission Implementing Decision (EU) 2022/484). More specifically, these derogations cover environmental measures that require farmers to, firstly, grow a minimum number of crops and, secondly, maintain at least 5% of the arable land on each holding larger than 15 hectares as an ecological focus area (EFA), where production is only permitted under certain conditions (Regulation (EU) No 1307/2013). Especially the latter has been acknowledged for its potential to benefit wildlife, by offering designated areas of land that are essentially taken out of production (fallow land) and subject to far less-intensive management practices than land not included in the EFAs (Pe'er et al. 2017).

Following the adoption of the Commission Decision, farmers were able to grow any crop on the EFAs and still receive full payments of their agricultural subsidies for the 2022 calendar year (Article 1). In addition, it authorizes MSs to allow farmers to derogate from conditions on pesticides and other plant protection products, the use of which is otherwise severely restricted in EFA's (Article 1). However, despite initial indications that these and other derogations from CAP rules would be limited to the 2022 claim year, the Commission adopted further legislation in July of that year to extend the possibility for MSs to allow farmers to derogate from environmental measures also in 2023.

Under Commission Implementing Regulation (EU) 2022/1317, MSs are able to allow farmers to derogate from additional obligations to keep agricultural land, including land that is not in production, in "good agricultural and environmental conditions," as laid out in Article 13 of Regulation (EU) 2021/2115 (Article 1). The Commission Regulation provides exceptions for two specific standards that recipients of CAP subsidies are otherwise required to observe concerning management conditions on crop rotation and the minimum share of agricultural land devoted to non-productive areas or features (Article 1). Respectively, these measures aim at preserving soil potential, as well as improving on-farm biodiversity, and thereby have the potential to impact productivity and the production decisions of farmers (Pe'er et al. 2017). In this light, the Preamble to the Commission Implementing Regulation emphasized that, given global food security concerns in 2022, it is "appropriate to exceptionally allow farmers flexibility to make use of their available arable land for food production, while mitigating the negative impacts of these choices on the environment and climate change" (Recital 4).

Furthermore, the Regulation stresses that MSs who choose to allow farmers to derogate from the obligation to keep land in "good agricultural and environmental conditions" are required to ensure that the production of food, rather than feed, crops are promoted (Article 1). Thus, MSs need to

> establish that the arable land which would not be devoted to non-productive areas as a result of the derogation . . . should not be used for growing maize and soya beans, considering that these crops are mostly not destined for food production.
>
> (Preamble: 4)

Productivity at the Expense of Ecological Integrity: The Long-Term Implications of Short-Term Gains

In their response to the 2022 food crisis, EU institutions and lawmakers have undoubtedly shown a great willingness to support increasing agricultural output as a primary tenet of their approach to contributing to "global food security." Not least, the Commission, under the direction of EU Commissioner for Agriculture Janusz Wojciechowski, has arguably played a leading role in formulating and implementing common measures to this effect. In its communication of 23 March 2022, the Commission set the tone of the initial EU reaction and clearly emphasized the need to ensure the productive capacity of agricultural sectors and their potential to increasing global food supplies (European Commission 2022a). In the case of wheat, for instance, the Commission argued that it is "fundamental from a geo-strategic point of view that the EU contribute to covering the production gap to address the expected global shortages" (EU Commission 2022a: 3).

However, while the EU is a net exporter and an important supplier of wheat on the international market, it is also a region with exceptionally high crop yields per hectare (Le Gouis et al. 2020). Indeed, being that it is one of the most intensely farmed areas in the world, these historically high levels of output have already come at the cost of ecological loss and destruction, with global implications (Jack 2016). This is highlighted within the context of the Planetary Boundaries framework, which affirms that intensive agricultural production and land use in industrialized countries have been a primary driver of changes to the Earth system (Stoate et al. 2009). With regards to the loss of biodiversity and disruption to the Earth's phosphorous and nitrogen cycles, in particular, EU agriculture has contributed significantly to reaching the tipping points identified in the literature (Steffen et al. 2015). Thus, the aim of maximizing agricultural productivity and output, which has historically been central to the CAP framework and enshrined in the Treaty on the Functioning of Europe (Articles 38–44), has ultimately served to undermine ecological integrity and long-term food security (Jack 2016).

Notwithstanding the high levels of productivity and considerable pressures already placed on agricultural systems, the Commission stressed that in order to address the market instability resulting from the 2022 war in Ukraine, "the Union agricultural production potential both for food and feed supply should be increased" (Preamble: 4). In addition, it argued that it was necessary to allow for derogations from environmental rules, to enable "an increase of the total arable land area available for the production of food and feed" (Decision (EU) 2022/484, Preamble: 6). The ability to derogate from the obligation to maintain EFAs on holdings larger than 15 hectares, as discussed earlier, was particularly instrumental in this respect, since it provides the ability to bring formally fallow land into production.

The EU's insistence on putting fallow land into production is problematic, however, given that farmers are likely to have already designated the least-productive areas of the agricultural holding for this purpose (Pe'er et al. 2017). Derogations from the crop rotation and EFA measures may, therefore, in many instances, be unlikely to meaningfully address food shortages, since the increases in output are expected to be only marginal (Pe'er et al. 2014). Importantly, increasing output on less-productive land may require greater use of inputs, such as fertilizers and pesticides, as well as other intensive land management practices. Thus, the decision to allow farmers to use pesticides and other plant protection products in EFAs underscores the potential for further ecological costs to be incurred, while the gains of bringing fallow land into production will remain limited. Not least, this undermines the benefits for biodiversity that EFAs are intended to provide and seriously casts doubt over the EU's long-term objective to reduce the ecological impacts of agriculture and contribute toward key environmental and climate targets.

Against this background, the measures introduced in response to the ongoing food crisis certainly have the potential to increase the gap between the EU targets specified for agricultural sectors and the actions taken to realize them (Benton et al. 2022). However, while the EU Commission has recognized that failure to implement the Green Deal targets would mean that food security "will be severely at risk in the medium and long-term, with irreversible impacts globally" (EU Commission 2022a: 9), it has arguably prioritized actions aimed at increasing productivity, without appropriate safeguards to ensure that ecological integrity is not further undermined. Although it is yet too early to estimate the impacts of these derogations on the production choices of EU farmers, a considerable number of MSs have expressed their intention to apply them for as long as it is possible to do so (Fortuna et al. 2022). This suggests that there is a reasonable possibility that environmental measures applicable to EU farmers could be further suspended to encourage them to maintain the highest possible levels of output.

Conclusions

The 2020 pandemic outbreak and subsequent supply chain disruptions caused severe volatility on global markets and primarily affected the ability of import-reliant countries to ensure food supplies due to rising production costs and food commodity prices. Thus, while the war in Ukraine has had a particularly dramatic effect on global food price inflation, these impacts were amplified by a number of pre-existing conditions and challenges to food security. Although European food supplies were not directly under threat, given that the EU is largely self-sufficient with regard to key agricultural commodities and staple crops, the 2022 crisis amplified the risks of growing uncertainty for producers and their ability to supply staple crops to international markets and import-reliant countries.

Responses to these crises by the EU indicate that short- and medium-term measures are at the forefront of policies aimed at ensuring food availability and addressing systematic challenges to productivity. In particular, the EU's readiness to allow for the suspension of environmental rules applicable to the farming sector as a means of securing increased productivity raises serious questions about its commitment and ability to attain important environmental objectives set to be achieved by 2030 and beyond. This is highly problematic, for several reasons, not least because the EU's ability to increase output is significantly limited, given the high levels of productivity already attained. Moreover, it signals a willingness to sacrifice environmental priorities for temporary gains in output at a time when ambition and actions need to be amplified in order to protect the integrity of the ecological systems that underpin food security.

The EU has also taken steps to extend the derogations from environmental standards beyond 2022, even though global food price inflation started to decline at the end of that year. There is therefore a risk that the crisis will continue to be used to justify the suspension of environmental safeguards and standards applicable to the farming sectors. Surely, this would further undermine the transition toward sustainable agricultural and food systems that are required in the years to come.

Reference List

Behnassi, M. & El Haiba, M. (2022) "Implications of the Russia–Ukraine war for global food security," *Nature Human Behaviour*, 6, pp. 754–755.

Béné, C., et al. (2021) "Global assessment of the impacts of COVID-19 on food security," *Global Food Security*, 31.

Ben Hassen, T. & El Bilali, H. (2022) "Impacts of the Russia-Ukraine war on global food security: Towards more sustainable and resilient food systems?" *Food*, 11(15).

Benton, T.G., Froggatt, A., & Wellesley, L. (2022) "The Ukraine war and threats to food and energy security: Cascading risks from rising prices and supply disruptions," *Research Paper*, London: Royal Institute of International Affairs. Available at: https://doi.org/10.55317/9781784135225.

Bureau of Labor Statistics (2022) "Consumer prices up 8.6 percent over year ended May 2022," *The Economics Daily: U.S. Department of Labor*. Available at: www.bls.gov/opub/ted/2022/consumer-prices-up-8-6-percent-over-year-ended-may-2022.htm.

Council of the European Union (2022a) *Special Meeting of the European Council (30 and 31 May 2022): Conclusions*, EUCO 21/22: General Secretariat of the Council.

EURACTIVE (2014) "Moscow retaliates with import ban on EU, US food," *EURACTIVE*, 7 August 2014. Available at: www.euractiv.com/section/agriculture-food/news/moscow-retaliates-with-import-ban-on-eu-us-food/.

European Commission (2020a) A Farm to Fork Strategy: For a fair, healthy and environmentally-friendly food system, COM (2020) 381 final.

European Commission (2020b) EU Biodiversity Strategy for 2030: Bringing nature back into our lives, COM (2020) 380 final.

European Commission (2022a) Safeguarding food security and reinforcing the resilience of food systems, COM (2022) 133 final.

European Commission (2022b) Ensuring availability and affordability of fertilisers, COM (2022) 590 final/2.

European Commission (2022c) An action plan for EU-Ukraine Solidarity Lanes to facilitate Ukraine's agricultural export and bilateral trade with the EU, COM/2022/217 final.

European Commission (2022d) Commission Decision of 12.5.2022 on the signing of a High Level Understanding between the European Union and Ukraine on indicative maps of the Trans-European transport network in Ukraine, being a high level agreement within the meaning of Article 49(6) of Regulation (EU) No 1315/2013 (TENT regulation), C (2022) 3204 final.

European Council (2022) *Informal Meeting of the Heads of State or Government: Versailles Declaration*, European Council.

European Parliament (2014) *The Russian Embargo: Impact on the Economic and Employment Situation in the EU. Policy Department A: Economy and Scientific Policy*, European Union.

European Parliament (2022a) *Russia's War on Ukraine: Impact on Food Security and EU Response*, European Parliamentary Research Service.

European Parliament (2022b) *The Ukraine War and Energy Supply*, Briefing: European Parliament.

European Parliament (2022c) *Russia's War on Ukraine: EU Food Policy Implications*, European Union.

Fink, M. (2022) "Naval blockade and the Russia–Ukraine conflict," *Netherlands International Law Review*, 69, pp. 411–437.

Food and Agricultural Organization [FAO] (2022a) *The Importance of Ukraine and the Russian Federation for Global Agricultural Markets and the Risks Associated with the Current Conflict*, Information Note, Rome: FAO.

Food and Agricultural Organization [FAO] (2022b) *Impact of the Ukraine-Russia Conflict on Global Food Security and Related Matters under the Mandate of the Food and Agriculture Organization of the UN*, Rome: FAO.

Food and Agricultural Organization, et al. (2021) *The State of Food Security and Nutrition in the World 2021: Transforming Food Systems for Food Security, Improved Nutrition and Affordable Healthy Diets for All*, Rome: FAO.

Fortuna, G. (2022) "Timmermans cautions against relaxing Green deal goals for sake of food security," *EURACTIVE*. Available at: www.euractiv.com/section/agriculture-food/news/timmermans-cautions-against-relaxing-green-deal-goals-for-sake-of-food-security/.

Fortuna, G., Dahm, J., & Foote, N. (2022) "Agrifood brief: Ch-ch-ch-ch-changes," *EURACTIVE*. Available at: www.euractiv.com/section/agriculture-food/news/agrifood-brief-ch-ch-ch-ch-changes/.

Glauben, T., et al. (2022) "The war in Ukraine, agricultural trade and risks to global food security," *Intereconomics*, 57(3).

Gualtieri, T. & Soto, A. (2022) "Spain's switch to Latin American grain offsets Ukraine war impact," *Bloomberg*. Available at: www.bloomberg.com/news/newsletters/2022-04-28/supply-chain-latest-spain-shields-farming-industry-with-south-american-grains.

Jack, B. (2016) *Agriculture and EU Environmental Law*, New York: Routledge.

Jagtap, S., et al. (2022) "The Russia-Ukraine conflict: Its implications for the global food supply chains," *Food*, 11(14).

Joint Research Centre (2022) *Potash: Impact Assessment for Supply Security*, European Union.

Laborde, D., et al. (2020) "COVID-19 risks to global food security: Economic fallout and food supply chain disruptions require attention from policy-makers," *Science*, 369(6503).

Le Gouis, J., et al. (2020) "How changes in climate and agricultural practices influenced wheat production in Western Europe," *Journal of Cereal Science*, 93.
Lin, F., et al. (2023) "The impact of Russia-Ukraine conflict on global food security," *Global Food Security*, 36.
Lonardo, L. (2022) "Law: The use of force, EU sanctions, and assistance to Ukraine," in: *Russia's 2022 War Against Ukraine and the Foreign Policy Reaction of the EU*. Global Foreign Policy Studies, Cham: Palgrave Macmillan. Available at: https://doi.org/10.1007/978-3-031-18694-3_4.
Okou, C., Spray, J., & Unsal, D.F. (2022) *Staple Food Prices in Sub-Saharan Africa: An Empirical Assessment*, IMF Working Papers, WP/22/135.
Organization for Security and Cooperation in Europe [OSCE] (2022) *Report on Violations of International Humanitarian and Human Rights Law, War Crimes and Crimes against Humanity (1 April–25 June 2022)*. Available at: www.osce.org/odihr/522616.
Pe'er, G., et al. (2014) "EU agricultural reform fails on biodiversity," *Science*, 344 (6188).
Pe'er, G., et al. (2017) "Adding some green to the greening: Improving the EU's ecological focus areas for biodiversity and farmers," *Conservation Letters*, 10(5).
Pereira, P., et al. (2022) "The Russian-Ukrainian armed conflict will push back the sustainable development goals," *Geography and Sustainability*, 3, p. 3.
Pinedo, E. (2022) "Spain set to approve emergency U.S., Argentina corn buying," *Reuters*. Available at: www.reuters.com/business/spain-set-approve-emergency-us-argentina-corn-buying-2022-03-14/.
Rankin, J. (2014) "Russia responds to sanctions by banning western food imports," *The Guardian*. Available at: www.theguardian.com/world/2014/aug/07/russia-retaliates-western-sanctions-ban-food-imports.
Santeramo, F.G. & Kang, M. (2022) "Food security threats and policy responses in EU and Africa," *Sustainable Horizons*, 4.
Sirengo, E., Mahdi, M., & Kwarkye, S. (2022) "Food scarcity could contribute to unrest and instability," *ISS Peace and Security Council Report*, 2022(152).
Steffen, W., et al. (2015) "Planetary boundaries: guiding human development on a changing planet," *Science*, 347.
Stoate, C., et al. (2009) "Ecological impacts of early 21st century agricultural change in Europe: a review," *Journal of Environmental Management*, 91(1).
Swinnen, J. & McDermott, J. (2020) "Covid-19 and global food security," *EuroChoices*, 19(3).
United Nations [UN] (2022) *Global Impact of the War in Ukraine: Billions of People Face the Greatest Cost-of-Living Crises in A Generation*, UN Global Crises Response Group on Food, Energy, and Finance. Brief No. 2.
Webber, J., et al. (2022) "Russia's war on Ukraine: EU food policy implications," *Financial Times*. Available at: www.ft.com/content/622b9762-cfc1-4258-8cd4-bd8a7efd53fc.
World Food Programme (2022) *Sri Lanka: Rising Prices Reduce Access to Food for Millions*. Available at: www.wfp.org/stories/sri-lanka-rising-prices-reduce-access-food-millions.

Chapter 6

Zero Tolerance for Environmental Crimes

Going Scorched Earth to Save the World

Joseph M. Gabris

Introduction

The mass annihilation of ecosystems across the globe signals the arrival of an unprecedented era of existential concern, making environmental justice a categorical imperative the world must collectively pursue. The International Criminal Court (ICC) – which was established by the Rome Statute of the International Criminal Court (Rome Statute) in 1998 – is the first and only permanent international court with jurisdiction to prosecute individuals for certain crimes. The crimes recognized by the ICC, as identified in the Rome Statute, are genocide, crimes against humanity, war crimes, and crimes of aggression. What all these crimes have in common is that they rise to the level of conduct that is most harmful and offensive to humanity as a whole. But virtually nowhere does the Rome Statute, in its original form, mention anything about environmental harms. Because the purpose of the Rome Statute and the ICC is to hold individuals internationally liable for the most egregious offenses known to man, and because mass environmental destruction can easily be characterized as such an offense, it logically follows that the Rome Statute should be amended to include some form of environmental crime in order to advance global environmental justice and save our world from climatic calamity.

International Environmental Law

Global discipline on environmental protection and restoration is constantly evolving. However, most of this growing body of international law – which includes a compendium of declarations, treaties, and rules – are non-binding instruments, meaning, they possess no legal force (UNOHCHR 2013). Many binding international environmental agreements exist; however, they are only legally binding for countries that have formally ratified them, and even then, countries are virtually free to withdraw from such agreements, absolving them of their obligations under that framework (Rogoff 1980: 266–269). An example of this was the United States' disappointing withdrawal from the Paris Agreement in 2020 (Greene 2020).

DOI: 10.4324/9781003440871-7

Aside from the lack of force such international environmental agreements possess, current bodies of international environmental law also lack effective enforcement mechanisms. As international environmental scholar Hilary French states, "[i]nternational law has the force of moral suasion, but few real teeth" (French 1994). Indeed, these bodies of law carry little administrative force and struggle to go beyond simply raising global awareness of environmental issues. Due to the principle of sovereignty, nation-states cannot be forced to participate in these international environmental agreements (Cryer 2005). Additionally, few provisions exist across this body of law that advance public participation in environmental protection. Further still, existing international laws addressing environmental protection and justice tend to narrowly focus on organizations and entities, such as corporations, that cause environmental harm, while ignoring individual liability of the same harms (French 1994).

Where do we stand today on global ecological integrity within the context of law? With the development of international environmental law addressing regulation and prevention of environmental harm caused by humans, we are moving in the right direction, but this is simply not enough. With respect to the global environment, according to the United Nations (UN), "[c]limate [c]hange is the defining issue of our time and we are at a defining moment" (UN 2022). Climate change, along with ozone depletion and mass extinction of wildlife, is only one of the major concerns involving widespread environmental degradation (Hunter 2021). While certainly laws exist both domestically and internationally to address these environmental harms, much more is needed in order to ensure posterity. State and global governance on environmental protection is inadequate, and current methodologies have proven insufficient in fully addressing environmental harm across the world. Indeed, domestic approaches are largely limited by their jurisdiction, and international approaches are complex and underdeveloped. Likewise, both state and international approaches tend to shy away from criminalization and typically focus on economic sanctions and other civil remedies.

A (Quick) History of Today's World

The environmental movement that seemed to capture the world's attention all at once began in the 1960s. Prior to the 1960s, there was "little environmental awareness and only a few isolated international environmental regulatory initiatives" (Greene 2020). However, public opinion became more aware of the dangers threating the planet in the 1960s. Some of the more memorable events that ushered in this era of environmentalism were the publication of the book *Silent Spring* by Rachel Carson in 1962, which documented the negative effects of pesticides on birds and the environment (Carson 1962), and the release of the image known as *Earthrise*, taken by astronaut William Anders in 1968 during the Apollo 8 mission.

In 1972, the UN hosted the first UN Conference on the Human Environment, held in Stockholm, Sweden, which produced the first international document to recognize the right to a healthy environment, known as the Stockholm Declaration. The Stockholm Declaration recognizes this right through 26 principles, most of which have played a critical role in the subsequent development of international environmental law. At the same conference, the UN General Assembly created the UN Environment Programme (UNEP), which is the central body in charge of international environmental affairs today (Greene 2020).

A little over a decade later, the UN created the World Commission on Environment and Development, known as the Brundtland Commission, in 1983, which focused on the relationship between human development and the environment. Four years later, 46 nations signed the Montreal Protocol to combat the depletion of the ozone layer. The Montreal Protocol is often considered a prime "example of successful international cooperation." Today, there are 197 signatories to the Montreal Protocol, which include all UN members, the European Union, and the Holy See (Greene 2020).

Following the Cold War, the UN held a major conference in Rio de Janeiro in 1992 to establish international cooperation on developing environmental issues affecting the planet. This conference is formally known as the UN Conference on Environment and Development (UNCED), also known as the "Rio Summit," or "Earth Summit." During this conference, two major environmental conventions were presented: (1) the Convention on Biological Diversity and (2) the UN Framework Convention on Climate Change (UNFCCC). The Rio Declaration was also established, which reinforced the environmental protection goals and objectives of the Stockholm Declaration. The central theme of the Rio Summit was sustainable development in the face of destructive human activities responsible for environmental harm. The Rio Declaration advanced two important concepts in international environmental law: (1) the precautionary principle, which dictates that "[w]here there are threats of serious or irreversible damage, lack of full scientific certainty shall not be used as a reason for postponing cost-effective measures to prevent environmental degradation," and (2) principle 10, which recognizes the right to information, participation, and justice in environmental matters (UN 1992; Greene 2020).

Following the Rio Summit, most major economic treaties began incorporating some sort of environmental protection within its language (Greene 2020). Since 1995, the signatories of the UNFCC have met annually at the Conference of Parties (COP) to address progress in dealing with climate change. The Kyoto Protocol – which was signed in 1997 and ran from 2005 to 2020 – was the first international agreement to establish legally binding obligations for developed countries. The Kyoto Protocol was eventually superseded by the Paris Agreement, which was adopted in 2015 and entered into force in 2016. Under the Paris Agreement, signatory countries commit themselves to prevent the average temperature of the planet from rising by 2°C to the greatest extent possible, compared to pre-industrial levels. It is noteworthy to point out that the relationship between climate change and human rights was recognized in its Preamble. Thus, the Paris Agreement reinforces the importance of sustainable development by acknowledging the need for sustainable economic growth with a focus on the poor and respect for human rights, which was laid out in the Millennium Declaration of 2000 (Greene 2020).

Two years after the Millennium Declaration, representatives from 190 countries attended the UN World Summit on Sustainable Development in Johannesburg in 2002 to follow up on the commitments of the Rio Summit. During this conference, the Declaration of Sustainable Development was adopted, with a focus on poverty eradication through "public–private partnerships." Ten years later, in 2012, the UN organized the third Conference on Sustainable Development, commonly known as "Rio +20," which produced a non-binding document referred to as "The Future We Want," where nation-states renewed their commitment to sustainable development (Greene 2020). Of course, while this is certainly not an exhaustive discussion on the modern history of international environmental law, it serves our purposes here as a reliable foundation to build the remainder of our discussion on.

International Accountability for Environmental Destruction

The critical question, then, is this: How can we advance international progress in environmental justice beyond the scope of existing legal frameworks? Given the noticeable lack of criminal, and even individual, liability globally regarding environmental harm, accordingly, some system of international criminalization of environmental harms must be established.

Criminalization of environmental harm is certainly not a novel concept. In fact, "in terms of scale, environmental crime is the fourth-largest criminal category, surpassed only by drugs, counterfeiting and human trafficking," and this number is expected to grow by "five to seven percent annually" (Chalk & Siegel 2021: 100). Examples of environmental crimes recognized by various international bodies include pollution crimes, wildlife crimes, and illegal mining (Nurse 2020). The G8, Interpol, European Union, UN Environment Programme, and UN Interregional Crime and Justice Research Institute all recognize some form of environmental crime (Black & Hayman 2002). Most countries also recognize environmental crimes domestically.

A noteworthy aspect about environmental crimes is their profitability. This has led to organized crime syndicates, or "eco mafias," being formed around the world (Walters 2013). The UNEP estimated that these groups were making between 20 and 30 billion USD per year around the early 2010s, and the number is expected to be higher now (Skinnider 2011). Similarly, Interpol estimated global wildlife crime to be worth $10 billion USD per year for the same early-2010s time period (Skinnider 2011). Often, when an entity is left with a choice to do something environmentally conscious as opposed to environmentally destructive, it is more efficient and cheaper for them to pick the latter, which includes having to deal with any possible monetary penalties. Indeed, most environmental crimes are economically motivated to some extent – an important detail which some governments, such as Finland, have based their approach on addressing such crimes (Paukku 2022).

With respect to the international theater, few legal mechanisms exist to enforce international environmental crimes. In 2008, the European Commission published a binding instrument within the European Union (EU) for the protection of the environment through criminal law (European Commission 2022). However, due to lack of enforcement and prosecution of environmental crimes, the European Commission proposed a new "Environmental Crime Directive" in 2021 by attempting to

> improve and add precision, legal certainty and effectiveness to the EU's legal framework on environmental crime so that it sanctions the activities most damaging to the environment, provide . . . better definitions, allow[ing] for a greater choice of adequate sanctions and a more targeted approach to applying them.

The ICC is the first and only permanent international court with jurisdiction to prosecute individuals for crimes. Only four crimes are recognized by the ICC through the Rome Statute: (1) genocide, (2) crimes against humanity, (3) war crimes, and (4) crimes of aggression (Rome Statute 1998). However, in 2016, the ICC widened its prosecutorial remit to allow for the prosecution of environmental harms, particularly "crimes that are committed by means of, or that result in, inter alia, the destruction of the environment, the illegal exploitation of natural resources or the illegal dispossession of land" (ICC 2016). While the ICC did not formally extend its jurisdiction, the court now assesses existing offenses, particularly crimes against humanity, in a broader context than it did before. The ICC will also

> seek to cooperate and provide assistance to States, upon request, with respect to conduct which constitutes a serious crime under national law, such as the illegal exploitation of natural resources, arms trafficking, human trafficking, terrorism, financial crimes, land grabbing or the destruction of the environment.
>
> (ICC 2016)

However, the ICC's 2016 prosecutorial remit has proven to be severely ineffective. Indeed, the effort has many problematic caveats. For one, without the creation of a separate crime specific to environmental harm, the ICC "will have to pursue environmental crimes in furtherance of other crimes currently within the scope of the Rome Statute," thus giving it less-than-first priority (Patel 2016: 182). Furthermore, the avenue the ICC must take in order to prosecute environmental crimes is quite difficult.

The ICC must first determine if sufficient evidence exists to warrant an investigation, which involves reviewing the gravity of the offense and the ability of the Court to provide justice to victims (Patel 2016: 183). Additionally, the ICC "operates under an inherently anthropocentric view of harm," and thus, crimes are evaluated to the extent they negatively impact people – "actions polluting or degrading the environment alone will not fit the remit of the ICC" (Durney 2018). So what, then, is the next step toward effective international enforcement of liability for environmental harms? The answer, to put it simply, is to amend the Rome Statute to recognize a fifth international crime: ecocide.

A Subject Matter Approach – Ecocide

Ecocide can be broadly defined as "mass damage or destruction of ecosystems, committed with knowledge of the risks" (Stop Ecocide Foundation 2022). Plant biologist and bioethicist Arthur W. Galston originally coined the term "ecocide" in a 1970 conference paper as "the willful and permanent destruction of the environment in which people can live in a manner of their own choosing" (Galston, cited in Zierler, 2011). This term was originally used to categorize the type of widespread harm created by the US use of Agent Orange during the Vietnam War – Agent Orange being a deadly herbicide that is extremely harmful to the environment and can be fatal to humans (Zierler 2011). An important aspect of ecocide is that intent is not necessarily relevant (Westing 1974). Indeed, ecocide is often the result of war, where destruction of the environment is not necessarily the underlying goal.

Today, at least nine countries have developed ecocide laws in their national jurisdictions (Ecocide Law 2022). The next step forward is for the ICC to develop laws recognizing ecocide as an international crime.

A Way Forward: Amending the Rome Statute to Include Ecocide as the Fifth Core Crime

The ICC, which was established by the Rome Statute, is the fundamental body that oversees international criminal law. Recognizing ecocide as an international crime under the Rome Statute would give much greater authority to the ICC to investigate and prosecute environmental crimes, thus advancing environmental justice around the globe. But what purpose does the ICC serve, and to what extent does the ICC serve that purpose today?

In 1998, a conference of 160 nation-states adopted a treaty, known as the Rome Statute of the International Criminal Court, establishing the ICC. The

ICC is a "permanent international court established to investigate, prosecute and try individuals accused of committing the most serious crimes of concern to the international community as a whole" (ICC 2022). The primary goal of the ICC "is to help put an end to impunity for the perpetrators of such crimes and contribute to their prevention" (ICC 2022).

The Rome Statute, which entered into force on 1 July 2002, established the four core international crimes previously mentioned. These crimes are not subject to any statute of limitations. Under the Rome Statute, the ICC can only investigate and prosecute these four crimes in situations where states are unable or unwilling to do so themselves, with the jurisdiction of the ICC being complementary to the jurisdiction of domestic courts. Additionally, the ICC only has jurisdiction over such crimes if they are committed within the territory of a party state or by a national of a party state – but an exception exists for crimes authorized by the UN Security Council (ICC 1998).

It is noteworthy to point out that, during the drafting of the Rome Statute, the crime of ecocide was considered to be included as another core international crime, in addition to the existing four. The crime of ecocide was considered by the UN International Law Commission as a "crime against peace" in the initial drafts of the Code of Crimes Against the Peace and Security of Mankind (which eventually developed into the Rome Statute); however, it was ultimately excluded from the final treaty, and reasons as to why it was not included in the Rome Statute remain unclear (Gupta 2021). Even so, one reference to environmental crimes exists in the 1989 version of the Rome Statute, under Article 8(2)(b)(iv) (prior to the 2016 remit), which relates to war crimes (Bilotta 2019). However, the harms identified under Article 8(2)(b)(iv) fall short of ecocide and are largely inadequate in addressing environmental harms, as such harms must be within the context of international armed conflict. Furthermore, the wording of Article 8(2)(b)(iv) is severely limiting – the environmental damage resulting from war must be "[w]idespread, long-term, ***and*** severe," thus creating a very high threshold to satisfy (Bilotta 2019).

Even in light of the ICC's 2016 prosecutorial remit, the Rome Statute as it exists today is largely ineffective in addressing international environmental crimes. Again, the 2016 remit did nothing to expand the ICC's jurisdiction and, to this day, has done virtually nothing in terms of advancing environmental justice globally. For instance, in 2021, the non-governmental organization All Rise formally communicated to the ICC, requesting an investigation into then–Brazilian President Bolsonaro's role in crimes against humanity stemming from activities resulting in the destruction of the Amazon rainforest (All Rise 2021). However, the inquiry is still pending at the time of this writing, and concern exists as to whether the ICC will actually open up an investigation. Additionally, prosecuting Bolsonaro's destructive activities as a crime against humanity will likely prove difficult, largely because crimes against humanity take an anthropocentric approach, which is focused on the element of human harm. Marcio Astrini, Executive Secretary of the Brazilian

Climate Observatory, who supported the investigation, stated that the destruction of the Amazon rainforest would almost certainly continue under Bolsonaro's regime, and that "the complaint is unlikely to change how Bolsonaro governs" (Astrini, cited in Teixeira 2021). However, if ecocide were included in the Rome Statute, many of these problems would be resolved or at least mitigated.

A New Instrument: Blueprint for an Amended Rome Statute

Amending the Rome Statute requires more than mere general interest; it requires popular demand. While any state party to the Rome Statute can *propose* an amendment, the amendment can only be adopted by a two-thirds majority vote, and amendments (except amendments to the list of existing crimes, that is, Articles 5, 6, 7, or 8) will only be entered into force upon ratification by seven-eighths of the state parties (ICC 1998).

In 2010, Polly Higgins, a Scottish barrister and the founder of the Stop Ecocide Foundation (SEF), submitted a draft law of ecocide to the UN Law Commission (Higgins 2016: 61–92). The proposed amendment to the Rome Statute defined the crime of *ecocide* as (Higgins 2016)

(1) acts or omissions committed in times of peace or conflict by any senior person within the course of State, corporate or any other entity's activity which cause, contribute to, or may be expected to cause or contribute to serious ecological, climate or cultural loss or damage to or destruction of ecosystem(s) of a given territory(ies), such that peaceful enjoyment by the inhabitants has been or will be severely diminished.
(2) To establish seriousness, impact(s) must be widespread, long-term, *OR* severe.
(3) [Definitions] [f]or the purposes of paragraph 1 [exist here].

The elements of ecocide as proposed by Higgins include: (1) "[t]he perpetrator's acts or omissions caused, contributed to, or may be expected to cause or contribute to [ecocide];" (2) "[t]he perpetrator's activity has or will severely diminish peaceful enjoyment by the inhabitants;" (3) "[t]he perpetrator had knowledge or ought to have had knowledge of the likelihood of ecological, climate or cultural harm;" and (4) "[t]he perpetrator was a senior person within the course of State, corporate or any other entity's activity in times of peace or conflict" (Higgins 2016).

Unfortunately, Higgins's proposal was rejected, but the fight for amending the Rome Statute to include ecocide has not stopped there. In 2021, the Independent Expert Panel for the Legal Definition of Ecocide proposed a 165-word draft definition of ecocide to be included in the Rome Statute (Stop Ecocide Foundation 2021). The draft defines *ecocide* as "unlawful or wanton acts committed with knowledge that there is a substantial likelihood of severe and either widespread or long-term damage to the environment being caused by those acts" (Stop Ecocide Foundation 2021).

If added to the Rome Statute, ecocide, unlike the four existing crimes, would be the only crime un which human harm is not a prerequisite for prosecution. Jojo Mehta, Chair and Cofounder of SEF, articulates that

> elements of human harm . . . can be included in [the definition of ecocide], but [ecocide] also extends to damage, per se, to ecosystems . . . [s]o effectively you're looking at something that has, at least in part, potential to be a crime against nature, not just a crime against people.
>
> (Mehta, cited in Fischels 2021)

Of course, several other compelling reasons to include ecocide under the Rome Statute exist as well.

Repercussions of a Percussive Instrument: Effect of Adding Ecocide to the Rome Statute

Aside from creating individual liability for environmental harm across the globe, criminalizing wholesale environmental destruction by adding ecocide to the list of core international crimes recognized by the Rome Statute would significantly heighten the seriousness and scope of environmental destruction across the globe. Indeed, criminalization is an effective deterrent, particularly where it threatens individuals with punitive consequences, where such people would otherwise be okay with paying a fine for engaging in illegal activity (Cohen 2000). In many cases, violators have found it more cost-effective to continue to pollute more than the law allows and simply pay any resulting fines and penalties.

Additionally, acknowledging ecocide as its own, separate crime under the Rome Statute would promote environmental justice from being a secondary consideration into a priority. Accountability for environmental destruction would go beyond purely wartime conflict, and the ICC will no longer have to pursue environmental crimes in furtherance of the four existing crimes recognized by the Rome Statute, thus making prosecution of such crimes more efficient. Likewise, environmental crimes would no longer require proof of human harm, thus moving away from the anthropocentric model supporting the current four international crimes.

Nevertheless, perhaps the most compelling reason that ecocide should be added to the Rome Statute is to help mitigate the progression of climate change (Bilotta 2019). Including ecocide in the Rome Statute would emphasize the types of measures required to combat climate change. Likewise, the addition would serve as a preventative measure against the other crimes acknowledged by the ICC (Bilotta 2019).

Of course, adding ecocide to the Rome Statute is not the end-all-be-all for the fight against climate change and other forms of environmental harm. For one, the ICC has limited jurisdiction in that it only regulates nations that have subscribed to the Rome Statute. Considering that the United States, Russia, China, and India – the top four global contributors to climate change, accounting for more than half of the world's CO_2 emissions – are not signatories to the Rome Statute, the Rome Statute's reach is currently very limited. Furthermore, the ICC can only prosecute individuals – it cannot hold countries or organizations criminally liable. Even so, all this is certainly not to say that adding ecocide to the Rome Statute is a fruitless endeavor; rather, these problems only serve to further highlight the complexity of addressing environmental harm on an international level, and that the fight for environmental justice will remain gradual.

Conclusions

More international effort is needed in furthering environmental justice. While the ICC may do an adequate job of investigating and prosecuting the four crimes currently recognized by the Rome Statute, environmental crimes are not recognized, and adding ecocide to the list would be highly beneficial to the protection of local, transnational, and global environments. By adding ecocide to the Rome Statute, the seriousness and scope of environmental harms are raised. Furthermore, criminalizing environmental harm would likely serve as a more effective deterrent than civil penalties. Likewise, global response to the international criminalization of environmental destruction will likely result in greater international cooperation to address these serious environmental harms, such as climate change. To be precise, while adding ecocide to the Rome Statute will not resolve or reverse the extent of environmental damage that exists today, it is certainly the next step in the right direction.

Reference List

All Rise (2021) *Communication under Article 15 of the Rome Statute of the International Criminal Court Regarding the Commission of Crimes Against Humanity Against Environmental Dependents and Defenders in the Brazilian Legal Amazon from January 2019 to Present, Perpetrated by Brazilian President Jair Messias Bolsonaro and Certain Former and Current Principal Actors of His Administration.* Available at: http://climatecasechart.com/climate-change-litigation/wp-content/uploads/sites/16/non-us-case-documents/2021/20211012_14633_na.pdf.

Bilotta, A. (2019) "Should the Rome Statute include the crime of ecocide?" *E-International Relations*. Available at: www.e-ir.info/2019/08/28/should-the-rome-statute-include-the-crime-of-ecocide/.

Black, D. & Hayman, G. (2002) "International environmental crime: The nature and control of environmental black markets," *Royal Institute of International Affairs*. Available at: https://www.chathamhouse.org/sites/default/files/public/Research/Energy,%20Environment%20and%20Development/environmental_crime_background_paper.pdf.

Carson, R. (1962) *Silent Spring*, 1st ed., Boston: Houghton Mifflin.

Chalk, R. & Siegel, P. (2021) "Ecocide and the threat of trade embargos and economic sanctions," *Business Law International*, 22(1), pp. 99–110.
Cohen, M. (2000) "Empirical research on the deterrent effect of environmental monitoring and enforcement," *The Environmental Law Reporter*, 30, pp. 10245–10252.
Cryer, R. (2005) "International criminal law vs state sovereignty: Another round?" *European Journal of International Law*, 16(5), pp. 979–1000.
Durney, J. (2018) "Crafting a standard: Environmental crimes as crimes against humanity under the international criminal court," *Hastings Environmental Law Journal*, 24(2).
Ecocide Law (2022) *Ecocide Law in National Jurisdictions*. Available at: https://ecocidelaw.com/existing-ecocide-laws/.
European Commission (2021) *Questions and Answers on the Revised EU Directive on Environmental Crime*. Available at: https://ec.europa.eu/commission/presscorner/detail/en/qanda_21_6745.
European Commission (2022) *Combatting Environmental Crime*. Available at: https://ec.europa.eu/environment/legal/crime/.
Fischels, J. (2021) "How 165 words could make mass environmental destruction an international crime," *National Public Radio*. Available at: www.npr.org/2021/06/27/1010402568/ecocide-environment-destruction-international-crime-criminal-court.
French, H. (1994) "Strengthening international governance," *Journal of Environment & Development*, 3(1).
Greene, F. (2020) "International environmental law: History and milestones," *Inter-american Association for Environmental Defense*. Available at: https://aida-americas.org/en/blog/international-environmental-law-history-and-milestones.
Gupta, S. (2021) "The proposed definition of 'ecocide': An attempt to constitute fifth international crime?" *Jurist*. Available at: www.jurist.org/commentary/2021/07/sarthak-gupta-ecocide-fifth-international-crime/.
Higgins, P. (2016) *Eradicating Ecocide: Law and Governance to Prevent the Destruction of Our Planet*, 2nd ed., London: Shepheard-Walwyn.
Hunter, D. (2021) "International environmental law," *American Bar Association*. Available at: www.americanbar.org/groups/publicf_education/publications/insights-on-law-and-society/volume-19/insights-vol-19-issue-1/international-environmental-law/#:~:text=International%20environmental%20law%20is%20the,and%20mass%20extinction%20of%20wildlife.
International Criminal Court [ICC] (1998) *Rome Statute*.
International Criminal Court [ICC] (2016) *Policy paper on case selection and prioritization*, Office of the Prosecutor. Available at: www.icc-cpi.int/itemsDocuments/20160915_OTPPolicy_Case-Selection_Eng.pdf.
International Criminal Court [ICC] (2022) *What is the ICC and what does it do?* Available at: www.icc-cpi.int/sites/default/files/iccdocs/get-involved/lesson1.pdf.
Nurse, A. (2020) "Contemporary perspectives on environmental enforcement," *International Journal of Offender Therapy and Comparative Criminology*, 66(4), pp. 327–344.
Patel, P. (2016) "Expanding past genocide, crimes against humanity, and war crimes: Can an ICC policy paper expand the court's mandate to prosecuting environmental crimes?," *Loyola University Chicago International Law Review*, 14(2).
Paukku, E. (2022) "Environmental crime in a welfare state – a case study on the prosecution of environmental crimes in Finland 2015–2020," *International Journal of Comparative and Applied Criminal Justice*. Available at: www.tandfonline.com/doi/full/10.1080/01924036.2022.2087701.

Rogoff, M. (1980) "The international legal obligations of signatories to an unratified treaty," *Maine Law Review*, 32. Available at: https://mainelaw.maine.edu/faculty/wp-content/uploads/sites/4/rogoff-mlr-32.pdf.

Skinnider, E. (2011) "Victims of environmental crime – mapping the issues," *The International Centre for Criminal Law Reform and Criminal Justice Policy*. Available at: https://globalinitiative.net/wp-content/uploads/2017/12/ICCLR-Victims-of-Environmental-Crime.pdf.

Stop Ecocide Foundation (2021) *Independent Expert Panel for the Legal Definition of Ecocide: Commentary and Core Text*. Available at: https://static1.squarespace.com/static/5ca2608ab914493c64ef1f6d/t/60d7479cf8e7e5461534dd07/1624721314430/SE+Foundation+Commentary+and+core+text+revised+%281%29.pdf.

Stop Ecocide Foundation (2022) *FAQs – Ecocide and the Law: What Is Ecocide?* Available at: www.stopecocide.earth/faqs-ecocide-the-law.

Teixeira, F. (2021) "Brazil's Bolsonaro accused of crimes against humanity as Amazon felled," *Thomson Reuters Foundation*. Available at: www.reuters.com/article/brazil-deforestation-lawsuit/brazils-bolsonaro-accused-of-crimes-against-humanity-as-amazon-felled-idUSL8N2R84SL.

United Nations [UN] (1992) *Report of the United Nations Conference on Environment and Development: Rio Declaration on Environment and Development*. Available at: www.un.org/en/development/desa/population/migration/generalassembly/docs/globalcompact/A_CONF.151_26_Vol.I_Declaration.pdf.

United Nations [UN] (2022) *Climate Change*. Available at: www.un.org/en/global-issues/climate-change#:~:text=Climate%20Change%20is%20the%20defining,scope%20and%20unprecedented%20in%20scale.

UN Office of the High Commissioner for Human Rights [UNOHCHR] (2013) *Mapping Human Rights Obligations Relating to the Enjoyment of a Safe, Clean, Healthy and Sustainable Environment: Individual Report on Non-Binding International Environmental Instruments*. Available at: https://www.ohchr.org/sites/default/files/Documents/Issues/Environment/Mappingreport/10.Non-binding-environmental-instruments-5Mar.docx.

Walters, R. (2013) "Eco mafia and environmental crime," *Crime Justice and Social Democracy*, pp. 281–294.

Westing, A. (1974) "Arms control and the environment: proscription of ecocide," *Bulletin of the Atomic Scientists*, 30(1), pp. 24–27.

Zierler, D. (2011) *The Invention of Ecocide: Agent Orange, Vietnam, and the Scientists Who Changed the Way We Think about the Environment*, 1st ed., Athens: University of Georgia Press.

Further Readings

Broswimmer, F. (2002) *Ecocide: A Short History of the Mass Extinction of Species*, London: Pluto Press.

Gillett, M. (2022) *Studies on International Courts and Tribunals: Prosecuting Environmental Harm Before the International Criminal Court*, Cambridge: Cambridge University Press.

Kreike, E. (2021) *Scorched Earth: Environmental Warfare as a Crime Against Humanity and Nature*, Princeton: Princeton University Press.

Section 2

Ecological Integrity and Climate Security

Chapter 7

The Political Morality of Adaptation

Recognition and Relationality in a Climate-Changed World

Bruce Jennings

Introduction

In *Adaptation to Climate Change*, geographer Mark Pelling sees a dual opportunity in the ways we reorganize power and transform law in order to adapt to serious disruptions brought about by climate change. The first opportunity is to adapt effectively so that many of our past values and ways of life can be preserved (Diamond 1988). The second is to resist the hegemony of neoliberalism when we bring about change in the globalized political economy. However, these opportunities will be difficult to seize.

During the past several decades, the deregulatory and anthropocentric orientation of neoliberal global capitalism has facilitated climate change denial and popularized a false and dangerous belief that there is no alternative open to contemporary societies and nation-states but to continue the extraction and burning of energy-rich fossil carbon, thereby exacerbating the worsening of global warming and other aspects of climate change. In virtually every country, those at or near the bottom of the pyramid of their society's political economy of wealth and power – those whose human rights and material interests are being undermined – throw up their hands in the face of climate mitigation and adaptation. Unfortunately, as Pelling points out, preserving just and progressive values and resisting the ideology of neoliberalism do not always accompany adaptive thinking and practice:

> Adaptation has been framed in terms of identifying what is to be preserved and what is expendable, rather than what can be reformed or gained. Dominant development discourses put the economy as first to be preserved, above cultural flourishing or ecological health. . . . Climate change is no longer an external threat to be managed "out there," but is an intimate element of human history – both the outcome and driver of development decisions for individuals, organizations, and governments. This requires a closer look at social relations and practices, even values, as sites for adaptation.
>
> (Pelling 2011: 1, 6)

DOI: 10.4324/9781003440871-9

Here, I want to build on the notion that relationality, social practices, and cultural norms and values are sites of climate adaptation. I believe this is especially the case when we focus on practices or patterns of interaction and interdependence of right relationality among persons and groups that can flow from cognate practices of right recognition. More specifically, my argument in this chapter is that in place of neoliberal market solutions, the emerging encounter of humanity with climate change adaptation may lead to a recrudescence of more localized economic institutions and a more participatory and deliberative mode of democratic governance (Orlov 2011; Howe 2023). This will most likely take place amid massive population relocations and the breakdown of agricultural production and freshwater availability in many regions. Most likely, the current operations of the financial and credit systems will be reduced and more regionalized in their functioning. Social precarity will fuel political unrest. The pressures placed on systems of social order and public health will be unprecedented. No doubt there will be a martial response as sovereignty is challenged in nation-states. But there will also need to be an ethical and a social psychological response. Speech acts and dialogic relationships will have to be deployed in the service of normative rebirth. Vision and conscience will have to be renewed; traditions of hospitality and decency will have to be tapped. Mistakes made in the past must be learned from and atoned for.

Against this backdrop, careful philosophical and phenomenological analysis of recognitive and relational practices is not merely an academic luxury but a practical political necessity. This is because we shall need to understand how practices of right recognition and right relationality can serve to support what we can appropriately refer to as a new, ecological "political morality" – to accompany a new ecological political economy – for flourishing and resiliency – for integral ecosystems of life – on a climate-changed planet (Jennings 2016).

The overarching theoretical schema of this chapter is that practices of right recognition will foster transformations in patterns of social and moral relationality in a society, which in turn will reinforce new modes of intentionality and agency, both at the collective and the individual levels. Let me explicate this terminology and conceptual framework more fully. According to the usage I employ, *recognition* is not primarily a passive, receptive mode of perception. Instead, it is a mode of active normative social and linguistic agency in its own right; it affords an open-ended occasion for recognizing persons, places, and things (Ricoeur 2005). Recognitions are formulated and articulated by subjects and enter a circuit of reciprocal exchange with others, thereby cocreating a meaningful connection and opening potential pathways for mutual cooperation and interaction (Jennings 2018). This is not limited solely to human communication but takes place among other species and in other realms of life, yet human language is the most fertile ground for recognitions of great plasticity and complexity that we know (MacIntyre 2007: 187–203; Bourdieu 1977, 1990).

Moreover, when they take on organizational form as widespread, rule-governed normative practices, right recognitions – such as solidarity, care, respect, concern, and accommodation, which is a recognitional practice of particular pertinence to this chapter – can orient individual and group agency by changing both cognitive normative commitments and also affective motivational contentments.

Part of what I am claiming is that active valuation through right recognition is a prerequisite for right relationality, for if we cannot perceive and affirm the moral standing of the Other, the inherent moral standing and worth of that person, it is less likely that ongoing right relationality among selves and others – relating practices of reciprocity, respect, trust, fidelity, and truthfulness and loyalty – will be sustainable, because they will never grow beyond situational advantage, what Kant referred to as humanity's "unsocial sociability" (Sonenscher 2023).

Similarly, the practice of right relationality can also facilitate right recognition that often needs prompting by a prior relational experience with the other person. So the dynamic between recognition and relationality is vital to the political morality of adaptive climate governance that I envision.

However, it is important not to rest content with only the procedural or constructivist dimension of an ecological political morality. There is also a substantive ontological aspect standing behind recognition and relationality that anchors these norms. This is the material and evolutionary basic and the fundamental norm, the *grundnorm*, of international environmental law and of the diverse embodiments of political morality in local politics and law. There are things that are real, and they possess inherent or intrinsic, not merely instrumental, value. Also, one might say that the grounding norm is not itself merely an inherent value, one among many. Instead, the grounding norm is best understood as a meta-value, the value of there being value at all. The grounding norm opens the possibility of any thing or creature possessing intrinsic value, which, so to say, is a reflection of such value from this source. This offers an account of the phenomenology of the lived experience of the normative in the human conscience and in the phenomenon of life as such. It is something so deeply embedded in the nature of life on Earth that without it, right recognition and right relationality would be emptied of moral worth so that, in time, the affirming and attentive practices of recognition and relationality, such as solidarity and care, would collapse (Jennings 2018).

Ecological Integrity as Grounding Norm

For many decades and in many international agreements and documents, the concept of ecological integrity has been used to denote such fundamental normative grounding. Klaus Bosselmann notes that a concept like sustainability is incomplete without the specification of what condition, properties, or functions are to be sustained. He argues that it is ontological integrity that supplies the content for sustainability practices. Bosselmann writes:

> The core idea is the preservation of the integrity of Earth's ecological system, an idea that has been expressed in a number of key international agreements albeit in a somewhat hidden manner. . . . Earth system integrity includes not just "natural" systems, but also "human" systems such as social or economic or financial systems. All systems possess a certain degree of integrity, i.e., the ability to maintain the functions of self organization, while also responding to external changes. In this way, it is possible to perceive integrity as an overarching policy concern for ecological, energy, financial, economic, and political systems.
>
> (2023)

In another sense, the notion of integrity (and resilience) does not simply denote the capability of systems to maintain their form and function in the face of outside perturbation. Integrity is integral to the creation of form. The account of creation that has come down to us in the book of Genesis speaks of a primal existence that is without form and void. The creative act achieved order in place of chaos, turning form into formlessness, and plenitude into emptiness. This instilled integrity into the world.

Now, amidst order, form, and plenitude that have been given, it falls to us to recognize that we have disrupted and imperiled the integrity, the inherent capability of all creatures to maintain their own form and function that is the essence of their being alive. When we adapt, we are not so much protecting ourselves or bringing living beings to heel to do our bidding as we are trying to recapture the relationality, the fit, between human and natural being, from which we have become estranged.

Ecological Political Morality

This opens a space for a new deliberative and enactive democratic citizenship (Di Paolo et al. 2018; Tronto 2013). Even as enactive citizenship is fostered by pre-existing institutions and cultural norms, it can also prompt new, transformative social institutions and the practices of recognition they foster. In this way, despite great adversity and, no doubt, recurrent emergency and disaster, our climate-adapted future does provide the opportunity and hope for practices of right relationality based on practices of greater equality, justice, empowerment, and new kinds of liberty and security (Orr et al. 2019). This is a vision of cognitive and agentic transformation that could be realized in many, if not all, societies today. It is one that should be further theorized in ecological law, political philosophy, and governance.

Can that theoretical potential be actualized in practical reasoning and in both collective action and individual conduct (Brandom 2019)? Capabilities are achieved when human potentialities are transformed into realized functionings by taking advantage of opportunities and resources that the social and natural environment affords. To sustain ecosystem integrity is also to sustain natural affordances for human flourishing. Within living memory, ecosystemic and social environments have provided people with readily accessible affordances, at least those residing in the Global North. Much of the political morality of the liberal democratic tradition (which incorporates neoliberalism but is much broader than that perspective) has been designed to function well in that climatic context. However, those historic patterns of accessibility are going to change, and indeed, that process has already begun (IPCC 2022). Therefore, the continuing viability of the received political morality is in serious question (Jennings 2015).

As philosopher Ronald Dworkin (2011) explicates the notion, a political morality addresses what the ends and means of governance should be, what constitutes a good social order to strive toward, how power should be distributed in society, and how it should be used in the regulation of social, economic, and personal conduct. A political morality also has ontological presuppositions. An ecological political morality will clarify how and why a new page must be turned in the long story of relationships between humans and nature. We won't be able to "face the planetary" unless we get many things right, not the least of which is a proper account of normativity (Connolly 2017; Chakrabarty 2021).

Built into the idea of recognition is a deep logic of hospitality – embracing difference and inviting inclusion. As such, it is a natural fit with the logic of adaptation and especially "accommodation" itself – making room for, taking in, making things fit, offering a place to dwell in the world. Recognition has to do with a just and morally right perception of what each of us owes to other human beings and the responsibility humanity must take for the larger living world. Right recognition attests to the moral standing of others as beings of inherent, not simply instrumental, worth (Honneth 2012). This has a direct bearing on the nature and goals of adaptive climate preparedness and planning, and indeed on adaptive climate governance generally.

Understood in this way, practices of recognition are dynamic and subject to change but nonetheless exhibit a certain endurance and continuity in a society, a culture, and in the personality and self-esteem of individuals. They generate action-guiding reasons and motivations. They affect authority, power, distribution, and exchange, and they can contribute to the reinterpretation of social rules, roles, and relationships, out of which large-scale social change emerges. As human beings aspire to be recognized, so do they strive to recognize.

One other general point should be mentioned. Adaptation is self-reflective and asks what constitutes its own effective and responsible practice. What is the best way to understand this aspect of the discourses and recognitions inherent in climate adaptation? The answer I suggest is fourfold. First, as a practice, adaptation involves right recognition of others, human and non-human alike. Second, adaptation involves right relationality in living mutually interdependent lives with others. Third, there are developmental aspects of moral and civic learning that come from engaging in adaptation as a practice of recognition. We learn to adapt, and we adapt to learn. Fourth, right recognition and civic learning can be institutionalized in an activity variously known as climate adaptation planning, emergency preparedness planning, or disaster planning. Community-based planning is one of the primary institutional settings within which the recognitional practice of adaptation can be pursued.

I am tempted to say that planning as such in its many forms – including urban planning, transportation planning, environmental planning, and the like – can be understood as a distinctive practice of recognition (see Schön 1984; Minteer 2006; Fischer 2009; and Forester 1999). For the purposes of this discussion, however, it is sufficient to treat planning primarily as an institutional setting within which the recognitional practice of adaptation or taking a stance of accommodation can be pursued. Of course, adapting to climate change is a societal activity in the broadest, planetary sense. It is not limited to settings and functions of governance alone but permeates the social, cultural, and psychological fabric of societies in which the norms of an egalitarian, pluralistic, inclusive, and participatory mutuality prevail.

No existing societies today fully express a political morality of right recognition and right relationality and its attendant norms of political morality. But when such norms have at least some, even if not predominant, social expression, then practices of recognition can move toward cultural and social changes that will be judged right and responsible by reasonable, well-informed persons of civic goodwill.

Adaptation as a Practice of Recognition

I turn now to consider climate change adaptation as a concept in order to lay the groundwork for understanding and directing it later as a practice. When we undertake to adapt, what is most fundamental to what we are doing?

All social practices, skills, habits, meanings, and in particular, practices of right recognition allow for the other to be encountered and made visible. We need to rearticulate and reinforce human agency in and through relationships of recognition, mutuality, respect and concern, need, and vulnerability. When such values and concerns fade from the moral fabric of our lives, we lose something, some part of our capability to see, to know, to feel. In this relational dynamic, the watchword is the *variety*, not the *uniformity*, of intentional and evaluative agency and interdependent positions and projects. A common element in recognitive practices is the establishment of cognitive receptivity and reciprocity among two or more conscious and aware subjects who engage in various forms of encounter in various settings

Adaptation may be said to be a dynamic of recognition that develops and transforms the kind and depth of recognition and the types and consequences of relationships among persons over time. This may hold a key to developing a moral pluralism without an attendant moral relativism or to affirm a social individuality without the accompanying atomistic ideology of individualism.

Adaptation as a Practice of Accommodative Relationality

Let me turn now to a dynamic of developing moral acuity that I interpret as inherent in the stance of accommodation with practices of adaptation.[1] My working hypothesis is that the trajectory from a sense of displacement to an affirmation of symbiosis is the key response to accommodation and is its moral goal. If this is plausible conceptually, socially, and psychologically, then we can devise some practical aspects of civic deliberation and preparedness to ease this trajectory and to avoid becoming hung up or fixated at earlier stages of the process.

In its everyday use, the word *adaptation* generally carries a passive or reactive connotation. To adapt is to respond to (new) demands (an incursion or perturbation) imposed by one's environment, not with the goal of eliminating or expelling those demands, but to achieve a state of equilibrium or *modus vivendi* with them.

I seek to move away from this analytic of passivity. I propose that what most basically constitutes the recognitive and relational posture of adaptation is *accommodation*. This concept contains the entire circuit beginning with right recognition (how we see the other), leading to an adjustment or affirmation of relationality (how we treat the other), and then culminating in intentional and transformative agency (how we live with the other inclusively together in mutually affirming interdependence).

When one adopts a posture and stance of accommodation toward others, one makes room for them, often by moving over or relocating oneself on behalf of the other. Adaptation is the accommodation of a changed relationality, either socially with other people or in relation to not human life in the natural world. Accommodation also partakes of the gestures and speech acts of hospitality. To engage with another hospitably is to invite someone into a space of inclusion, thereby allowing them to vacate a space of exclusion – and providing refuge from the precarity and anxiety of exclusion in the process. To accommodate is to adjust your own physical, preferred, or relational surround to ease the new "implacement" of another (Casey 1997). This is not a passive process, even when coercively imposed, and certainly not when affirmatively embraced. This new implacement involves making the needs and

1 The following discussion draws on a more detailed treatment presented in Jennings (2022).

characteristics of the other as subject to moral consideration; it provides the embodied presence of the other with moral "considerability" (Goodpaster 1978). One of the most difficult challenges in a climate-changed world will be to construct a practical, new social imaginary in which the ecological forces at work changing the conditions of life on Earth are viewed as occasions for enhancement of human and natural flourishing and not simply instances of its demise (Weintrobe 2021).

In these senses, the metaphors of adaptation and accommodation point us in the direction of embodiment and presence. Having been invited into a new placement, a new abode, the newcomers must respect certain explicit and tacit limits on their behavioral and emotional repertoire, but in return, they are placed under the canopy of rights, protections, and equalities of mutual respect and concern.

Here I offer one heuristic description of a pathway of accommodation in response to climate change:

- *Accommodation as loss and displacement*. In this phase, accommodation is undertaken while perceived as an intrusion to be endured and begrudged. Here, making room for is interpreted and experienced as being put upon. All parties may be conscious of this, but they tolerate it and perhaps view it as the best arrangement that can be struck under the circumstances. It is a thin accommodation, an expedient only, hopefully temporary. In some cases, it can be a temporary phase that later develops into a fuller and more complex type of relationality.
- *Accommodation as a gesture of authentic self-realization*. Even deeper kinds of making room for can fall prey to a merely instrumental recognition of the other who is being accommodated or welcomed into place. The psychological and relational dynamics are subtle. The emotional tone may be kind and solicitous, but if the accommodative situation is tacitly seen as a means to an end of gaining something useful – such as the favorable perceptions of third parties or enhanced self-esteem – then the full range of the moral possibilities of acceptance and accommodation will go unfulfilled.
- *Accommodation as power*. Here, being confronted by a change in one's lifeworld, or by a new claim on one's time and attention, may be welcomed as an effective way to eventually capture control over the person or situation that is being adapted to. This phase of accommodative practice must be on guard against the fallacy of seeing adaptation as simply a sorting process of retaining some aspects of social and ecological life and abandoning others. The narratives of adaptation are more ambivalent, and the possibilities are more radical and transformative, than this kind of border and boundary construction notion of adaptive governance suggests.
- *Accommodation as welcoming, embracing hospitality*. This phase of attitude and perception stands at the other end of the continuum from

enduring and begrudging. It is an unreserved welcoming accepted and embraced primarily for its own sake. Hospitality and welcoming, making room for, here are viewed as intrinsically valuable, regardless of what other kinds of instrumental value the posture and stance of inclusion may have in the eyes of one or more of the parties involved.

- *Accommodation as symbiosis.* This is the final phase of accommodation as a recognitive practice that I will mention here (additional phases may be identified as the analysis is fine-tuned). A new phase of the recognitive relationship between self and other is reached when the interaction is phenomenologically and hermeneutically revealed as a form of symbiotic living, against the backdrop of a symbiotic ontology of being alive. This "re-grounds" the interpretation of adaptation as a practice of recognition, places it on a different philosophical footing. The other is now understood as an authentic symbiont (companion), and the accommodative relationship is subsumed (for the first time, perhaps) into a fundamental way of being in the world, most especially for those who feel the work of adaptation to climate change to be a special vocation or calling. This does not entail a loss of embodied boundaries or loss of a distinctive sense of self, but it does involve a kind of kinship recognition. Individuals are not atoms but kindred selves, not only when blood relatives, but also as closely tied members of the social and natural fabric of life (Van Horn et al. 2021). That fabric is the basic thing that the parties to an adaptation or accommodation need to have in common.

I regard symbiotic affirmation as a world-historical opportunity. The adaptation to and accommodation of symbiotic being and its constraints are a covenantal moment – a moral and scientific transformative promise, not only as a theological one. It then becomes a duty to publicize and share it for the enlightenment of others. This is one of the things that an adaptation plan does. It is the performative function of a plan as an act of public speech (Searle 1969). The force of public speech is one of the ways social imaginaries, worldviews, and political moralities undergo development and change.

Moreover, recognitive accommodation suggests a number of broader features of adaptive ecological governance itself. The general concept and goal of adaptation can be a discourse of legitimation, particularly in the domains of health and environmental policy and planning. An ethos and philosophy of adaptation can displace the neoliberal and global capitalistic philosophy of growth. It can also supply a critique of the moral psychology imputed to the character of *Homo economicus* – the calculating, strategic, efficiency, and utility maximizer of instrumental value. Having been told so often in the Western capitalist tradition that growth and possessive individualism are inherent in human nature, mounting such a critique is a tall order. Yet I think this is possible and realistic, because the conceptual logic of norms of growth and greed is fundamentally incompatible with the constitutive norms and practices of adaptation as I am construing them here.

I hope and believe that climate adaptation planning, when it is better institutionalized and more visible, can foster – and be facilitated by, in turn – an ongoing cultural and normative reorientation in the deliberative sphere of governance and in the experience of social respect, voice, mutuality, and belonging so that more people come to experience in their everyday lives. This normative transition at the level of civil society will emerge from new articulations of traditional values in speech actions.

To be sure, adaptive governance can evolve in either a technocratic or a deliberative and participatory direction. If we want to explore the deliberative pathway, it is important to develop more fully from both a philosophical and a practical point of view, "civic placemaking" and "civic learning" (Kaebnick et al. 2021; Eckenwiler 2013, 2016).

The kind of democratic deliberation and citizenship required by a political morality of accommodation needs to be rooted in the political culture and the everyday lived experience of large numbers of people. One way to strengthen these cultural and experiential foundations of participatory democracy is to ensure that there are venues to enable, enact, and nurture civic learning. *Civic learning* simply is the internalization of the ideals of mutual right recognition and reciprocal right relationship.

Adaptation and the Human Prospect

Adaptation is not simply about using natural resources differently and curtailing harmful and wasteful behaviors. Adaptation will be aided by, but is not reducible to, technological innovations. Adaptation, as I have argued, is normatively a transformational developmental process concerning the human capability to live in new quarters and in new ways – to live, that is to say, better relational lives. It starts from an accommodation of necessity, hopefully temporary. Yet accommodation can develop beyond that into an accommodation of intrinsic value in others that expresses hospitality, refuge, and welcome.

Many anomalous weather events and trends are being felt by ordinary people, and they are now being understood in the context of climate change. As everyday life increases awareness of these realities, the motivation to support climate action policies and measures politically will become stronger eventually.

When political support for climate action eventually does build and bubble over in a destressed society – as the history of large-scale social changes in the past suggests it eventually will – the hard work of adaptation and governance will only have begun. Then we shall need to channel climate action–oriented political power toward constructive cultural and social change. In part, this will involve developing new values, new institutions, and new ways of living creatively, constructively, justly, and humanely. Adaptive governance and planning are fundamentally about nurturing this type of civic and moral learning, in addition to the necessary and practical changes in the material resources and aspirations of everyday life. The interesting complexity and challenge visible in the prospect of civic and moral learning is that many traditional values, many mentioned in this presentation, will need to be reaffirmed and transformed at the same time. This dialectical learning process of continuity within transformation should be leisurely, and its argumentation serene. Alas, thanks to our monumental procrastination, the best we can hope for is that climate change adaptation will stay at least one step ahead of what is coming after us.

Reference List

Bosselmann, K. (2023) "Collapse or sustainability? Ecological integrity as a fundamental norm of law," in: P. Burdon & J. Martel (eds) *Research Handbook for Ecological Law in the Anthropocene*, pp. 117–127, London: Routledge.

Bourdieu, P. (1977) *Outline of a Theory of Practice*, Cambridge: Cambridge University Press.

Bourdieu, P. (1990) *The Logic of Practice*, Redwood City: Stanford University Press.

Brandom, R.B. (2019) *A Spirit of Trust: A Reading of Hegel's Phenomenology*, Cambridge, MA: Harvard University Press.

Casey, E.S. (1997) *The Fate of Place: A Philosophical History*, Berkeley: University of California Press.

Chakrabarty, D. (2021) *The Climate of History in a Planetary Age*, Chicago: University of Chicago Press.

Connolly, W.E. (2017) *Facing the Planetary: Entangled Humanism and the Politics of Swarming*, Durham: Duke University Press.

Diamond, C. (1988) "Losing your concepts," *Ethics*, 98, pp. 255–277.

Di Paolo, E.A., Cuffari, E.C., & De Jaegher, H. (2018) *Linguistic Bodies: The Continuity between Life and Language*, Boston: MIT Press.

Dworkin, R. (2011) *Justice for Hedgehogs*, Cambridge, MA: Harvard University Press.

Eckenwiler, L. (2013) "Ecological selves as citizens and governance as place-making," *Minding Nature*, 6(1), pp. 18–25.

Eckenwiler, L. (2016) "Defining ethical placemaking for place-based interventions," *AJPH*, 106(11), pp. 1944–46.

Fischer, F. (2009) *Democracy and Expertise: Reorienting Policy Inquiry*, New York: Oxford University Press.

Forester, J. (1999) *The Deliberative Practitioner: Encouraging Participatory Planning Processes*, Boston: MIT Press.

Goodpaster, K. (1978) "On being morally considerable," *Journal of Philosophy*, 75(6), pp. 308–325.

Honneth, A. (2012) *The I in We: Studies in the Theory of Recognition*, Cambridge: Polity Press.

Howe, N. (2023) *The Fourth Turning Is Here*, New York: Simon and Schuster.

IPCC (2022) *Climate Change 2022: Impacts, Adaptation, and Vulnerability*. Contribution of Working Group II to the Sixth Assessment Report of the Intergovernmental Panel on Climate Change, H.-O. Pörtner, D.C. Roberts, M. Tignor et al. (eds), Cambridge: Cambridge University Press.

Jennings, B. (2015) "Ecological political economy and liberty," in: P.G. Brown & P. Timmerman (eds) *Ecological Economics for the Anthropocene: An Emerging Paradigm*, pp. 272–317, New York: Columbia University Press.

Jennings, B. (2016) "Putting the bios back into bioethics: Prospects for health and climate justice," in: C. Cox Macpherson (ed) *Climate Change and Health: Bioethical Insights into Values and Policy*, pp. 11–37, New York: Springer Press.

Jennings, B. (2018) "Solidarity and care as relational practices," *Bioethics*, 32, pp. 553–561.

Jennings, B. (2022) "The ordeal of adaptation: Recognition and relationality in a climate changed world," *Journal of Theoretical and Philosophical Psychology*, 42(3), pp. 177–188.

Kaebnick, G.E., Gusmano, M., Jennings, B., Neuhaus, C.P., & Solomon, M.Z. (eds) (2021) "Democracy in crisis: Civic learning and the reconstruction of common purpose," *Hastings Center Report*, 51(1), Special Report. Available at: https://onlinelibrary.wiley.com/toc/1552146x/2021/51/S1.

MacIntyre, A. (2007) *After Virtue: A Study in Moral Theory*, 3rd ed., South Bend: University of Notre Dame Press.

Minteer, B.A. (2006) *The Landscape of Reform: Civic Pragmatism and Environmental Thought in America*, Boston: MIT Press.

Orlov, D. (2011) *Reinventing Collapse*, Gabriola Island: New Society.

Orr, C.J., Kish, K., & Jennings, B. (eds) (2019) *Liberty and the Ecological Crisis: Freedom on a Finite Planet*, New York: Routledge.

Pelling, M. (2011) *Adaptation to Climate Change: From Resilience to Transformation*, New York: Routledge.

Ricoeur, P. (2005) *The Course of Recognition*, Cambridge, MA: Harvard University Press.

Schön, D.A. (1984) *The Reflective Practitioner*, New York: Basic Books.

Searle, J.R. (1969) *Speech Acts: An Essay in the Philosophy of Language*, Cambridge: Cambridge University Press.

Sonenscher, M. (2023) *After Kant: The Romans, the Germans, and the Moderns in the History of Political Thought*, Princeton: Princeton University Press.

Tronto, J. (2013) *Caring Democracy: Markets, Equity, and Justice*, New York: New York University Press.

Van Horn, G., Kimmerer, R.W., & Hausdoerffer, J. (eds) (2021) *Kinship: Belonging in a World of Relations*, Libertyville: Center for Humans and Nature Press.

Weintrobe, S. (2021) *Psychological Roots of the Climate Crisis: Neoliberal Exceptionalism and the Culture of Uncare*, New York: Bloomsbury Academic.

Chapter 8

Ecological Integrity and Indigenous Farming Practices in Africa

Balancing Traditional Knowledge and Scientific Methods in Climate Adaptation

Ngozi Finette Unuigbe

Introduction

In the pursuit of maintaining ecological integrity and ensuring sustainable agricultural practices, it is crucial to recognize the intricate relationship between Indigenous farming methods and the natural environment. The majority of Indigenous farmers in Africa, deeply connected to their land and reliant on its ecological balance, depend on rain-fed agricultural production for their livelihoods. In this context, reliable Indigenous weather forecasts become essential tools in guiding rural farmers' decisions, especially in adapting to the challenges posed by climate change.

However, much of the literature reveals that an exclusive reliance on Western scientific knowledge often proves counterproductive at the rural level. This is particularly true in regions where traditional knowledge systems have evolved in tandem with local ecological conditions, enabling farmers to make informed decisions that respect and uphold ecological integrity. These Indigenous practices, often overlooked or undervalued, offer invaluable insights into sustainable resource management and climate adaptation strategies.

To foster a harmonious relationship between humanity and nature, it is imperative to integrate these traditional methods with modern scientific approaches. This integration not only enhances the resilience of agricultural systems against the impacts of climate change but also ensures the preservation of ecological integrity, which is vital for global security, public health, and lasting peace. Therefore, in the quest for sustainable solutions to the challenges faced by Indigenous farmers in Africa, a balanced and respectful synthesis of Indigenous knowledge and Western science stands as a cornerstone in maintaining ecological integrity and advancing global environmental justice.

In sub-Saharan Africa (SSA), pastoral communities represent a significant portion of the population, with their livelihoods deeply intertwined with the environment. These communities, comprising about 50 million people, or more than 10% of the SSA population, have developed a profound understanding of their natural surroundings. Their observational skills and the

DOI: 10.4324/9781003440871-10

detailed environmental knowledge encoded in their languages enable them to monitor biodiversity changes and predict weather patterns over various timescales. This Indigenous knowledge system is a critical asset in the face of climate variability and change.

However, the very nature of pastoralism, reliant on natural resources and predictable rainfall, renders these communities particularly vulnerable to the impacts of climate change. This vulnerability is exacerbated by the increasing unpredictability of weather and climate conditions in arid and semi-arid areas, as reported in the 5th Assessment Report of the Intergovernmental Panel on Climate Change (IPCC). Despite their inherent adaptability, with a nomadic or semi-nomadic lifestyle designed to withstand environmental shocks, pastoralists face increasing pressures due to climate change.

Unfortunately, Western scientific meteorological forecasts, often criticized for their lack of accuracy and utility at the local scale, fail to adequately serve these Indigenous communities. The technical and systematic nature of this climatic information, as noted by Braman et al. (2013) and Kom et al. (2022), is often poorly interpreted by Indigenous farmers. This gap in effective communication and understanding highlights the limitations of relying solely on Western scientific knowledge to combat climatic risks. Finucane (2009) and Mugambiwa and Rukema (2019) have documented the need to integrate local Indigenous knowledge with scientific approaches to effectively address these challenges.

The Paris Agreement under the UN Framework Convention on Climate Change (UNFCCC), specifically in Article 7(5), underscores the importance of including Indigenous and local knowledge in understanding climate change and developing adaptation strategies. This inclusion marks a significant recognition of the value that Indigenous knowledge systems bring to the table, especially in devising effective and culturally appropriate responses to climate-related challenges. Despite this, pastoralists often find themselves marginalized from the political economy, with their valuable knowledge and practices overlooked in policymaking processes. This marginalization not only undermines the effectiveness of climate change adaptation strategies but also exacerbates the vulnerability of these communities.

This chapter will delve into how Indigenous climate knowledge has been instrumental over the past century in assisting rural farmers' households in SSA to tackle climate stressors, thereby enhancing decision-making for adaptation. It will argue for the advancement of assessment processes to ensure that Indigenous knowledge systems receive adequate recognition and credit. Furthermore, the chapter will explore how policymakers in SSA can enhance and utilize Indigenous climate knowledge among pastoralists in the region, addressing climatic stressors and potentially increasing crop production. This approach not only aligns with the goals of the Paris Agreement but also offers a pathway toward more resilient and sustainable agricultural practices in the face of climate change.

The Nature of Pastoralist Climate Forecasting in Sub-Saharan Africa: A Complement for Western Science

Pastoralism is a way of life centered on the raising and herding of livestock, such as cattle, goats, sheep, and camels. It is a livelihood strategy adapted to diverse environmental conditions and is particularly prevalent in regions where farming is limited due to harsh climates, such as arid and semi-arid lands. Pastoralists are often nomadic or semi-nomadic, moving their herds in search of pasture and water in response to seasonal changes. This lifestyle requires a deep understanding of the environment, including knowledge of grazing patterns, water sources, and the health of their animals.

Central to pastoralism is the use of Indigenous or traditional knowledge – a cumulative body of knowledge, skills, and practices passed down through generations. This knowledge encompasses an understanding of weather patterns, animal behavior, vegetation cycles, and natural resource management. It enables pastoralists to make informed decisions about migration, grazing, and animal care, thereby ensuring their survival and livelihood in often-unpredictable environments.

Indigenous knowledge plays a major role in local livelihoods and is crucial to supporting local efforts to forecast and make sense of seasonal climate situations at local levels. However, the progressive loss of Indigenous knowledge threatens the ability of pastoralists and farmers to cope with and adapt to climate change. The challenge ahead is finding ways of integrating Indigenous weather forecasting with scientific weather forecasting systems. Such integration can potentially create more accurate, relevant, and useful weather information that caters to the specific needs of pastoralist communities, enhancing their resilience and adaptive capacity in the face of climatic changes. Pastoralists are among those who are most vulnerable to drought. Drought events will increase by more than 20% in most of the world by 2080, with the number of people affected by droughts rising by 9–17% in 2030 and 50–90% in 2080 (Guye et al. 2022). Several pieces of evidence indicate that Africa has warmed in recent decades, which is coherent with anthropogenic climate change and has a significant impact on climate-sensitive sectors (UNICEF 2012; Balehegn et al. 2019). Drought, unlike other climatic hazards, has a slow onset and a long duration, making effective prediction difficult (Blauhut et al. 2016). In pastoral areas where access to technology is limited and the use of scientific weather prediction is unlikely, taking early action to reduce the risk is seemingly challenging. Climate change is likely to increase the frequency of extreme events, such as droughts and floods, contributing to livestock death and crop production decline (Arnold et al. 2015). There are significant disparities in the availability of timely, accurate, and user-friendly climate forecast information to farmers and pastoralists (Radeny et al. 2017). Furthermore, scientific weather forecasts are not accessible or usable by pastoralists, making them less effective at the district level (Adu et al. 2018).

Pastoralism is one of the human activities in the African savannahs that has long demonstrated resilience to a variable climate. In the arid and semi-arid savannah regions of Africa, a land marked by its variable climate, the practices of pastoralism have long been shaped by environmental extremes, from droughts to floods and famine. This unpredictability of weather patterns necessitates a keen awareness of climatic conditions, as the livelihoods of pastoralists and farmers are intimately tied to these environmental factors.

Historically, national meteorological services across these regions have provided daily weather forecasts. These forecasts are crucial for day-to-day decision-making, offering insights into immediate weather conditions that can affect daily activities. However, the complex and often extreme nature of the climate in these regions demands a more comprehensive approach to weather prediction.

In response to this need, meteorological services have expanded their scope to include seasonal forecasts. As noted by Osman et al. (2023), these forecasts represent a significant advancement in meteorological services. They provide critical long-term weather information, essential for strategic planning and decision-making in agriculture and pastoralism.

Seasonal forecasts offer predictions about weather patterns for the upcoming months. This information typically includes projections about rainfall levels and temperature trends, which are vital for determining the feasibility and timing of agricultural activities. For instance, accurate forecasts of rainfall can indicate a favorable season for planting, suggesting that crops planted during such a period are more likely to thrive and result in successful harvests. Conversely, predictions of a dry season might prompt farmers to adjust their planting strategies, perhaps by selecting drought-resistant crops or altering their planting schedules to maximize yield despite reduced rainfall.

For pastoralists, these seasonal forecasts inform crucial decisions about grazing and migration. Knowledge of anticipated rainfall patterns can guide pastoralists to areas with better prospects for grazing, ensuring adequate food for their livestock. In periods of forecasted dry conditions, pastoralists might plan migrations earlier than usual to secure adequate grazing for their herds.

Additionally, seasonal forecasts play a critical role in hazard preparedness. They provide early warnings about potential extreme weather events, such as droughts or floods. Armed with this knowledge, communities can take preemptive actions to mitigate the impacts of these events. For example, in anticipation of drought, communities might focus on conserving water and food supplies, while preparations for potential floods could include reinforcing infrastructures and developing evacuation plans.

In summary, the provision of seasonal forecasts by national meteorological services is a critical development for communities in Africa's arid and semi-arid regions. These forecasts offer essential information for long-term planning and decision-making, enhancing the resilience of pastoral and agricultural communities to the challenges posed by a variable and often harsh climate. This advancement in weather services, enabling communities to plan for optimum productivity and prepare for potential hazards, is a significant step toward mitigating the impacts of climate variability in these regions. One example of this information gap concerns rainfall. Pastoralists' lives are completely reliant on rain, because the availability and spatial distribution of pasture and water are determined by rainfall (Napier & Desta 2011). However, not only is there a shortage of information on climatic risks, but the dissemination and interventions are also inconsistent. Pastoralists suffer difficulties as a result of this (Martin et al. 2016).

Most pastoralists rely on Indigenous knowledge and local observations to analyze, anticipate, and interpret seasonal weather patterns and climate (Radeny et al. 2017). Indigenous knowledge can be described as institutionalized local, place-based knowledge rooted in local cultures that is frequently passed down through oral traditions from generation to generation (Selemani 2020; Mugi-Ngenga et al. 2021). Many Indigenous communities worldwide rely on Indigenous weather forecasting to help them cope with or adapt to the impact of climate change–induced extremes in weather events (Nassef & Belayhun 2012). Following several years of inspection, Indigenous weather forecasting (IWF) is now embedded and entrenched in many African cultures and communities, with various societies and cultures using various indicators to forecast future weather conditions (Khan et al. 2016). This necessitates the use of effective early warning systems to address the impacts from the drought cycle and reduce the negative repercussions of droughts (Filho et al. 2021). In pastoral areas with limited access to modern drought forecasting and information dissemination via radio, television, newspapers, and social media, the role of Indigenous weather forecasting becomes crucial for predicting future climate conditions and sustaining livelihoods. In sub-Saharan Africa, established forms of pastoralist weather forecasting include observations of and inferences drawn from animal behavior, cloud type and cover, and specific vegetation phenomena. The early warning alert signs and indicators have been useful in specific contexts for predicting weather conditions and/or climatic impacts, including the onset of rains and rainfall yield (i.e., high or low amount of rainfall). Pastoralists forecast local weather patterns, such as the onset, cessation, intensity, and distribution of rainfall, as well as the incidences and magnitude of drought and flood events, by observing living organism behavior, wind direction, and cloud types (Radeny et al. 2017).

Meteorological and astrological Indigenous weather forecasting indicators include direction and strength of winds, star–moon alignment, apparent movement of stars, direction of the moon crescent, types of clouds, temperature conditions, lightning and thunder, color of the sky, and rainbow to forecast the next rainy season. In Uganda, Indigenous knowledge forecasters associate the onset of rainfall with the appearance of clouds. The appearance of nimbostratus and cumulonimbus clouds indicates a high probability of rainfall. While in Ethiopia, the appearance of white feather-like column (vertically standing) cloud in the sky is an indication that rain is about to come. A sky dominantly covered with light cloud indicates drought. Astrological indicators were more pronounced among pastoralists (Guye et al. 2022: 45).

Biological indicators focus on the behavior and activities of domestic and wild animals, insects, and different species of plants for weather forecasting. For instance, in Uganda, the *Mvule* tree indicates onset of the rainy season. While in Ethiopia, the intestines of cattle, sheep, and goats are used to forecast about the magnitude, severity, and duration of drought, drought-affected places, disease outbreak, the prospect of peace, and/or conflict (Desalegn et al. 2017). In Tanzania, the occurrence of large flocks of swallows and swans roaming from the south to the north during the months of September to November is an indication of onset of short rains (Desalegn et al. 2017).

In Kenya, the flowering of coffee trees is an indication that the rainy season is near, while the blossoming of peach trees in Botswana, South Africa, Tanzania, and Zimbabwe is linked to the beginning of the rainy season. Soil fertility management is another area in which Indigenous local knowledge is used, whereby Indigenous farmers classify soils according to their potential for crop production and crop yield. For many decades, ILK land management practices have also been shown to conserve carbon in soils (Filho et al. 2023).

Despite its benefits, IWF has been criticized for its lack of precision in predicting the magnitude of the effect and the duration of the climatic shock. However, IWF is decisive in assisting endeavors to advance access to climate services in many African communities, particularly where meteorological forecasts are not accessible. There is therefore a growing interest in documenting ILK for its intrinsic heritage value and its practical use in potentially supporting relevant and locally acceptable climate change adaptation (Filho et al. 2023). However, though a great deal of empirical research has been devoted to ILK, this type of knowledge has not yet been adequately articulated in adaptation and mitigation planning, especially in the Global South. The main criticism of Indigenous and local knowledge is in its locally specific influence, endemic to a social group, which may limit its scalability.

Despite being acknowledged as a valuable resource for climate adaptation, current national adaptation policies on the African continent still show serious gaps in effectively integrating ILK systems within the legal frameworks to reduce vulnerability. ILK should be better integrated with modern climate change adaptation strategies to anticipate more effective responses. Both rural communities and relevant government agencies should complement the use of ILK with climate change strategies so as to maximize its contribution to the effective implementation of climate change policies.

Integrating Indigenous and Western Scientific Knowledge to Adapt to Climate Risks

Several studies have indicated that "[I]ndigenous knowledge method tends to be qualitative, and they develop a diachronic databank, that is, a result of observations from local environment over an extended time period" (Berkes 2012). Indigenous people are the observers, so their decision-making is directly linked to the quality and reliability of their social and environmental observations. On the other hand, Western science observations created by a small group of scientists tend to be quantitative and frequently represent synchronic information observations from a wide range of sites, with lack of longtime epoch use by the Indigenous people (Jury 2013). Local knowledge, however, offers "observations and interpretations at a much finer spatial scale with considerable temporal depth by highlighting elements that may not be considered by climate scientists" (IPCC 2018: 10).

Indigenous and local knowledge systems are distinct from scientific knowledge systems, with different intellectual heritages, contexts, and applications, and sustained by substantially different types of institutions. However, far from being opposed to each other, Indigenous and local knowledge systems, like the African pastoralists' knowledge, and Western scientific knowledge systems may share important features regarding the centrality of observation in producing knowledge, the compiling of reliable insights on the basis of long-term verified patterns, and according to some theorists, how hypotheses and models are built and tested, and with complementary results.

Tracker and scientist Louis Liebenberg proposes that Western science shares its roots with the Indigenous art of tracking. In *The Art of Tracking: The Origin of Science*, he argues that the two systems share a fundamental "hypothetico-deductive reasoning" that, as a system, permits moving between inductive reasoning based in observation and evidence and the capacity to develop hypotheses and apply deductive reasoning, a process that is continually tested and adjusted (Liebenberg 2021: 9). Despite their differences of origin, practices, and focus, these knowledge systems rely on some shared principles, including observation, hypothesis building, cross-referencing of various indicators, and systems of adaptation and modification. As the systems are based on observing different phenomena at different scales, with different approaches, they provide important opportunities for complementarity, particularly with regards to climate adaptation, which involves biological and atmospheric factors, impacts, and options for behavior modification.

Given the seriousness of the global climate crisis and the vulnerability of rural communities, particularly in the least developed countries (LDCs), the potential complementarity and synergies between mainstream Western science and Indigenous knowledge systems invite us to investigate and recognize the unique characteristics and explanatory potential of ILK to enrich our understanding of weather and climate and, moreover, about ecosystems and biodiversity (IPCC 2013). These descriptions are valuable right now as the phenomena occur. With the benefit of observations by elders over a lifetime, and through their unique languages, which evolved in situ and contain detailed taxonomies of biodiversity, seasonal cycles, precipitation patterns, and atmospheric phenomena, we can also gain transgenerational information and baselines from earlier times, before there was scientific observation present in these territories. Taking time for dialogue, investigation, and exploring such synergies and complementarity may reinforce Indigenous peoples' resilience, fundamentally change current approaches to climate adaptation and resilience building, and potentially create opportunities for innovations in the coproduction of knowledge about weather, climate, and nature.

Pastoralist Knowledge Research: Toward a Synthesis

Worldwide, pastoralism sustains approximately 200 million households and nearly a billion head of animals, including goats, cattle, and camels, occupying a quarter of the world's land area, predominantly arid and semi-arid, with 60% of these pastoral households located in Africa (Ahmed & Bihi 2019). As accurate, relevant, and accessible climate information becomes more urgent for achieving successful adaptation and sustainable development, international attention is focusing on how to embed meteorological and climate services and information in national development strategies, particularly adaptation planning for national adaptation plans (NAPs) (Ahmed & Bihi 2019).

Different Indigenous weather forecasting indicators have varying levels of acceptance, depending on their precision. For instance, in some places, star–moon alignment is the most dependable weather information source, while cloud color is the preferred indicator in other sites. The challenges facing IWF include insufficient documentation of the knowledge, a poor knowledge transfer system, lack of coordinated research to investigate its accuracy and reliability, death of forecast experts, and influence of religion and modern education.

There is potential in increasing the interaction between the Indigenous knowledge systems and meteorological and other environmental sciences to generate improved modelling of climate change, predictive capacity in forecasting, and an overall approach to knowledge-based climate adaptation. A dialogue between knowledge systems could help better understand rainfall, rain quality, and wind patterns, especially in the arid and semi-arid lands of Africa, to generate more accurate data and understanding of how the atmosphere, terrestrial biodiversity, and ecosystems are changing – allowing us a greater time depth of observation and details of current changes from year to year. The interaction of Indigenous knowledge and Western science could increase resilience options, help focus policy interventions, and generally improve the capacity of all actors to make better predictions and respond appropriately (Jiri et al. 2016).

Research shows that pastoralists use a number of means to form projections on when it is going to rain, or to know if they are going to face conditions such as high winds or low moisture, how long these trends are likely to last, and what their intensity might be. They can also gather information on where good grazing and watering will likely occur within a large territory. A number of signs are used, including the flowering of certain trees, the behavior of certain animals and insects, and changes in winds and clouds. Even the milk that cows produce can provide information about trends in the natural world. These tools, referring to traditional methods and Indigenous knowledge used by herders, are crucial for their survival and productivity in arid and semi-arid lands. They provide herders with the ability to interpret environmental cues, predict weather patterns, and make informed decisions about grazing, migration, and resource management. However, these forecasting methods have limitations, particularly as the climate grows increasingly unpredictable and extreme. Traditional knowledge, while invaluable, may not always accurately predict the more erratic and severe weather events brought about by climate change, such as prolonged droughts, unexpected heavy rains, or unseasonal temperature shifts. As a result, herders face greater challenges in sustaining their livelihoods and maintaining the health of their livestock under these changing conditions. This challenge is shared by Western meteorological science, again reinforcing the principle that combining different types of knowledge and analysis may improve our overall capacity and understanding. The gaps that these two knowledge systems have remind us of the importance of working together and finding new synergies in forecasting, observation, and climate adaptation responses. Combining certain traditional weather forecasting with meteorology may make forecasts stronger and more useful for pastoralists, as well as neighboring communities and urban centers.

Pastoralists need climate and weather information at the level of the ecosystem or landscape in which they operate. Dialogue around maps can be used as the basis for generating downscaled information. Participatory mapping allows pastoralists to document and make visible how their nomadic or semi-nomadic land use operates. Far from being random, traditional transhumance is designed from a perspective that can be useful for adaptation as well as sustaining the ecosystems which support pastoralism.

In terms of adaptation to climate change, pastoralist practices are based on long experience of interaction with the landscape and numerous environmental variables and include the deployment of practices to cope with the risks existing in their environments, where both floods and droughts can be regular occurrences. The emphasis on a transdisciplinary approach in climate adaptation marks a significant shift from a purely pragmatic strategy to one that is decolonial, non-hierarchical, and inclusive. This approach represents a transformative way of addressing environmental challenges by recognizing Indigenous knowledge and scientific knowledge as equally valuable and complementary. It moves away from the traditional dominance of Western scientific paradigms, instead advocating for a more holistic and inclusive understanding that respects and integrates diverse knowledge systems. By doing so, it not only enhances our collective capacity to respond to climate challenges but also respects and uplifts the voices and rights of Indigenous communities, ensuring their involvement in the creation of sustainable and effective solutions.

Pastoralist Knowledge Research: Blueprints for Coproduction

One matter that remains unclear is how the integration of Indigenous local knowledge (ILK) into transformative forms or national adaptation planning can happen in practice. To effectively bridge knowledge systems, Nalau et al. (2018) both highlighted and criticized an inherent bias of adaptation that needs to be dissolved: Current models of ILK integration as well as scientific knowledge scrutinize neither Western thinking nor the current notions of development or adaptation options. Scholarly driven research design would preclude from appreciating the central importance of creative and coproduced learning environments over more rigorous research design, analysis, and evaluation, which, while allowing for certain types of learning opportunities, block others. Indigenous communities may, in turn, perceive the integration of knowledge systems as a different kind of colonization (Nalau et al. 2018). There are ways and means to overcome such barriers and go beyond local, place-based impacts, as highlighted in Tengö et al.'s (2017) review. Two key aspects are:

- *Engagement of knowledge carriers and institutions.* Adaptation involving ILK represents a collaborative, iterative, and inclusive process in which everyone participating has stakes in decision-making, although participation does not guarantee, in and of itself, that ILK is adequately and appropriately built into adaptation (UNFCCC 2013: 3). Moreover, collaboration is likely to fail if the relationship among the involved actors is not based on trust, if power asymmetries exist, if marginalized groups are overlooked, and if it does not follow cultural, ethical, and legal protocols (Makondo & Thomas 2018).
- *Building trust and communication.* local knowledge is generally context-specific and manifests differently across place and time, so ILK may be contradictory to scientific and technological understandings of universal knowledge and development. According to Tengö et al. (2017), the difference in understanding of ILK, that is, differing worldviews and experiences, can be a potential source of disagreement. Language and literacy may also represent barriers for understanding and participation (UNFCCC 2013). There is a growing consensus among climate scholars and development practitioners that research on climate change adaptation should not rely on Western knowledge alone, and the engagement of diverse knowledge systems is crucial for effective and context-specific climate adaptation planning. If there is a common agreement regarding the benefits of integrating ILK into formal climate change adaptation strategies, there is,

however, relatively limited empirical evidence portraying how it can take place in practice. From our perspective, this paradigm shift would require a range of actions, and we outline three shifts that could help transform climate research approaches to be more inclusive.

First, a change of the programmatic mindset that guides most scientific work is required. Scientists only trust evidence generated through a "rigorous" Western, science-based protocol with statistically analyzed data that demonstrates acceptable levels of reliability and validity. This mental shift would create space for the decolonization of climate science that is heavily guided by colonial legacies, such as languages and approaches (Trisos et al. 2022). The change would also allow scientists to reconsider what "science" means and what can be labeled as scientific outcomes. Climate research is exclusionary, framed by Western, science-based approaches, and though traditional knowledge does follow the same path, this knowledge system is grounded on careful observation of environmental changes that Indigenous communities have relied on for many centuries to cope with climate challenges.

Second, limited funding flows to climate research in general, and more specifically to local knowledge evidence in the Global South, is a major barrier. A recent study on thematic prioritization of funding for climate change research in Africa shows that while institutions from the Global North – principally Europe and North America – received 78% of funding for climate research, African institutions received only 14.5% (Overland et al. 2021). Most local and Indigenous knowledge is generated in the Global South – Africa, Latin America, and Asia – and this funding discrepancy means that climate-related local knowledge is not fully captured in climate research to create space for the integration. This suggests that funding bodies need to set more dedicated allocations for documenting the practical integration of local knowledge. Tengö et al. (2017) call for substantial investments of both time and funds to support knowledge-holders and institutions, underscoring the large efforts needed to properly consider the diverse contexts in which ILK exists.

Finally, the integration of local knowledge/Indigenous knowledge and Western, scientific knowledge should not just be seen as a combination of the two knowledge systems. The integration should be part of a more complex coproduction process where climate scholars and local/Indigenous communities interact effectively and are interested in questioning how local knowledge resonates in the scientific framing or vice versa (Adu et al. 2018). It is about co-generation of knowledge, not a category of stakeholder, often researchers, pretending to know enough about local needs and aspirations to be able to offer solutions that are sufficiently finished to require little or no subsequent local adaptation. This cooperation calls for an ethical shift, humility from both "scientists" and local communities for mutual learning with no "asymmetry of power" that gives one group more power to determine the direction that the cooperation takes (Liboiron 2021). Here we challenge climate research to be more flexible with regard to its approach that should not only be grounded in the Western perspective of "science."

The knowledge dialogues between scientists and herders can address climate science gaps and build capacities of national authorities to monitor, predict, and respond to slow-onset and extreme climatic events. By combining knowledge systems, such projects could aim to refine assessments, to identify complementarity in forecasting and use of biotic and abiotic indicators, and to broaden the understanding of the impacts of climate change at the local level, while helping to identify adaptation opportunities that match the needs and aspirations of these vulnerable communities.

Conclusion

As climate science becomes increasingly influential in policy formation, its integration with Indigenous weather forecasting in sub-Saharan Africa highlights the crucial role of ecological integrity in this process. This integration signifies that policies should extend beyond simply mitigating climate change effects to actively nurturing the resilience and diversity of ecosystems. It is about envisioning a future where environmental guardianship is as fundamental to policy decisions as economic or political factors. This shift, underpinned by ethical mindfulness, is not just advantageous but vital for the planet's health and sustainability.

Indigenous local knowledge (ILK), particularly Indigenous weather forecasting methods in sub-Saharan Africa, offers a rich and diverse understanding of the natural world. These traditional forecasting techniques, honed over generations, provide insights into seasonal and climatic variations that are deeply embedded in the sociocultural fabric of various communities. Integrating this knowledge with contemporary climate science can lead to more nuanced, adaptable, and regionally relevant strategies, enhancing the effectiveness of policy decisions.

The amalgamation of ILK with scientific methods can spark groundbreaking advancements in climate adaptation strategies. This fusion is not merely about preserving ancient wisdom; it is an expansion of scientific inquiry that benefits all life on Earth. Indigenous weather forecasting challenges and refines scientific assumptions, contributing to more robust and comprehensive climate solutions.

To realize the full potential of this integration, capacity building within rural communities and government agencies is crucial. This ensures that the application of ILK in climate policies is not an isolated practice but is interwoven with broader climate change strategies. When governments and scientific institutions acknowledge and incorporate ILK, a more profound understanding of environmental dynamics emerges, significantly contributing to global climate change mitigation and adaptation efforts.

In summary, the integration of Indigenous wisdom, particularly weather forecasting in sub-Saharan Africa, with modern climate science is imperative for advancing global environmental science and policy. It represents a collaborative path toward a sustainable future, where the wealth of diverse knowledge systems is fully utilized for the benefit of all life on Earth. This approach not only enriches science but also ensures that decision-making is rooted in a deep understanding of and respect for the natural rhythms and wisdom of our world.

Reference List

Adu, D.T., Kuwornu, J.K.M., Anim-Somuah, H., & Sasaki, N. (2018) "Application of livelihood vulnerability index in assessing smallholder maize farming households' vulnerability to climate change in Brong-Ahafo region of Ghana," *Kasetsart Journal of Social Sciences*, 39(1), pp. 22–32.

Ahmed, M. & Bihi, M. (2019) "Indigenous knowledge for resilience and adaptation in pastoral production system of Somali regional state in Ethiopia," *Interdisciplinary Description of Complex Systems*, 17(4), pp. 723–737.

Arnold, N., Branson, M., Kuang, Z., Randall, D., & Tziperman, E. (2015) "MJO intensification with warming in the superparameterized CESM," *Journal of Climate*, 28, pp. 2706–2724.

Balehegn, M., Balehey, S., Fu, C., & W. Liang, W. (2019) "Indigenous weather and climate forecasting knowledge among Afar pastoralists of northeastern Ethiopia: Role in adaptation to weather and climate variability," *Pastoralism*, 9(8), pp. 1–14.

Berkes, F. (2012) *Sacred Ecology*, 3rd ed., New York: Routledge.

Blauhut, V., Stahl, K., Stagge, J.H., Tallaksen, L.M., De Stefano, L., & Vogt, J. (2016) "Estimating drought risk across Europe from reported drought impacts, drought indices, and vulnerability factors," *Hydrology and Earth System Sciences*, 20(7), pp. 2779–2800.

Braman, L.M., Van Aalst, M.K., Mason, S.J., Suarez, P., Ait-Chellouche, Y., & Tall, A. (2013) "Climate forecasts in disaster management: Red cross flood operations in West Africa (2008)," *Disasters*, 37, pp. 144–164.

Desalegn, A., Radeny, M., & Mungai, C. (2017) *Indigenous Knowledge in Weather Forecasting: Lessons to Build Climate Resilience in East Africa*. Available at: www.preventionweb.net/news/indigenous-knowledge-weather-forecasting-lessons-build-climate-resilience-east-africa.

Filho, W.L., Matandirotya, N.R., Lütz, J.M., Alemu, E.A., Brearley, F.Q., Baidoo, A.A., Kateka, A., Ogendi, G.M., Adane, G.B., Emiru, N., & Mbih, R.A. (2021) "Impacts of climate change to African indigenous communities and examples of adaptation responses," *Nature Communications*, 12(1), pp 13–16.Filho, W.L., Wolf, F., Totin, E., Zvobgo, L., Simpson, N.P., Musiyiwa, K., Kalangu, J.W., Sanni, M., Adelekan, I., Efitre, J., Donkor, F.K., Balogun, L., Rui Mucova, S.A., & Ayal, D.Y. (2023) "Is indigenous knowledge serving climate adaptation? Evidence from various African regions," *Development Policy Review*, 41(2), p. e12664. Available at: https://doi.org/10.1111/dpr.12664.

Finucane, M. (2009) "Why science alone won't solve climate crisis: Managing the climate risks in the Pacific," *Asia Pacific Issues*, 89, pp. 1–8.

Guye, M., Legesse, A., & Mohammed, Y. (2022) "Indigenous weather forecasting among Gujii pastoralists in southern Ethiopia: Towards monitoring drought," *Pastoralism*, 12, pp. 43–54.

Intergovernmental Panel on Climate Change [IPCC] (2013) *Climate Change 2013: The Physical Science Basis*. Contribution of Working Group I to the Fifth Assessment Report of the Intergovernmental Panel on Climate Change, T.F. Stocker, D. Qin, G.-K. Plattner, M. Tignor, S.K. Allen, J. Boschung, A. Nauels, Y. Xia, V. Bex & P.M. Midgley (eds), Cambridge: Cambridge University Press.

Intergovernmental Panel on Climate Change [IPCC] (2018) *Summary for Policymakers*. Global Warming of 1.5C. An IPCC Special Report on the Impacts of Global Warming of 1.5°C above Pre-Industrial Levels and Related Global Greenhouse Gas Emission Pathways, in the Context of Strengthening the Global Response to the Threat of Climate Change, Sustainable Development, and Efforts to Eradicate Poverty, V. Masson-Delmotte, P. Zhai, H.O. Pörtner, D. Roberts, J. Skea, P.R. Shukla, A. Pirani, W. Moufouma-Okia, C. Péan, R. Pidcock, S. Connors, J.B.R. Matthews, Y. Chen, X. Zhou, M.I. Gomis, E. Lonnoy, T. Maycock, M. Tignor, & T. Waterfield (eds), Geneva: World Meteorological Organization.

Jiri, O, Mafongoya, P., & Mubaya, C. (2016) "Seasonal climate prediction and adaptation using indigenous knowledge systems in agriculture systems in Southern Africa: A review," *Journal of Agricultural Science*, 8, p. 156.

Jury, M.R. (2013) "Climate trends in Southern Africa," *South African Journal of Science*, 109(1/2), pp. 1–11.

Khan, A.S.M.M.R., Eyasmin, R., Rashid, M.H., Ishtiaque, S., & Chaki, A.K. (2016) "Variability, heritability, character association, path analysis and morphological diversity in snake gourd," *Agriculture and Natural Resources*, 50(6), pp. 483–489.

Kom, Z., Nethengwe, N.S., Mpandeli, S., & Chikoore, H. (2022) "Indigenous knowledge indicators employed by farmers for adaptation to climate change in rural South Africa," *Journal of Environmental Planning and Management*, 66(13), pp. 2778–2793. Available at: https://doi.org/10.1080/09640568.2022.2086854.

Liboiron, M. (2021) "Decolonizing geoscience requires more than equity and inclusion," *Nature Geoscience*, 14, pp. 876–877.

Liebenberg, L., Ao, A., Lombard, M., Shermer, M., Xhukwe, U., Biesele, M., Xao, D., Carruthers, P., Kxao, O., Hansson, S.O., Langwane, H., Elbroch, L.M., Ui, N., Keeping, D., Humphrey, G., Gaq'o, U., Steventon, J.K.N., & Voysey, M. (2021) "Tracking science: An alternative for those excluded by citizen science," *Citizen Science: Theory and Practice*, 6(1): 6, pp. 1–16.

Makondo, C.C. & Thomas, D.S.G. (2018) "Climate change adaptation: Linking indigenous knowledge with western science for effective adaptation," *Environmental Science & Policy*, 88, pp. 83–91.

Martin, R., Linstädter, A., Frank, K., & Müller, B. (2016) "Livelihood security in face of drought – assessing the vulnerability of pastoral households," *Environmental Modelling and Software*, 75(5), pp. 414–423.

Mugambiwa, S.S. & Rukema, J.R. (2019) "Rethinking indigenous climate governance through climate change and variability discourse by a Zimbabwean rural community," *International Journal of Climate Change Strategies and Management*, 11(5), pp. 730–743.

Mugi-Ngenga, E.W., Kiboi, M.N., Mucheru-Muna, M.W., Mugwe, J.N., Mairura, F.S., Mugendi, D.N., & Ngetich. F.K. (2021) "Indigenous and conventional climate-knowledge for enhanced farmers' adaptation to climate variability in the semi-arid agro-ecologies of Kenya," *Environmental Challenges*, 5, p. 100355.

Nalau, J., Becken, S., Schliephack, J., Parsons, M., Brown, C., & Mackey, B. (2018) "The role of indigenous and traditional knowledge in ecosystem-based adaptation: A review of the literature and case studies from the Pacific Islands," *Weather, Climate, and Society*, 10(4), pp. 851–865.

Napier, A. & Desta, S. (2011) *Review of Pastoral Rangeland Enclosures in Ethiopia*, Pastoralist Livelihoods Initiative Policy Project. Available at: www.fic.tufts.edu/assets/Tufts-Range-Enclosure-ReviewPLI.pdf.

Nassef, M. & Belayhun, M. (2012) *Water Development in Ethiopia's Pastoral Areas. A Synthesis of Existing Knowledge and Experience*, Save the Children USA and Overseas Development Institute. Available at: https://pdf.usaid.gov/pdf_docs/PA00KQ4P.pdf.

Osman, M., Beeril, R., Bueler, D., & Grams, C. (2023) "Multi-model assessment of sub-seasonal predictive skill for year-round Atlantic-European weather regimes," *Quarterly Journal of the Royal Meteorological Society*, 18(775, Part B), pp. 2386–2408.

Overland, I., Fossum Sagbakken, H., Isataeva, A., Kolodzinskaia, G., Simpson, N.P., Trisos, C., & Vakulchuk, R. (2021) "Funding flows for climate change research on Africa: Where do they come from and where do they go?" *Climate and Development*, 14(8), pp. 705–724.

Radeny, M., Desalegn, A., & Mungai, C. (2017) *Indigenous Knowledge in Weather Forecasting: Lessons to Build Climate Resilience in East Africa*. Available at: https://ccafs.cgiar.org/news/indigenous-knowledge-weather-forecasting-lessons-build-climate-resil ience-east-africa-0.

Selemani, I.S. (2020) "Indigenous knowledge and rangelands' biodiversity conservation in Tanzania: Success and failure," *Biodiversity and Conservation*, 29(14), pp. 3863–3876.

Tengö, M., Hill, R., Malmer, P., Raymond, C.M., Spierenburg, M., Danielsen, F., Elmqvist, T., & Folke, C. (2017) "Weaving knowledge systems in IPBES, CBD and beyond – lessons learned for sustainability," *Current Opinion in Environmental Sustainability*, 26–27, pp. 17–25.

Trisos, C.H., Adelekan, I.O., Totin, E., Ayanlade, A., Efitre, J., Gemeda, A., Kalaba, K., Lennard, C., Masao, C., Mgaya, Y., Ngaruiya, G., Olago, D., Simpson, N.P., & Zakieldeen, S. (2022) "Africa," in: H.-O. Pörtner, D.C. Roberts, M. Tignor, E.S. Poloczanska, K. Mintenbeck, A. Alegría, M. Craig, S. Langsdorf, S. Löschke, V. Möller, A. Okem, & B. Rama (eds) *Climate Change 2022: Impacts, Adaptation, and Vulnerability*. Contribution of Working Group II to the Sixth Assessment Report of the Intergovernmental Panel on Climate Change, Cambridge: Cambridge University Press.

UN Children's Fund [UNICEF] (2012) *Climate Change and Environmental Education: A Companion to the Child Friendly Schools Manual*, UNICEF's Division of Communication.

UN Framework Convention on Climate Change [UNFCCC] (2013) *Best Practices and Available Tools for the Use of Indigenous and Traditional Knowledge and Practices for Adaptation, and the Application of Gender-Sensitive Approaches and Tools for Understanding and Assessing Impacts, Vulnerability and Adaptation to Climate Change*, UNFCCC Technical Paper FCCC/TP/2013/11.

Chapter 9

A Healthy Environment, a Healthy Climate, and the Right to Health

European Union and International Developments

Ottavio Quirico

Introduction

In July 2022, the UN General Assembly adopted a Resolution that recognizes the human right to a "safe, clean, healthy and sustainable environment," with particular regard to the problem of climate change (UN General Assembly 2022). The Resolution is the outcome of a long process carried forward by the independent expert and special rapporteurs on human rights and the environment. It is certainly a meaningful step toward the achievement of environmental and climate protection. Taking into account the different adjectives featured in the Resolution, this chapter focuses specifically on the question of a "healthy" environment and climate, through the lens of the EU contribution to international developments.

EU law has indeed provided meaningful input to the development of the right to a healthy environment in international law, particularly through the pivotal role of the Charter of Fundamental Rights of the European Union (EUCFR) (EU 2000). This is notably the case in light of EUCFR Article 37 on environmental protection. Besides this normative evolution, significant developments are also taking place with respect to the possible recognition of the fundamental right to a healthy climate, as the EU is adopting a progressive policy in the area, notably via Resolution 2020/2134 (INI) on the Effects of Climate Change on Human Rights and the Role of Environmental Defenders on This Matter (European Parliament 2021).

The analysis proceeds in two parts. The first part outlines the contribution of EU law to the international development of a human right to healthy environment and, within this context, to the more specific human right to a healthy climate. Taking stock of this assessment, the second part considers essential implications of the recognition of these different rights for the right to health. The research unfolds against the background of the fundamental issue of ecological integrity and its relationship to fundamental rights (Bosselmann 2017: 129–130), at the crossroads between the possibility of greening human rights thus far followed by most jurisdictions, particularly the European Court of Human Rights (Pirjatanniemi 2016), and the idea

DOI: 10.4324/9781003440871-11

of affording the environment and climate human rights protection per se, which progressively shifts the focus from an anthropocentric vision to an eco-centric view of fundamental rights and the right to health (Boyd 2017).

The Right to a Healthy Environment and the Right to a Healthy Climate

International Evolution

Following the suggestions of the UN Human Rights Committee (HRC 2021: 3, para. 1), in July 2022, the UN General Assembly recognized that "the right to a safe, clean, healthy and sustainable environment" is "a human right" via Resolution A/76/L.75 (UN General Assembly 2022). These developments are the result of a trajectory that started in 1972, with Principle 1 of the Stockholm Declaration on the Human Environment acknowledging a "fundamental" right to an "environment of quality," conceived of as an *erga omnes* commitment vis-à-vis the international community as a whole, including future generations (UNCHE 1972). This approach was confirmed 20 years later via Principle 1 of the Rio Declaration, solemnly stating that human beings are at the center of sustainable development and entitled to a healthy life in harmony with nature, at the crossroads between environmental protection, life, and health (UNCHE 1992). In 1994, Principles 2 and 4 of the UN Draft Principles on Human Rights and the Environment acknowledged the intergenerational right to a "secure, healthy, and ecologically sound environment," interwoven with all human rights (Ksentini 1994). Another key step in this thread is Article 1 of the Bizkaia Declaration on the Right to the Environment, stating that everyone has a right to "a healthy and ecologically balanced environment" (UNESCO 1999).

This trajectory indicates that there is a clear evolution in international law toward the recognition of a fundamental right to a healthy environment (European Parliament 2021), which is nonetheless not yet acknowledged as binding law. Indeed, despite the recent adoption of UN Resolution A/76/L.75, it cannot yet be said that the right to a healthy environment is an enforceable commitment, as a Resolution of the UN, although authoritative, is not an international treaty and does not, per se, create compelling duties.

Besides these main developments, the idea has sporadically emerged of a more specific human right to a healthy climate. This is particularly the case of *Juliana v. US*, a dispute where the US district court for the District of Oregon affirmed the existence of a fundamental right to "a climate system capable of sustaining human life" (D. Or. 2016: para. 82). Along these lines, the UN Special Rapporteur on Human Rights and the Environment has acknowledged that states have an obligation to respect, protect, and fulfil "the right to a safe climate," particularly via the adoption of adequate mitigation and adaptation measures (Boyd 2019: 18, para. 65). On this basis, the

Declaration on Human Rights and Climate Change adopted by the Global Network on Human Rights and the Environment (GNHRE) articulates a universal human right to "the highest attainable standard of health" that is "free from dangerous anthropogenic interference with the climate system such that rising global temperatures are kept well below 2 degrees centigrade above preindustrial levels" (Davies et al. 2017: 250; GNHRE 2021).

The European Union Trajectory: Progressive Developments

The EU has followed a trajectory similar to international regulation in the matter of environmental protection, influenced by and influencing the global normative environment. Indeed, although its founding treaties did not initially embed environmental protection, the EU has progressively developed an advanced normative system in this respect, particularly via the innovative case law of the European Court of Justice (ECJ) and the Court of Justice of the European Union (CJEU). Thus, among other norms, Article 3(3) of the current Treaty on European Union (TEU) establishes sustainable development and a high level of environmental protection as fundamental principles of the Union. Correspondingly, TEU Article 21(2)(f)-(g) provides that preserving environmental quality and sustainability is a core tenet of the EU external relations. These principles are further spelled out in detail in Title XX of the Treaty on the Functioning of the European Union (TFEU) on Environment. Building on this ground and on the text of TEU Article 11, EUCFR Article 37, on Environmental Protection, establishes the obligation for the Union to integrate a high level of environmental protection and the improvement of environmental quality into all its policies within the remit of sustainable development as a human rights matter.[1]

Within such an advanced environmental framework, the Union has also made significant progress as concerns the recognition of a more specific human right to a healthy climate. This is particularly witnessed by the adoption of Resolution 2020/2134 (INI) (European Parliament 2021). This instrument fosters the introduction of "a right to a safe and healthy environment" in the EUCFR (Principle 8) and commits the EU to supporting the UN Special Rapporteur on Human Rights and the Environment in the global recognition of the fundamental right to a safe, clean, healthy, and sustainable environment (Principle 7). In light of this premise, the Resolution underscores the impact of climate change on the effective enjoyment of the human rights to life, food security, safe drinking water and sanitation, health, and housing (Preamble, B). It also highlights the importance of climate change litigation in holding states and corporations accountable to allowing a "healthy life for present and future generations" (Preamble, D). The Resolution further

1 EUCFR Article 37 essentially replicates the content of TEU Article 11.

recalls the numbers provided by the World Health Organization (WHO) on the impact of climate change in terms of yearly deaths owing to malnutrition, malaria, and heat stress, as well as the effect of climate shocks reported by the World Food Programme, exposing around 750 million people, that is, one in ten persons worldwide, to severe food insecurity (Preamble, U). On this basis, the Resolution acknowledges the necessity of positing a universal "fundamental right to a safe, clean, healthy and sustainable environment and to a stable climate" that is "fully enforceable through the justice system at all levels" (Principle 9). This normative environment clearly affirms the human right to a safe (and healthy)[2] climate as a component of the human right to a healthy environment.

Similar developments are taking place in the Council of Europe system. Thus, in September 2021, the Parliamentary Assembly of the Council of Europe adopted Recommendation 2211/2021, focusing on the necessity of Anchoring the Right to a Healthy Environment, asking the Committee of Ministers to adopt an additional protocol to the European Convention on Human Rights (ECHR) and another to European Social Charter (ESC) on the right to a safe, clean, healthy, and sustainable environment, particularly to ensure a life of dignity and good health (Parliamentary Assembly of the Council of Europe 2021). Furthermore, the Recommendation requested the preparation of a feasibility study for a convention on environmental threats and technological hazards threatening human health, dignity, and life, as well as the revision of Recommendation CM/Rec (2016: 3) on human rights and business, aiming to foster corporate environmental responsibility with respect to the human right to a safe, clean, healthy, and sustainable environment. As a follow-up, the Committee has adopted a Recommendation requesting states parties to the Council of Europe to implement the right to a healthy environment via domestic legislation (Committee of Ministers of the Council of Europe 2022). These developments are in line with international developments, particularly at the UN level, and are meaningful to the EU system, in light of the foreseeable accession of the Union to the ECHR envisaged in TEU Article 6(2). It is clear, though, that the EU has adopted a more advanced policy as compared to both the ECHR and general international law and is thus fostering progressive developments as concerns the recognition of a fundamental right to a healthy climate in the wider context of the human right to a healthy environment.

2 On the link between the right to a "stable" or "safe" climate and the right to health, see De Vido (2020): 102.

Implications for the Right to Health

Right to Health, Environmental Protection, and Climate Change

Climate change affects the right to health in several ways. It is considered that rising temperatures will entail approximately 250,000 additional deaths per year between 2030 and 2050 because of their impact on phenomena such as malnutrition and heat stress (WHO 2020: 4, paras 6–7; WHO 2021), whereby vulnerable groups are particularly affected (McIver et al. 2016: 1707–1714; Special Rapporteur on the Right to Health 2016: para. 20). Climate change is indeed aggravating infectious disease transmission, mental illness triggered by heat as well as air pollution, extreme weather conditions, flooding, and drought affecting health (Crimmins et al. 2016: 25–42; Bell & Greenberg 2018: S54).

Besides being recognized in about 120 constitutions around the world via different formulations, the right to health is acknowledged as an extensive fundamental claim in Article 12 of the International Covenant on Economic, Social, and Cultural Rights (ICESCR), beyond mere access to health care, essentially creating the conditions necessary to a healthy life. Article 12(1) establishes the state duty to recognize the "right of everyone" to the "highest attainable standard of physical and mental health," which is conducive to a life of dignity (CteESCR 2000: para. 1; UNOHCHR and WTO 2008: 3). Article 12(2) sets out the corresponding state duty to take the necessary steps to achieve the full realization of such a right, including those necessary for, inter alia, the improvement of all facets of environmental and industrial hygiene, as well reduction of the stillbirth rate and of infant mortality, healthy development of infants, control of all forms of diseases, and the establishment of conditions necessary to provide medical assistance in the event of sickness. This has created a basis for the Committee on Economic, Social, and Cultural Rights (CteESCR) to point out that the "underlying determinants of health" include issues such as access to safe drinking water, food, housing, and environmental conditions (CteESCR 2000: para. 4).[3]

ICESCR Article 12 must be read in conjunction with Article 2, which provides that states, together with relevant international organizations, such as the WHO (CteESCR 2000: paras 64–65; Meier et al. 2022: 26 ff.), must make all possible efforts, to the maximum of available resources, to promptly take

3 Meaningfully, the Constitution of Ecuador (1998), Chapter IV (Economic, Social, and Cultural Rights), Article 42, acknowledges that "the State guarantees the right to health, its promotion and protection, through the development of food security, the provision of drinking water and basic sanitation, the promotion of a healthy family, work and community environment, and the possibility of permanent and uninterrupted access to health services, in conformity with the principles of equity, universality, solidarity, quality, and efficiency."

action and progressively fulfil the right to health, particularly via legislative measures.[4] Corporations are also committed to providing a significant contribution, according to the concept of responsible business conduct (Esty & Bell 2018: S80). This entails as a minimum the implementation of a national strategy aimed at ensuring the enjoyment of the right to health, according to adequately monitored indicators and standards (CteESCR 2000: paras 2, 30, 36), involving the minimum core obligation to guarantee access to essential food, safe drinking water, shelter, housing, and sanitation (CteESCR 2000: paras 15, 43).[5] There is therefore a strict interrelation with the state obligation to adopt adequate climate mitigation and adaptation policies, given that, for instance, storms, droughts, and floods affect the availability of food, water, and housing, thus negatively affecting the right to health (Parker-Flyn 2018: S58; WHO 2020: 4, para. 7; Wu 2021: 124). Such an obligation is further strengthened by the duty to protect the right to health, which entails that a state prevents excessive greenhouse gas (GHG) emissions from private energy companies and remedies consequences via adequate adaptation policies (CteESCR 2000: para. 33; Special Rapporteur on the Right to Health 2016: para. 35). This is in line with the obligation to enact and enforce laws that prevent the pollution of water, air, and soil by extractive and manufacturing industries (CteESCR 2000: para. 51). Furthermore, given that part of GHG emissions are directly produced by the state, notably because companies in the energy sector are often either operated or participated by the state, respect for the right to health reinforces the duty to implement mitigation policies that prevent excessive GHG emissions from private energy companies and remedy consequences via targeted adaptation policies (CteESCR 2000: para. 33). This is in line with the duty of a state to refrain from polluting air, water, and soil through industrial waste outlined by the CteESCR (CteESCR 2000: para. 34).

A direct link between the right to health and climate change was established in the framework of the Council of Europe by the European Committee on Social Rights (ECteSR) in the case of *Marangopoulos* (see further Glicksman et al. 2018: S104). In this dispute, the Committee was requested to decide whether Greece had failed to comply with the duty to protect public health against air pollution under ESC Article 11 (ECteSR 2005). Greece allowed the operation of lignite mines and power stations fueled by lignite as a key source of energy at the time of the complaint, causing, inter alia, GHG

4 Consistently, the Constitution of South Africa (1996) Chapter II, Section 27 (Health Care, Food, Water, and Social Security) provides that "(1) [e]veryone has the right to have access to (a) health-care services, including reproductive health care; (b) sufficient food and water" and "(2) [t]he State must take reasonable legislative and other measures, within its available resources, to achieve the progressive realization of each of these rights."

5 Along these lines, the Constitution of India (1950), Part IV, Article 47, provides that "[t]he State shall regard the raising of the level of nutrition and the standard of living of its people and the improvement of public health as among its primary duties."

emissions, without taking sufficient action to reduce their environmental impact: emitted particulates exceeded EU limits and induced respiratory diseases, inconsistent with the Kyoto Protocol (ECteSR 2005: para. 34). Based on various regulatory instruments, including the UNFCCC, Kyoto Protocol, TEU and TEC environmental norms, and the EU ETS, the ECteSR considered that the right to health under ESC Article 11 embeds the "right to a healthy environment" (ECteSR 2005: para. 160). In light of this, the Committee held Greece in violation of ESC Article 11(1), establishing a duty to "remove as far as possible the causes of ill-health," because its energy policy grounded in lignite exceeded binding CO_2 emission limits without adequate monitoring and safety standards. Furthermore, the Committee considered Greece in breach of the duty to prevent diseases under ESC Article 11(3) because of the absence of regular health assessments and of the obligation to promote health under ESC Article 11(2), in light of the failure to implement adequate environmental impact assessments (ECteSR 2005: paras 222 ff.).

As the right to health is interdependent, indivisible, and interrelated with other human rights, violations of the right to health impair other fundamental duties, such as the rights to housing, food, water, adequate standards of living, and life, and vice versa. Thus, the Appeals Court in *Urgenda* (*The State of the Netherlands* v. *Urgenda Foundation* 2019) upheld the argument of the cross-claimant against the Netherlands, considering that Articles 2 and 8 of the European Convention on Human Rights (ECHR) impose on the state a duty of care to take action and prevent foreseeable breaches of the rights to life and to private and family life caused by climate change, whereby the claims to life and to private and family life cover "environment-related situations" and the "right to a healthy environment" (ECtHR 2009: paras 107 and 112), including a "state duty to protect" against "a real threat of dangerous climate change." Similarly, recent claims in the ECtHR are based on the right to health as part of an extensive interpretation of the rights to life and to private and family life under ECHR Articles 2 and 8. For instance, in *Mex M. v. Austria* (ECtHR 2021), a citizen affected by Uhthoff's syndrome claims that Austria is in breach of the right to private and family life under ECHR Article 8, extensively interpreted so as to protect the right to a healthy environment – a component of the right to health – for failing to adopt effective climate policies, arguing that excessive temperatures impair the ability of the claimant to walk. Furthermore, increased temperature would expose the claimant to the risk of premature death, in violation of the right to life under ECHR Article 2.

Right to a Healthy Environment, Right to a Healthy Climate, and Right to Health

Within the outlined framework, the question arises whether acknowledging a human right to a healthy environment rather than a human right to a healthy climate, along the lines of the developments that are taking place in the progressive evolution of EU law, might make a difference. The right to a healthy climate is indeed narrower and more specific than the claim to a healthy environment, with relevant ramifications.

First, the right to a healthy climate arguably allows consideration of the duty to stabilize temperature increase within sustainable limits as a more stringent obligation to achieve the result rather than a simple obligation of conduct (Fitzmaurice 2010: 107). It is indeed easier to assume that states and other legal persons have an obligation to achieve the aim of stabilizing GHG emissions rather than the aim of achieving environmental protection at large.

Second, it is simpler to assume that the right to healthy climate is a human right, as unlike the right to a healthy environment, this would not affect other obligations, such as the human rights nature of claims relating to the disposal of hazardous waste. This could contribute to raising the minimum standard states have to respect in the progressive fulfilment of the right to health via "concrete and targeted steps," as underscored by the Special Rapporteur on the right at issue (Special Rapporteur on the Right to Health 2016: para. 26). Concretely, the human right to a healthy climate would provide a specific lens for monitoring how governments generate or miss opportunities to allocate budgets and funds to implement the right to health to the maximum of available resources (Special Rapporteur on the Right to Health 2016: para. 26; Tait et al. 2018: S64-65). Indeed, in its current status, it is uncertain what obligations precisely the right to health imposes on the state (Goodman 2005: 643).

Third, it is easier to posit that the right to a healthy climate is an *erga omnes* duty, as, unlike the right to a healthy environment, it does not compel assuming the *erga omnes* nature of unrelated claims, such as those concerning the disposal of hazardous waste. In other words, the right to a healthy climate entails a narrower overlap with the right to health, as compared to the holistic right to a healthy environment, which allows a more flexible approach to the uncertainties surrounding the nature and structure of the right to health. That the right to health should, in principle, be recognized as an *erga omnes* duty is witnessed by the fact that it is embedded in the 1946 Universal Declaration on Human Rights as a "right of everyone," as well as in the 1962 Covenant on Civil and Political Rights, which has a universal scope of application (European Parliament 2022: 45). Furthermore, the universal nature of the right to health has been recognized by the CteESCR in the context of the international duty of cooperation, compelling states toward respecting and protecting the enjoyment of the right to health in third world countries (CteESCR 2000: para. 39; Meier et al. 2022: 17 ff.).

On the other hand, specific protection offered by the human right to a healthy climate is naturally more limited than that afforded by the right to a healthy environment. The first can only reinforce the right to health via-à-vis the threat posed by GHG emissions, whilst the second strengthens protection from a wide set of threats, spanning water and air pollution, as well as hazardous waste and substances, also including GHG emissions (WHO 2020: 28, para 44; Boyd et al. 2021: 27 ff.). For instance, only the right to a healthy environment has a sufficient scope to cover exposure to "harmful substances," such as radiation and chemicals, and "other detrimental environmental conditions" directly or indirectly affecting human health, which the CteESCR has indicated as essential components of the right to health (CteESCR 2000: para. 15). A wide perspective is in line with the broad approach outlined by the WHO to tackle the relationship between health and climate change, in light of the 2030 Agenda for Sustainable Development, aiming to create "safe and healthy environments" and to "improve people's lives today and in the future" (UN 2015; WHO 2020: 8, para. 18). The WHO, indeed, envisages the adoption of "systematic consideration of health" and "comprehensive evaluation" in the elaboration of health policies beyond the health sector, spanning urban planning, energy, transport, housing, food systems, agriculture, water, and sanitation (WHO 2020: 9, para. 19).

Conclusion

International law is progressively evolving toward the recognition of a human right to healthy environment, particularly under the impulse of the Special Rapporteur on the Environment and Human Rights. However, despite the recent adoption of Resolution A/76/L.75 on the matter by the UN General Assembly, it cannot be considered that the right to a healthy environment is yet established as a binding international obligation. Within this framework, it is possible to trace an evolution in EU law toward the recognition of a more specific human right to a healthy climate, fostering progressive international law development. This raises the question as to what the implications would be for the right to health of the recognition of a fundamental right to a healthy climate rather than a healthy environment.

On the one hand, the right to a healthy environment is a holistic concept and covers a broad range of threats, spanning radiations, chemicals, and any sorts of harmful substances, in line with the interrelationship between the environment and health outlined by the CteESCR. On the other hand, the right to a healthy climate is more targeted and could thus more easily contribute to establishing the right to health as an *erga omnes* human right and duty of result rather than one of conduct, with respect to climate-related threats.

Arguably, these developments should not be seen as inconsistent but, rather, as complementary. Establishing a human right to a healthy climate in the context of the right to a healthy environment, possibly seen as a fundamental right, would afford maximally improved respect, protection, and fulfilment for the right to health vis-à-vis environmental threats in general and climate change in particular. Indeed, UN Resolution A/76/L.75 and EU Resolution 2020/2134 (INI) acknowledge the human right to healthy environment, particularly in light of climate change.

Acknowledgments

The research for this chapter was done with the support of the Erasmus+ program of the European Commission. (Projects ESEU 101176249 and GLOBEU 101127105) Content reflects the view of the author only.

Reference List

Bell, M. & Greenberg, M. (2018) "Climate change and human health: Links between history, policy, and science," *American Journal of Public Health*, 108(S2), pp. S54-S55.

Bosselmann, K. (2017) *The Principle of Sustainability Transforming Law and Governance*, 2nd ed., London: Routledge.

Boyd, D. (2017) *The Rights of Nature: A Legal Revolution That Could Save the World*, Toronto: ECW Press.

Boyd, D. (2019) *Report of the Special Rapporteur on the Issue of Human Rights Obligations Relating to the Enjoyment of a Safe, Clean, Healthy and Sustainable Environment*, UN Doc. A/74/161.

Boyd, D., Knox, J., & Limon, D. (2021) *The Case for Universal Recognition of the Right to a Safe, Clean, Healthy and Sustainable Environment*, Universal Rights Group.

Committee of Ministers of the Council of Europe (2022) *Recommendation to Member States on Human Rights and the Protection of the Environment*, CM/Rec (2022)20.

Crimmins, A., et al. (2016) *The Impacts of Climate Change on Human Health in the United States: A Scientific Assessment*, Washington, DC: US Global Change Research Program. Available at: http://doi.org/10.7930/J0R49NQX.

CteESCR (2000) *The Right to the Highest Attainable Standard of Health (Art. 12)*, General Comment No. 14.

D. Or. (2016) *Juliana v. United States*, 217 F Supp (3d) 1224.

Davies, K., et al. (2017) "The declaration on human rights and climate change: A new legal tool for global policy change," *Journal of Human Rights and the Environment*, 8(2), p. 250.

De Vido, S. (2020) "Climate stability as a common good: A strategy for the European union," in: S. Baldin & S. De Vido (eds) *Environmental Sustainability in the European Union: Socio-Legal Perspectives*, pp. 101–120, Trieste: EUT.

ECteSR (2005) *Marangopoulos Foundation for Human Rights (MFHR) v. Greece*, Collective Complaint No. 30.

EctHR (2009) *Tatar v. Romania*, Appl. No. 67021/01, Judgment.

EctHR (2021) *Mex M. v. Austria*, Complaint.

Esty, D.C. & Bell, M.L. (2018) "Business leadership in global climate change responses," *American Journal of Public Health*, 108(Suppl 2), pp. S80-S84.

EU (2000) *Charter of Fundamental Rights*, 43 OJ C 364/1.

European Parliament (2020) *Resolution on the Effects of Climate Change on Human Rights and the Role of Environmental Defenders on This Matter*, 2020/2134(INI).

European Parliament (2021) *A Universal Right to a Healthy Environment*, Doc. PE 698.846.

European Parliament (2022) *Right to Health – A Comparative Perspective: Canada*, Doc. PE 729.444.

Fitzmaurice, M. (2010) "Responsibility and climate change," *German Yearbook of International Law*, 53, pp. 89–138.

Glicksman, R., Paddock, L., & Kim, D. (2018) "The role of health in climate litigation," *American Journal of Public Health*, 108(S2), pp. S104–S108.

Global Network on Human Rights and the Environment [GNHRE] (2021) *Declaration on Human Rights and Climate Change*. Available at: https://gnhre.org/declaration-human-rights-climate-change.

Goodman, T. (2005) "Is there a right to health?" *Journal of Medicine and Philosophy*, 30(6), p. 643.

Ksentini, F.Z. (1994) *Human Rights and the Environment, Review of Further Developments in Fields with Which the Sub-Commission Has Been Concerned, Final Report of the UN Sub-Commission on Prevention of Discrimination and Protection of Minorities*, United Nations Economic and Social Council, UN Doc. E.CN.4/Sub.2/1994/9.

McIver, L., et al. (2016) "Health impacts of climate change in pacific island countries: A regional assessment of vulnerabilities and adaptation priorities," *Environmental Health Perspectives*, 124/11, pp. 1707–1714.

Meier, M., de Mesquita Benjamin, J.B., & Williams, C.R. (2022) "Global obligations to ensure the right to health: Strengthening global health governance to realise human rights in global health," *Yearbook of International Disaster Law*, 3(1), pp. 3–34.

Parker-Flynn, J.E. (2018) "Linking mitigation and adaptation to climate change," *American Journal of Public Health*, 108(S2), pp. S58–S59.

Parliamentary Assembly of the Council of Europe (2021) *Anchoring the Right to a Healthy Environment*, Recommendation 2211.

Pirjatanniemi, E. (2016) "Greening human rights law: A focus on the European convention on human rights," in: J. Bos & M. Düwell (eds) *Human Rights and Sustainability: Moral Responsibilities for the Future*, pp. 19–31, London: Routledge.

Tait, M.E. (2018) "Building a culture of health in our changing climate," *American Journal of Public Health*, 108(S2), pp. S64–S65.

UN (2015) *Transforming Our World: The 2030 Agenda for Sustainable Development*. Available at: https://sdgs.un.org/2030agenda.

UN Conference on Human Environment (1972) Report, Doc. A/Conf.48/14/Rev.1.

UN Conference on Human Environment (1992) Report, Doc. A/CONF.151/26.

UN Educational, Scientific and Cultural Organization [UNESCO] (1999) *Bizkaia Declaration on the Right to Environment*, Doc. 30 C/INF.11.

UN General Assembly (2022) *The Human Right to a Clean, Healthy and Sustainable Environment, Resolution*, Doc. A/76/L.75.

UN Human Rights Council [HRC] (2021) *The Human Right to a Safe, Clean, Healthy and Sustainable Environment*, UN Doc. A/HRC/48/L.23/Rev.1.

UN Office of the High Commissioner on Human Rights [OHCHR] and World Trade Organization [WTO] (2008) *The Right to Health*, Factsheet n. 31.

UN Special Rapporteur on the Right to Health (2016) *Right of Everyone to the Enjoyment of the Highest Attainable Standard of Physical and Mental Health*, UN Doc. A/71/304.

World Health Organization [WHO] (2020) *WHO Global Strategy on Health, Environment and Climate Change: The Transformation Needed to Improve Lives and Wellbeing Sustainably Through Healthy Environments*. Available at: https://www.who.int/publications/i/item/9789240000377.

World Health Organization [WHO] (2021) *Climate Change and Health*. Available at: www.who.int/ news-room/fact-sheets/detail/climate-change-and-health.

Wu, C.-F. (2021) "Challenges to protecting the right to health under the climate change regime," *Health and Human Rights Journal*, 23(2), pp. 121–138.

Chapter 10

Lessons From World War II

Mobilizing Canada to Fight the Climate Crisis

Anne Venton

Introduction

Since 1990, Canada has failed to address climate change, as its greenhouse gas emissions have increased by 21%. As a result, Canada faces a climate crisis that requires a 50% reduction in its 2018 greenhouse gas emissions by 2030. This chapter will argue that to address this climate emergency, Canada must now mobilize to fight the climate crisis by utilizing lessons learned from its highly successful mobilization efforts for World War II. To do so, it will begin with a short history of Canada's present climate crisis based on reports from climate policy practitioners and scholars Elizabeth May, Gordon Laxer, William Carroll, Martin Lukacs, and Seth Klein. They explain how federal government policies for reducing fossil fuel emissions to the target of net-zero by 2050 are being obstructed in order to preserve the business-as-usual models of corporate companies in the fossil fuel and energy sectors in Canada's neoliberal capitalist economy, leading to a lack of political will to protect the ecological integrity of the country's land and its species. The chapter will conclude that Canada needs to mobilize to fight the climate crisis by utilizing ten lessons based on those learned in Prime Minister Mackenzie King's administration, particularly by C. D. Howe, when Canada "hit above its weight" in successfully mobilizing for World War II.

A Short History of Canada's Climate Crisis

On 18 May 2021, Elizabeth May, environmentalist and former leader of the Green Party of Canada (2006–2019), gave a virtual presentation to the Toronto Branch of the World Federalist Movement Canada (WFMC) titled "A Global Accountability Act: What Would it Look Like?" Her presentation included a brief historical review of Canadian government policy on restricting greenhouse gas emissions (May 2021).

May's presentation was about a tale of two treaties. The first treaty was the 1987 Montreal Protocol limiting threats to the Earth's ozone layer from chlorofluorocarbons in refrigeration appliances and products like aerosol

DOI: 10.4324/9781003440871-12

cans. It was a treaty which worked because it incorporated a clear, effective enforcement mechanism of foreign trade sanctions against delinquent countries and made its most stringent requirements on wealthy countries.

The second treaty was the 1997 Kyoto Protocol that was made under the 1992 Rio Summit's United Nations Framework Convention on Climate Change (UNFCCC). That framework committed states to reduce greenhouse gas emissions, based on the scientific consensus that human-made CO_2 emissions are driving global warming. The Kyoto Protocol was ineffective, in part, because it did not apply the trade sanctions architecture of the Montreal Protocol. In 1994, the World Trade Organization (WTO) had agreed on Trade-Related Aspects of Intellectual Property Rights (TRIPS). This removed the option for Kyoto to include trade sanctions which would have violated protection of intellectual property.

The alternatives to trade sanctions include group shaming during global assessments. Such general approbation is ineffective because ministers of finance and civil servants are unprepared to meet targets when there are no immediate term penalties. Canada's Bill C-12 (an Act respecting transparency and accountability in Canada's efforts to achieve net-zero greenhouse gas emissions by 2050) was adopted by the House of Commons on 22 June 2021. It is a weak law because it only involves reporting of the country's greenhouse gas emissions relative to target (i.e., public shaming). C-12 does not have enforcement penalties and procedures.

In response to questions from participants at the conference, May replied that the Canadian government had exhibited appalling incrementalism with respect to climate change. She added:

> [W]hile I love the Prime Minister Justin Trudeau, there is cognitive dissonance on his part when he says you can build the $17 billion Transmountain Pipeline and reduce greenhouse gas emissions. He exhibits no political will to enforce the Paris agreement targets. We are apparently distracted from meeting targets because we don't want to upset Alberta.
>
> (May 2021, *in synopsis*)

May concluded that there is no political will in Canada to enforce the Paris Agreement targets. The biosphere is at great risk and will be unlivable if the 1.5°C target is not met, yet we are currently on track for 3.0°C to 4.0°C. There is a very significant risk, almost a certainty, that human civilization will not be able to withstand the number of changes that will occur in our environment.

She added that for effective greenhouse gas management and accountability at the world level, there need to be enforcement penalties and procedures, such as dismantling the Trade-Related Aspects of Intellectual Property Rights (TRIPS) and reconfiguring the World Trade Organization (WTO) as the World Trade and Environment Organization (WTEO). Its purpose would be to stop the consumption of fossil fuels to enable a shared responsibility for a future that we can live with. May concluded with the remark that such action would bring an end to support for the profit and greed of multinational corporations at the expense of the environment and excessive climate change.

How Big Foreign Oil Companies Have Captured Canadian Energy and Climate Policy

In December 2021, the Council of Canadians (COC) and the Canadian Centre for Policy Alternatives (CCPA) co-published a groundbreaking report, "Posing as Canadian: How Big Foreign Oil captures Canadian energy and climate policy," authored by political economist and Professor *Emeritus* of the University of Alberta Gordon Laxer (Laxer 2021). Laxer is also the author of two award-winning books, *Open for Business: The Roots of Foreign Ownership in Canada* and *After the Sands: Energy and Ecological Security for Canadians*.

The two highlights of the report are that the Canadian Association of Petroleum Producers (CAPP) acts as an extensive lobby group that meets with politicians and high-ranking bureaucrats in the federal and Alberta governments who are responsible for the development of policies and regulations for reducing the production and consumption of greenhouse gases, and that 97% of oil produced by corporations on CAPP's board in 2018 came from fully or majority-owned foreign corporations, and most of that oil is produced by big corporations. Presumably, these facts would end the demonization of Canadian environmentalist critics of oil as anti-Canadian. The effects of lobbying are to let Big Foreign Oil (BFO) increase its production and emissions with few restrictions and maintain its high profits.

The report contends that to accrue social license, BFO and its supporters go to great pains to deepen the notion that the identity and economic well-being of Albertans and Canadians are inextricably linked to oil. One of their main strategies in gaining social legitimacy is through astroturfing: funding from groups that masquerade as grassroots citizen support for oil and gas extraction.

CAPP has influenced individual voters in provincial and federal elections through micro-targeting surveys and massive advertising spending. Despite being dominated by BFO, the lobby group has benefited from a loophole that allows it to circumvent rules preventing foreign money interference in Canadian elections. As a well-funded registered third party, it has been able to command the attention of voters and political leaders during election campaigns. In addition, the executives of oil corporations on CAPP's board of governors have used various techniques to circumvent bans on corporations to make sizable contributions to politicians favorable to their agenda.

Breaking Through Big (Foreign) Oil's "Regime of Obstruction"

In October 2020, the CCPA interviewed William Carroll, a critical sociologist at the University of Victoria, "Breaking through Big Oil's 'Regime of Obstruction'" (Carroll 2020a). The interview was about Carroll's edited book *Regime of Obstruction: How Corporate Power Blocks Energy Democracy* (2020b). The book is an anthology of 32 contributors who demonstrate how corporations secure popular consent and co-opt, disorganize, or marginalize dissenting perspectives and position the fossil fuel industry as a national public good.

In his interview, Carroll says that the new denialism does not deny science. It accepts that there is a climate crisis, but it offers up solutions that are obviously inadequate. For example, rather than making dramatic changes to reduce the production and use of fossil fuels, the new denialism proposes a very slow, incremental pace of change that enables companies to carry on business as usual without endangering their profits or public investments in their capital. More specifically, it does this by steering policies toward market decisions such as implementing carbon taxes and technological innovations that make carbon extraction less intensive in terms of emissions, without acknowledging the logical contradiction of endless economic growth in a finite planet. These inadequate policies will clearly not address the climate crisis that the government is expected to address. So the industry constructs new "denialist narratives" that will influence government policymakers and the public to accept them (Carroll 2020a).

He continued that the fossil fuel sector also offers partnerships and revenue sharing with Indigenous communities to support pipeline projects. These revenue sharing agreements are a financial benefit held out to Indigenous communities, and understandably, a number have agreed. The divide-and-rule approach is front and center both for the industry and the Canadian government in trying to get things built and keep a lid on dissent. Carroll sees fossil fuel capital influencing public opinion at various "scales" or targets, such as municipal and provincial governments (especially Alberta), as well as the federal government and global and policy-planning organizations, plus knowledge production institutions (e.g., think tanks) and the consumer capitalist culture of everyday life, including global and policy-planning organizations (Carroll 2020a: 40).

Carroll also raised that in March 2022, during the energy crisis in Alberta, CAPP asked the federal government to subsidize the fossil fuel sector with an ask of between $27 billion and $30 billion in bailout money (Carroll 2020a). The federal government was actually more targeted. In April, the government announced $2.45 billion in aid for workers, $1.7 billion to clean up abandoned oil and gas wells, $750 million on emissions-reducing activities, and the availability of higher risk in financing in the form of Export

Development Canada commercial loans. This package of industry bailouts was done in a way that is not problematic in terms of public concern about the climate crisis. For example, the package excludes the value of the loans from Export Development Canada that could be very large or the probability that they would never be repaid. Therefore, the cost of the package is likely understated. Carroll expects there will be maneuvering going on to package industry bailouts in ways that are optically not problematic in terms of public concern.

In addition, he noted that Canada already is paying the fossil fuel industry enormous amounts of public money. An International Monetary Fund study in 2019 found post-tax subsidies to be in the $43 billion range in 2015–2016. That was almost 20% of the federal budget in that year. Other countries massively subsidize their energy sectors as well. Canada is not really an exception. But it is quite remarkable the extent to which the profitability of this industry is almost entirely dependent on state subsidy (Carroll 2020a: 40).

The capitalized market value of $43 billion in profits is $645 billion, assuming a price-to-earnings ratio of 15. Therefore, phasing out of the industry would save the government a great deal of money, but it would reduce the financial wealth of shareholders and pension plans of others to the extent of the fraction of those investments in fossil fuel companies. Also, as he noted, it would reduce the value of investments in fossil fuel companies by the five big banks, which are heavily invested in the industry, as they own significant blocks of shares in the major fossil fuel corporations (Carroll 2020a).

As an estimated 80% of all energy in the global economy is generated from carbon, he contends that the entire capitalist system is based on cheap energy from burning fossil fuels (Carroll 2020a: 38). Without it, prices will rise until there is an adequate amount of green energy, which is described as a declining-cost industry over time and scale. The question is how much time it will take for the green economy to fully develop.

To fight back, it is necessary for civilians to pressure governments. Changing attitudes and divestment politics are also important. Carroll believes, "We need to think about the fronts that need to be opened up, from everyday life right up to local and national politics to the global" (2020a: 38). What is needed to confront the climate crisis is a just transition from fossil fuel capitalism to something different that would incorporate social justice.

The Trudeau Formula for Business as Usual

The phenomenon of the new denialism in Canada's fossil fuel industry is a symptom of a change in Western democracies from Keynesian economics, progressive income taxation, and the welfare state that developed in the period from 1945 to 1970 to neoliberal capitalism that has emerged over the 50 years since 1970. Neoliberal capitalism features policies of corporate income tax cuts, reduction in the progressivity of income taxation, preferential lower tax rates on capital gains than income. It also features the privatization of government programs and a partial withdrawal from the post–World War II welfare state plus deregulation of markets for goods and services and particularly finance. Extreme inequality of wealth and income and climate breakdown are the largest consequences of neoliberal capitalism.

In Canada, the politics of denial of neoliberal capitalism are best explained in Martin Lukacs's book *The Trudeau Formula: Seduction and Betrayal in an Age of Disconnect* (Lukacs 2019). In a *Canadian Centre for Policy Alternatives (CCPA) Monitor* article, Lukacs – who is also an investigative journalist with the CCPA – provides an analysis of how Canada's prime minister Justin Trudeau has seduced the public with "sunny ways and real change" (Lukacs 2020).

"Business as usual" refers to the operations of capitalist businesses in the neoliberal economic system that is supported by most governments that have adopted the paramount policy goal of ever-increasing economic production or "output," measured by gross domestic product (GDP). Not surprisingly the goal of GDP growth is compatible with the capitalist goal of maximizing profits. However, the increased production in GDP is heavily dependent upon burning fossil fuels to provide the energy for production – energy which is overly cheap due to mispricing in so-called free markets and large government subsidies of oil and gas companies. The tax policies of neoliberal capitalism over the last 50 years have also resulted in a huge increase in the inequality of income and wealth – most notably publicized in the Occupy movement of 2011 that followed the crash in the US real estate market of 2007. Preferential tax policies on capital gains that have deregulated finance have fueled the hyper-financialization of the economy, as evidenced by the increase since 1970 in private wealth, the number of billionaires and incomes of corporate CEOs relative to national averages.

In general, the formula for climate denial involves accepting its problem as a fact but then offering in reality a token or partial solution which is implicitly presented as the full solution. Martin Lukacs illustrates the Trudeau formula for denying neoliberalism with reference to Trudeau's 2015 speech to Bay Street financiers. Trudeau's speech was in response to the establishment media that had howled about Trudeau's redistributive dogma. In his speech to the financiers, Trudeau stated how the "sense of fairness in Canadian society had evaporated as a result of growing inequality, stagnating wages, soaring personal family debt, and an increasingly insecure safety net" and therefore was not sustainable; if a government could not put forward an agenda aimed squarely at restoring the sense of fairness (implicitly to be restored by his recent modest tax rate hike on the top 1% of income earners), Trudeau warned that Canadians will eventually entertain more radical options (Luckas 2020: 18).

The Trudeau formula revolves around making a close study of the values and aspirations of the country's progressive majority and co-opting its protests, language, and demands. Selective concessions to the left may follow, but the progressive aspirations behind these policies are frequently hollowed out as part of a quiet compact with the corporate elite, whose policies are channeled, repackaged, and advertised as a captivating option. Demobilization and integrating progressive challenges in this way enable the government to manufacture consent for the prevailing social order and capture the voting blocs necessary to win or remain in office (Lukacs 2020: 18).

Lukacs believes that the politics of "changeless change" features "a spectacle of splashy announcements and bold initiatives that appear to disrupt business-as-usual but, in fact, mostly shore up prevailing disparities of wealth and power" (Lukacs 2020: 18). He also notes that

> [e]ven the government's new pledge to reach Canada's 2030 target and get to net-zero sounds like having your cake and eating it too further noting that "Net zero" is a shifty concept, long pushed by the global fossil fuel industry to allow them to keep extracting oil and gas with the dim hope of technologies emerging to suck carbon out of the atmosphere or to spray the stratosphere with potentially very dangerous solar reflecting chemicals.
> (Lukacs 2020: 19)

To conclude, the goal of net-zero will never be achieved by the Liberal Party operating with the Trudeau formula in which the leader, through a politics of changeless change, seamlessly collaborates with the wealthy and powerful to enable them to conduct business as usual, which is powered by burning fossil fuels that generate greenhouse gas emissions. The forecasts of climate temperature for the business-as-usual model are stark. A war on climate change in Canada is now in order.

A Good War: Mobilizing Canada for the Climate Emergency

Canadian economist Seth Klein's 2020 book *A Good War: Mobilizing Canada for the Climate Emergency* chronicles how Canada "hit above its weight" in terms of mobilizing the media, industry, and all its citizenry to face the biggest challenge of its time: World War II (Klein 2020a). He developed a battle plan of 14 key lessons from Canada's war experience to guide Canadians now through the stages of climate mobilization to successfully reduce the nation's greenhouse gas emissions by 50% from their 2018 level by 2030 (Klein 2020a: 184). He summarized these lessons in a short article, "Mobilize like we mean it," in the *CCPA Monitor* (Klein 2020b: 31–35). The ten most important are:

1. *Adopt an emergency wartime mindset, and be prepared to do what it takes to win* (Klein 2020b: 32). "To win" means to reduce emissions by 50% from their 2018 level by 2030 (Klein 2020a: 184).

2. *Rally the public at every turn.* The public needs to be rallied through news media, educational institutions, and better inclusion of the public in decision-making through the use of citizen assemblies and other means of democratic engagement. The purpose of the rally is to build climate literacy for all, plus knowledge about practicing democratic deliberation about policies so that all the public are politically engaged and involved in decision-making on refining climate policies to achieve the 2030 target (Klein 2020b: 32).
3. *Lessen inequality, which is toxic to social solidarity and mass mobilization.* In World War II, inequality was dramatically reduced by limiting excess profits and introducing other redistributive policies, which were done to avoid World War I experiences in which inequality undermined war efforts (Klein 2020b: 32).
4. *Embrace economic planning and create economic institutions needed to get the job done.* Government economic planning in World War II utilized private sector extensively, but the allocation of scarce resources decisions was not given to corporations or the market. C. D. Howe's department undertook detailed economic planning to ensure wartime production and conducted a national inventory of wartime supply needed and production capacity and coordinated the supply chains of all core war production inputs, making tools, rubber, metals, timber, coal, oil, and more. Institutions of 28 Crown corporations were created to execute and implement the plan to achieve the targets. The climate emergency demands similar approaches to economic planning. We must again conduct an inventory of conversion needs, determining how many heat pumps, solar arrays, wind farms, electric buses, etc. we will need to electrify virtually everything and end our reliance on fossil fuels. We will need a new generation of Crown corporations to then ensure those items are manufactured and deployed at the requisite scale. We will require huge public investments in green and social infrastructure to expedite the transformation of our economy and communities. And as we did in the war, we will need to mobilize labor to get this job done, banishing unemployment for years to come (Klein 2020b: 33).
5. *Spend what it takes to win.* Federal government spending for the success of winning World War II was enormous. The federal government issued new public victory bonds, and new forms of progressive income taxation were instituted (Klein 2020b: 33). At war's end in 1945, federal debt was 108% of its annual GDP – way above the 40% average ratio of year-end debt to GDP over the 65-year period from 1950 to 2015 (Klein 2020a: 184).
6. *Leave no one behind.* The ambition of plans for returning soldiers should inspire us to develop robust programs for all workers whose economic and employment security is currently tied to the fossil fuel economy, with a special focus on those provinces and regions most reliant on oil and gas production (Klein 2020b: 33).

7. *Reject the straitjacket of neoliberal economic thinking.* The previous six lessons all share a common thread, the casting off of free market economic ideas that have kept us from doing what we need to in the face of the climate emergency. During World War II, given the urgency and scale, both the general public and the private sector leaders understood that the economic transformation had to be state-led (Klein 202b: 34).
8. *Transform government.* We need a Climate Emergency War Cabinet Committee plus a Climate Emergency Secretariat in the Prime Minister's Office and in each Premier's Office, all of whom are coordinating the emergency response as a "whole of government" approach. A new federal–provincial cost shared program focused on addressing the climate crisis, including a new federal Climate Emergency Just Transition Transfer, to collaboratively fund new green infrastructure and job training initiatives, with funding going disproportionately to the provinces with the most "heavy lifting" to do in the transition, is needed. We need to breathe a new ambitious spirit into the civil service. The challenge now is to transform the public service – to recruit and promote people willing and able to make bold things happen quickly. We need visionary and creative people in key leadership positions in the civil service and to bring in outside experts, civil society leaders, and entrepreneurs (Klein 2020b: 34).
9. *Everyone has to do their part.* Households will need to shift their consumption, their transportation, and how they heat their homes. All companies and institutions, public and private, need transition plans. Social movements are needed to keep governments' feet to the fire at every stage (Klein 2020b: 34).
10. *Know thine enemy.* We face numerous barriers to change, particularly a fossil fuel industry that has done much to block climate action. One of the most insidious barriers is a dynamic of climate denialism which currently dominates our politics and is the new modus operandi of the fossil fuel industry (Klein 2020b: 35).

Conclusions

Canada's overarching need is to find bold leaders for its political system who will reject neoliberal thinking and thereby curb the economic power of the Big Foreign Oil enemy, to lessen inequality across economic classes and leave no one behind. The purpose of these is to achieve social solidarity among all the citizenry who are to be enlisted in the economic planning and work of mobilization through Crown corporations and institutions to reach the goals of reduced greenhouse gas emissions by 2030 and net-zero by 2050. Canada needs a new bold leader like Mackenzie King or Franklin Delano Roosevelt to deliver us from our climate emergency to a green economy that is free of fossil fuel emissions and simultaneously deliver us from our neoliberal capitalist economy to a social democratic economy for the common public good.

Reference List

Carroll, W. (2020a) "Breaking through big oil's regime of obstruction," *CCPA Monitor*, 27(4), pp. 38–42.

Carroll, W. (ed) (2020b) *Regime of Obstruction: How Corporate Power Blocks Energy Democracy*, Athabasca: Athabasca University Press.

Klein, S. (2020a) *A Good War: Mobilizing Canada for the Climate Emergency*, Toronto: ECW Press.

Klein, S. (2020b) "Mobilize like we need it," *CCPA Monitor*, 27(3), pp. 31–45.

Laxer, G. (2021) *Posing as Canadian: How Big Foreign Oil Captures Canadian Energy and Climate Policy*, Council of Canadians & Canadian Center for Policy Alternatives Saskatchewan and British Columbia Offices. Available at: https://canadians.org/wp-content/uploads/2022/05/Posing-as-Canadian-Gordon-Laxer.pdf.

Lukacs, M. (2019) *The Trudeau Formula: Seduction and Betrayal in an Age of Discontent*, Montreal: Black Rose Books.

Lukacs, M. (2020) "Is it still business as usual?" *CCPA Monitor*, 26(5), pp. 17–20.

May, E. (2021) *A Global Climate Accountability Act: What Would It Look Like?* Presentation to the Toronto Branch of the World Federalist Movement Canada (WFMC).

Chapter 11

The Next Green Frontier

Legal Readiness for Climate Finance

Virginia Zambrano

Introduction

How to protect and restore the integrity of the ecological systems? Is any sustainable development possible that can guarantee the preservation of biological diversity and adequate living conditions for people, avoiding the impacts of climate change, including drought and desertification, which jeopardize access to water, food, and land and contribute to armed conflicts and political instability around the world? The solution is apparently simple. It is a question of changing production and consumption systems to safeguard the regenerative capacity of the Earth, human rights, and the well-being of future generations. However, due to the lack of adequate economic resources, public policies developed by states are unable to adequately pursue these goals. In this situation, the financial market has been able to intercept the challenge of sustainability and has developed a series of financial instruments that can provide liquidity – and thus encourage investment – on the move toward a sustainable economy (i.e., green finance products). The "sustainability" challenge is thus intercepted by a market that has increasingly created innovative financial products. But as genuine as the interest in the environment may be, the model offered to investors risks to be centered only on the usual earning of profits, unless a strict regulation of these "financial green products" is provided.

Finance Intercepting Sustainability

Dostojesky, in *The Brothers Karamazzov*, says, "Each of us is guilty of everything and for everyone on Earth, this is beyond doubt, not only because of the original common guilt, but each individually for all men and for every man on Earth." The message is simple and goes in the direction of considering the consequences of human activities as time-projected. The various IPCC Reports sound like criticism of the current production system (Albrizio et al. 2014). The risks are not only for health and social stability but also for an economy subjected to the challenges of climate change (ACPR 2020; Bank of England 2019).

DOI: 10.4324/9781003440871-13

Vulnerable under these premises, sustainability enters the official agendas and drives the choices of public actors and economic operators, enriching National Recovery and Resilience Plans (PNRR), action plans, taxonomy strategies, etc. with content and dreams. In spite of the declamations, legal discourse does not move away from the stylistic features of market exchange and denounces the difficulty of departing from cognitive structures and ways of thinking that are designed for another type of relationship (Alessi et al. 2019a; Pindyck 2013). Even legal language accounts for this. The use of "natural capital" appears in the European Commission Communication COM (2011), 3 May 2011, no. 244 ("Our life insurance, our natural capital: an EU biodiversity strategy to 2020"), but also in the report of the Italian Comitato per il Capitale Naturale (CCN), which defines it as

> the entire stock of natural assets – living organisms, air, water, soil and geological resources – that contribute to providing goods and services of direct or indirect value to humans and that are necessary for the survival of the environment from which they are generated.

If this is the case, it is a question of understanding how to make the most of natural capital and resources by meeting the challenge of sustainability, which, as the various reports of the Global Sustainable Investment Alliance (GSIA) tell us, raises the interest of the financial markets.

The capital market is the main channel for corporate financing. A market that is characterized by a large number of financial innovations and operates without limits, also intercepting the challenges coming from environmental and social governance (ESG). The goal of the financial markets is, in short, to shift investments toward sectors that are less exposed to ESG risks and more aligned with SDG goals (Zerbib 2019). Having turned sustainability into an asset, it is a question of identifying tools and synergies to mobilize capital, by setting clear targets to achieve the famous "climate change mitigation and adaptation."

The "sustainability" challenge is thus intercepted by a market that has been increasingly creating innovative financial transactions. But as genuine as the interest in the environment may be, the model offered to investors remains centered on the prospect of earning profits, much to the detriment of medium- and long-term expectations toward the environment and other stakeholders (Van der Ploeg & Rezai 2020). On the other hand – but only if we want to see the glass half-full – it is also true that green investments can be an opportunity and be strategically used to mitigate/adapt to adverse events produced by climate change, as well as produce profits (OECD 2020). The IPCC report itself reminds us that this is the case. It is clear that the use of capital is not dictated by altruistic reasons. Finance realizes that incorrect handling of the climate issue can have devastating effects. In this direction, the Taskforce on Climate-Related Disclosure and the Network for Greening the Financial System identifies three areas for action: (1) risk management framework, (2) sustainability reporting, and (3) ESG investments (NGFS 2019a).

The recipe seems simple. Having set the objectives, which are those of the Paris Agreement and the United Nations on the SDGs, developed countries should contribute to climate finance. But what "sustainability" consists of is not easy to determine.

The Lack of an Adequate Regulatory Framework

COP26 of the UNFCCC in Glasgow supported climate finance and the opportunities it could offer and recommended (1) mobilizing \$100 billion per year by 2030 for poor countries and (2) collaboration between public and private finance. Excellent ideas with a fragile recipe, made evident when the COVID-19 pandemic highlighted the inability of developed countries to move in tune with Paris. But there is more. Glasgow said "go" but did not show the way.

For its part, the market has long since intercepted the challenge (Banque de France 2018). Not to mention debt-for-nature swaps traded on international markets, and the more recent experimentation with climate impact investing instruments and funds, the focus on blue bonds, social bonds, SDG bonds, and green bonds tells of the system's tendency to dress up the concept of "sustainability" financially (Ceci 2016; European Bank Authority – EBA 2019). In the 2010–2021 period, the supply of sustainable funds increased fivefold, testifying a genuine interest in the environment and society (USD 523 billion in 2021) (Clark et al. 2015). In short, the investor market is reacting well (Gianfrate & Peri 2019).

Where, then, is the problem? The problem is not only in the dilution of the "ecological cause" that occurs whenever the loosening of the relationship with the underlying asset increases the speculative component, causing that misalignment between the genuine interest of stakeholders and the way in which capital is employed to ensure ESG, but also in the difficulty of constructing an adequate regulatory framework capable of making the whole coherent. Even when it has been decided (with the proposed EuGB Regulation) by calling on the legislator to outline the regulatory framework, as it happened in Europe, the system does not appear to be without gaps, testifying that complex systems require complex interventions. And there is complexity here. Suffice it to say that it is not easy to understand what a green project is, what it consists of, and what rules and measurement criteria it adheres to. Not to mention the fact that the investment universe must always be measured according to sustainability indices.

In the absence of a global standard for certifying a particular bond as "green," there are guidelines drawn up by the International Capital Market Association (ICMA). There are four principles developed by ICMA, and they are simple: (1) There must be clear identification of the destination of the proceeds, (2) the projects selected must fit into a precise list of categories, (3) utmost transparency must be ensured as to how the proceeds are used, and (4) solid reporting must be in place to show investors how their money is being used. For the rest, the market is free to develop its own practices and make the best use of the contractual instrument. But even then, all is not well.

The lack of unambiguous standards makes ESG rating methodology and outcomes opaque (Berg et al. 2019). Their inconsistency and divergence contaminate the development of the performance indicator framework (Lamperti et al. 2019). The system has then "invented" not-for-profit charities, such as the Carbon Disclosure Project, which holds the gold standard of environmental reporting with the richest and most comprehensive dataset on corporate and citizen action, or the Climate Bond Initiative and Science-Based Target, which produces a labeling scheme for bonds as a user-friendly tool for investors and governments.

Here, self-reported data from companies is collected to enable investors, companies, and regulators to make informed decisions on the move toward a sustainable economy (Ehlers et al. 2020). However, the data is provided by private individuals, managed by private individuals, independent rating agencies, and auditors (Polbennikov et al. 2016). It is difficult to say how reliable this system is (Hausfather & Peters 2020). Sometimes opacity is also linked to intellectual property protection issues. There is more. Even in those areas where a standard exists, for example, CO_2 emissions, the absence of standardized reporting and the fact that we are dealing with self-reported data make the results very heterogeneous. And in this context, it is clear that if the material data is unreliable, each risk assessment can be interpreted differently and opens the door to assessment bias. The inconsistency is evident since it triggers a decision-making subcircuit that substitutes objective analysis with absolutely subjective assessments (Di Giulio & Migliavacca 2020).

The outcome is greenwashing (in the presence of high expectations) or green-hushing (i.e., the tendency to communicate nothing in order to avoid falling into greenwashing). It is clear that falsehoods and omissions in declarations are punished on both sides of the ocean, and the system is well stocked with control structures (e.g., in Italy, Testo Unico delle Disposizioni in Materia di Intermediazione F – TUF, Decreto Legislativo del 24 febbraio 1998, n. 58, and in the United States, the Securities Exchange Act of 1934).

The US Approach

In the matter of "green projects," however, the discourse is more complex. At issue is the adequacy of commitments to reduce environmental impact and the circumstance that, even without a genuine commitment, something can come along and thwart the realization of the project. In the United States, the Securities Exchange Act 1934 (15 USC § 78a), § 21E(i)(1)(B) calls them "forward-looking statements" or "plans and objectives . . . for future transactions."

It is clear that even when they assume contractual implications, it is difficult for the investor to prove the company's liability in breach of its obligation. If one sets aside the practice of including specific disclaimer clauses in the contract, the investor only proves liability if it is shown that this "forward-looking statement" was adopted despite having "actual knowledge . . . that the statement was false or misleading" (Securities Exchange Act 1934).

When the statement refers to future events, it must be shown that, given company policy, the unfeasibility of the project was already evident. However, it is clear that a company's promise to reduce its carbon emissions by 50% within 2040 is itself subject to fluctuations in the market (Mondial Bank 2020; Campiglio et al. 2018). Investments in renewable energy infrastructure, such as batteries, solar panels, and wind farms, are subject, for example, to the risks associated with the speed of policy change regarding carbon taxes. And in the United States, the safe harbor doctrine exempts from liability the company that can prove it believed in the viability of the project.

Another reason that, in the absence of a set of rules to follow, forward statements are unlikely to generate credible commitments for companies concerns the question of the "green project relevance." In the Powering Progress in the United States campaign, Shell thunderously declared that its commitment to the environment is, after all, comparable to the promises each of us made in Christmas letters as children to our parents (see, generally, Shell 2024). The disclaimer used by Shell is significant as the company said that the commitment may "differ materially from those expressed or implied in these statements," which is standard boilerplate language used by companies for forward-looking statements. Lobbying does the rest, as evidenced by the efforts of carbon-intensive companies to cast doubts, both on politicians and the public, on the importance of climate science: a disinformation strategy that echoes that of tobacco companies.

The misrepresentation, on the other hand, to be material must be such as to alter, for a reasonable investor (significantly) "the total mix of information made available" (Securities Exchange Act 1934). In *Singh v. Cigna Corp*. 918 3d 57, 64 (2d Cir. 2019), district court judges in a case alleging violations of Sections 10(b) and 20(a) of the Securities Exchange Act of 1934 and the Securities and Exchange Commission (SEC) Rule 10b-5 regarding fraud, said that the information was generically disclosed in an attempt to highlight the company's environmentally friendly efforts, but that "the statements at issue in Cigna's Code of Ethics were a textbook example of puffery, and a reasonable investor would not rely on it."

In *James v. Exxon Mobil Corporation* (65 Misc. 3d 1233 (NY Sup. Ct. 2019)), the State of New York brought an action for fraud against ExxonMobil because the company was claimed to have falsified its statements on accounting for the climate change regulation costs. The case drew global media attention. The New York Supreme Court stated that there was no mistake, and that even if there was a mistake, it was not relevant because "no reasonable investor . . . would make investment decisions based on speculative assumptions of costs that might be incurred 20 or 30 years in the future."

There is another aspect to consider. Damages in a security fraud claim should compensate investors for the losses incurred, which are related to the over- or undervaluation of the company's shares. However, the companies that fail to do so will not necessarily suffer an immediate drop in share prices (NGFS 2020a). They may also opportunistically exploit short-term fluctuations in climate policy to generate profits. And in this context, one has to wonder as to the consequences if profits are still generated despite the failure of the green project (Alessi et al. 2019b). Of course, in the case of false declarations, the issue of criminal sanctions remains open, in Italy as in America. Nevertheless, it is not clear how to graduate the intensity of this application in proportion to the violation of environmental commitments.

The other route goes through regulation. But even here one has to understand the political dimensions. In 2011, for example, the Clean Energy Legislative Package was introduced in Australia. The law was welcomed because it created confidence in green investments. However, it was swept away by a political coalition that cancelled the carbon tax, the introduction of which is now considered as anathema.

The European Approach

In the European Union, the regulatory demand aims to harmonize and avert these dangers. The EU approach moves in line with the Sustainable Finance Action Plan 2018 and within the framework offered by the Taxonomy Regulation 2020/852. In the wake of a long-term corporate policy that is inclusive and manages economic, environmental, and social risks, the EU has the Sustainability Reporting Directive (CSRD), the proposed Corporate Sustainability Due Diligence Directive (CSDDD), and the EU Low-Carbon Benchmark Regulation 2019/2089, as well as efforts to introduce a golden rule in relation to European green bonds (EuGBs). The draft was presented in 2021, and the agreement between the European Parliament and the Council was finally reached on 23 February 2023.

The aim is to lower the cost of gathering information from investors, stimulate product comparisons, and reduce uncertainties in what is green in order to "play" on the reputational component so as to reduce the risks of greenwashing. Thus increasing information flows, increasing the number of projects to be financed and the ability of the company to finance itself at a lower cost with ESG benefits.

The key elements here are: (1) EuGB designation only for green projects, and if it is clear that it is green, then the project moves in accordance with the Taxonomy Regulation; (2) transparency on how the proceeds are allocated; and (3) the issue must be examined by an external auditor registered and supervised by the European Securities and Markets Authority (ESMA).

Intervention is important, and in the enduring public/private chase, the axis shifts in favor of the former. But this is not all it is about. In parallel is the publication of EU models for issuers of other bonds with environmental objectives that do not refer to the European green bond standard. The implementation of such models will, however, remain strictly voluntary.

Let us put aside any consideration on the grandiosity of the project. Article 15 of the proposed regulation says that ESMA registers the auditor who "has experience in quality assurance, quality control, conducting pre- and post-issuance reviews, and financial services." In other words, there is little about green because the message is that scientific data, GHG emissions, biodiversity loss, etc., are handled admittedly with care, but by people with expertise in finance, business, social sciences, corporate management, marketing, etc., that is, with no expertise in environmental matters. And so the doubt that greenwashing already begins at the level of competences is strong.

In addition, the EuGBR (European Green Bonds Regulation) proposal creates three levels of external control. First, it assigns the supervision of the issuer's substantive compliance with the EuGB, in particular the taxonomy requirements, to private auditors. Second, it prescribes public supervision of issuers' compliance with the audit and disclosure requirements under the EuGBS. And third, it establishes a procedure for the public registration and supervision of external auditors. In short, it defines a public–private axis.

But a substantive auditing is *de facto* outsourced to private entities. Under the EuGBR proposal, the national competent authorities (NCAs) do not have any power to take over the substantive audit and, therefore, cannot adequately supervise in the event of malpractice, not least because ESMA has only limited supervisory powers. Secondly, external auditors have an inherent conflict of interest because they are paid by the issuer for their review and are therefore motivated to issue positive reviews and not negative reviews because they are financially dependent on the benevolence of issuers, at least for future issues.

The internal compliance regime is then substantially similar to that for credit rating agencies under Regulation (EC) No. 1060/2009 (Credit Rating Agencies Regulation, as amended). In short, the European legislator has come up with a very complex system that seems to fit into the groove traced by MiFID II (Markets in Financial Instruments Directive 2014/65/EU; Solvency II Directive 2009/138/EU; Capital Requirement Directive 2013/36/EU). The payment of an external auditor is also a cost for the company issuing the bond, and this means that only large companies will be in a better position to cope with these increased costs. All this without considering that the required organizational standards for being auditors are so stringent that only very large private entities will be able to perform external audits. The risk of monopolies being created for large, established accounting firms or other auditing companies is clear.

In short, the EU seems determined to go down the path of standardization without leaving anything to private regulation. And indeed, the proposed regulation in some ways marks the departure from the light-touch policy of the Commission. The concern about the need for clarity and uniformity is high. The disclosure requirements, by reducing the costs of gathering information, should also solve the problem of greenwashing. This explains the call for taxonomy and the resort to technical standards as tools to certify green activity even in an area – such as green bonds – that does not show an excessive proliferation of standards. Not necessarily, this is the best solution (Hachenberg & Schiereck 2018).

Between Public and Private Regulation

Private regulation has its own limitations, such as fragmentation and the proliferation of inconsistent standards, but it has the advantage of flexibility and adaptability in an area where it is difficult to achieve a shared framework for both climate risks and the following green objectives to be met. At issue is not only a regulatory framework capable of attracting and mobilizing finance.

A good number of public actors are also called upon to "play the game," such as judges and administrative authorities operating alongside private actors, such as banks, pension funds, or large commercial companies. Among all these actors, however, the courts, by strengthening environmental disclosure and accountability, appear to be the best equipped to give substance to the concept of "legal readiness for climate change." At issue is not only the assessment of a regulatory framework created by the lawmaker but also its validation through the intervention of judges and courts. The aim is to create the conditions for what could be called a true ecosystem of innovative legal instruments. In short, it is a matter – and paraphrasing Aronofsky's film *The Black Swan* – of supplying the "green swan" with adequate tools to avoid being overwhelmed by its dark side (climate change), tools capable of strengthening environmental policies and improving corporations' conduct.

Reference List

ACPR (2020) *Climate-Related Scenarios for Financial Stability Assessment: An Application to France*, Working Papers, Banque de France, No. 774, July 2020. Available at: https://publications.banque-france.fr/sites/default/files/medias/documents/wp774.pdf.

Albrizio, S., Botta, E., Kozluck, T. & Zipperer, V. (2014) *Do Environmental Policies Matter for Productivity Growth? Insights from New Cross-Country Measures of Environmental Policies OECD*, Economic Department Working Papers n. 1176.

Alessi, L., Battiston, S., Melo, A.S. & Roncoroni, A. (2019a) *The EU Sustainability Taxonomy: A Financial Impact Assessment*, Publications Office of the European Union, Luxembourg. Available at: https://doi.org/10.2760/347810.

Alessi, L., Ossola, E., & Panzica, R. (2019b) *The Greenium Matters: Evidence on the Pricing of Climate Risk*, Management: Statistics Working Paper 418. Department of Economics, University Milan Bicocca.

Bank of England (2019) *The 2021 Biennial Exploratory Scenario on the Financial Risks from Climate Change*. Available at: www.bankofengland.co.uk/-/

media/boe/files/paper/2019/the-2021-biennial-exploratory-scenario-on-the-financial-risks-from-climate-change.pdf?la=en&hash=73D06B913C73472D0DF21F18DB71C2F454148C80.

Banque de France (2018) *Responsible Investment Report*. Available at: www.banque-france.fr/sites/default/files/media/2019/03/26/banque-de-france-responsible-investment-report-20180.pdf.

Berg, F., Koelbel, J., & Rigobon, R. (2019) *Aggregate Confusion: The Divergence of ESG Ratings*, MIT Sloan School Working Papers.

Campiglio, E., Yannis, D., Monnin, P., Ryan-Collins, J., Schotten, G., & Tanaka, M. (2018) "Climate change challenges for central banks and financial regulators," *Nature Climate Change*, 8(6), pp. 462–468.

Ceci, M. (2016) *Green, Social and Sustainability Bonds*, J.P. Morgan.

Clark, G.L., Feiner, A., & Viehs, M. (2015) *From the Stockholder to the Stakeholder: How Sustainability Can Drive Financial Outperformance*. Available at: https://ssrn.com/abstract=2508281.

Di Giulio, E. & Migliavacca, S. (2020) "Quanto è green la finanza mondiale?" *Energia*, 3(20), pp. 24–35.

Ehlers, T., Mojon, B., & Packer, F. (2020) "Green bonds and carbon emissions: Exploring the case for a rating system at the firm level," *BIS Quarterly Review*, pp. 31–47. Available at: https://ssrn.com/abstract=3748440.

European Banking Authority [EBA] (2019) *Action Plan on Sustainable Finance*, Paris: European Banking Authority.

Gianfrate, G. & Peri, M. (2019) "The green advantage: Exploring the convenience of issuing green bonds," *Journal of Cleaner Production*, 219, pp. 127–135.

Hachenberg, B. & Schiereck, D. (2018) "Are green bonds priced differently from conventional bonds?" *Journal of Asset Management*, 17(1).

Hausfather, Z. & Peters, G. (2020) "Emissions – the 'business as usual' story is misleading," *Nature Communication*, 577.

James v. Exxon Mobil Corporation, 65 Misc. 3d 1233 (NY Sup. Ct. 2019).

Lamperti, F., Bosetti, V., Roventini, A., & Tavoni, M. (2019) "The public costs of climate-induced financial instability," *Nature Climate Change*, 9, pp. 829–833.

Mondial Bank (2020) *State and Trends of Carbon Pricing 2020*. Available at: https://openknowledge.worldbank.org/handle/10986/33809.

Network for Greening the Financial System [NGFS] (2019a) *A Call for Action, Climate Change as a Source of Financial Risk, First Comprehensive Report*. Available at: www.ngfs.net/sites/default/files/medias/documents/ngfs_first_comprehensive_report_-_17042019_0.pdf.

Network for Greening the Financial System [NGFS] (2020a) *Status Report on Financial Institutions' Practices with Respect to Risk Differential between Green, Non-Green and Brown Financial Assets and a Potential Risk Differential*. Available at: www.ngfs.net/sites/default/files/medias/documents/ngfs_status_report.pdf.

Organization on Economic Cooperation and Development [OECD] (2020) *ESG Investing: Environmental Pillar Scoring Reporting*. Available at: www.oecd.org/finance/ESG-Investing-Environmental-Pillar-Scoring-Reporting.pdf.

Pindyck, R.S. (2013) "Climate change policy: What do the models tell us," *Journal of Economic Literature*, 51(3), pp. 860–872. Available at: http://doi.org/10.1257/jel.51.3.860.

Polbennikov, S., Desclee, A., Dynkin, L., & Maitra, A. (2016) "ESG ratings and performance of corporate bonds," *The Journal of Fixed Income*, 26(1), pp. 21–41.

Securities Exchange Act 1934 (15 U.S.C. § 78a).

Shell (2024) *Powering Progress in the United States*. Available at: www.shell.us/about-us/powering-progress-usa.html.

Singh v. Cigna Corp., 918 3d 57, 64 (2d Cir. 2019).
Van der Ploeg, F. & Rezai, A. (2020) *Stranded Assets in the Transition to a Carbon-Free Economy*, Annual Review of Resource Economics.
Zerbib, O.D. (2019) "The effect of pro-environmental preferences on bond prices: Evidence from green bonds," *Journal of Banking and Finance*, 98, pp. 39–60.

Chapter 12

A Nun, Some Children, Climate Change, and a Novel Duty of Care – *Sharma* Considered

Janice Gray

Introduction

Australia does not have an enviable record on climate change (CSIRO 2022). It has been (and continues to be) responsible for high levels of emissions (CSIRO 2021), yet Australian government policy has, over many years, demonstrated a reluctance to introduce more robust climate change prevention, adaptation, mitigation, and recovery measures. Indeed, Australia has more than once been the subject of criticism or ridicule on the world stage because of its cavalier approach to climate change issues (Schultz 2022: 11).

Government policy and, arguably, law have not kept pace with an initially slow but now more rapidly developing public consciousness that calls into question the idea that the Earth is so forgiving it can be harmed by human beings, without consequence – that it will simply regenerate or replenish irrespective of what is done to it. Instead, the "business-as-usual" model of environmental policy, law, and regulation that has permitted harmful conduct to continue is increasingly understood to be an irresponsible and unsatisfactory approach to environmental governance and management (IPCC 2022; *Sharma* case [289]).[1] Such an approach is subject to criticism for a range of reasons, including its contribution to global, long-term warming leading to sea level rises, land and marine biodiversity loss/extinction, negative ecosystem impacts, climate-related risks to health, negative livelihood impacts, and threats to food security, water supply, human security, and economic growth (IPCC 2022; Vohra et al. 2021; Bauman et al. 2022; CSIRO 2022; Longden et al. 2020). The business-as-usual approach is also criticized for undermining intergenerational equity and the right to a safe climate future, as well as pushing planetary boundaries to the limit (IPCC 2022); *Sharma* case: 51). Such outcomes are increasingly seen as unacceptable.

1 *Sharma v. Minister for the Environment (2021) FCA 560* (first instance judgment), referred to herewith as the *Sharma* case. *Minister for the Environment v. Sharma* (2022) FCAFC 35, referred to herewith as the *Sharma* appeal.

DOI: 10.4324/9781003440871-14

Although slow to embrace the need for climate change action, a significant percentage of the Australian public (83%) has come to the realization that climate change is a concerning issue, requiring urgent action (IPSOS 2022). Interestingly, a critical aspect of the call for climate change action has been the role of children and teenagers. Their role in "schooling" Australian adults has been profound (Huntley 2020: 2). While Swedish teenager Greta Thunberg's personal strike for climate action may have led the charge internationally, Australian teenagers also embraced the cause, organizing the students' School Strike 4 Climate (School Strike 4 Climate 2023) and joining Thunberg and others in storming the stage at the 2019 Madrid Climate Talks (Huntley 2020: 46) to express dissatisfaction with the pace of world leaders' efforts to address climate change.

However, another group of Australian children[2] ultimately chose litigation as their key tool to progress action on climate change. It is this latter group, of eight children, led by Anjali Sharma, with the assistance of an 86-year-old Brigidine nun, Sister Brigid Arthur, as a litigation guardian, that forms the focus of this chapter. Sharma and the other applicants brought a class action in relation to Whitehaven's application to expand its Vickery coal mine in the New South Wales town of Gunnedah. Although the first instance decision of Justice Bromberg was overturned on appeal, his decision and Justice Beach's reasons for judgment in the appeal case arguably lay the ground for further attempts to recognize a novel duty of care or fashion alternative, new legal tools that could assist future climate change litigation (*Sharma* case; *Sharma* appeal).

Background

To contextualize the *Sharma* children's decision to pursue litigation, it is helpful to consider the policy and political climate preceding their decision. It influenced them.

> The children observed a weak history of climate action and noted that politicians, should be of sufficient competence to realise the seriousness of climate change and the future harm it will cause to children through worsening extreme weather, and [that politicians] should take action to reduce carbon emissions without being forced [to do so] through a court case.
>
> The applicants also noted that climate change, is not something that should be fought by children in an institution like the Federal Court. Children should be able to be children without having such a deep concern about their futures. However, having tried every other avenue available to us, including asking, striking from school and protesting, we took on this legal sparring match nonetheless.
>
> (Equity Generation Lawyers 2022)

2 The term "children" is used because it is consistent with the language of the judgments.

In 2007, Prime Minister Rudd gave climate activists and others some optimism that action would follow when he observed that climate change represented one of "the greatest moral challenges of our time" (Australian Government 2009). Discussion ensued as to whether that challenge might be best met by the introduction of an emissions trading scheme or a carbon tax, the former effectively being a license to pollute up to a certain level, with government deciding how much carbon dioxide (CO_2) emission to permit in the first place and then allowing low emitters to sell their excess to higher emitters, while the latter involves a dollar tax per ton being placed on an emission of CO_2 (such as on a liter of petrol or a kilowatt-hour of electricity). It was thought that both approaches would raise the price of carbon-intensive goods, incentivizing producers and consumers to cut their greenhouse gas (GHG) emissions. However, neither scheme was implemented until 1 July 2012, when Prime Minister Gillard's Clean Energy Act (2011) came into effect. It introduced a carbon pricing scheme. However, the scheme was repealed in 2014, by the newly installed Coalition (Liberal and National Parties) government, under the leadership of then–Prime Minister Abbott. Under a succession of different Prime Ministers (Abbott, Turnbull, and Morrison), the Coalition government remained in power for eight years – eight years that are commonly seen as a climate change reform wilderness (Climate Council 2022).

Had the Abbott government not repealed the key Gillard-era reform of putting a price on carbon, 72 million tons of emissions would have been saved between 2015 and 2020 (Australia Institute 2020), while emissions in 2020 would have been more than "25 million tonnes CO2-e lower" (Grudnoff 2020: 2). The repeal was, therefore, an environmentally retrograde step.

Between 2013 and 2016, the Coalition government also attempted to abolish three complementary climate change measures: the Renewable Energy Target (RET), the Clean Energy Finance Corporation (CEFC), and the Australia Renewable Energy Agency (ARENA). However, cross-bench Senate opposition ultimately prevented those abolitions (Grudnoff 2020: 4). Given retention of the RET, CEFC, and ARENA has been found to have safeguarded \$23.4 billion worth of investment in renewable energy from 2013 to 2018 (Browne et al. 2018: 1) and resulted in emissions reduction of at least 334 million tons of CO_2 (Grudnoff 2020: 5), the prevention was a welcome relief. It is worth noting that the 344 million ton reduction figure was predicted to be more than the Coalition government's Central Climate Policy Emissions Reductions Fund would deliver over its six years of operation between 2014 and 2020 (Grudnoff 2020: 5, relying on Browne et al.).

Other Coalition climate policies also reflected a timidity in addressing predicted climate crises. For example, in terms of the second commitment period of the Kyoto Protocol, in 2013, Australia negotiated a target to reduce emissions by 5% below 2000 levels, by 2020. This was the weakest target in the world (Climate Council 2022: 6). In 2015, Australia set a 2030 emission reduction target of 26–28% below 2005 levels, making the target one of the weakest in developed countries (Climate Council 2022: 6).

In 2015, Australia also adopted the Paris Agreement, a legally binding multilateral agreement on climate change, whose goal is to limit global warming to well below 2° (preferably 1.5°C), compared with pre-industrial levels. This objective is to be achieved by (1) economic and social transformation that is based on the best available science and (2) a five-year cycle of increasingly ambitious climate action, undertaken by each signatory country. However, in 2018, Australia announced that it would use "carryover" credits from the Kyoto Protocol to meet its commitments under the Paris Agreement. That approach drew opprobrium from the international community. Further, in 2019, Australia was singled out as one of only a few countries contributing to the delay in the completion of the Paris Agreement's rule book, specifically Article 6, covering the international trade in emissions reductions (Climate Council 2022: 6).

In terms of health, between 2006 and 2017, there were around 36,000 premature deaths in Australia caused by extreme heat (Longden et al. 2020). Those death rates are likely to rise (IPCC 2022), but Australia's climate change policies have not adequately addressed related concerns.

Meanwhile, in 2019–2020, Australia saw the physical, emotional, and economic impact of ineffective climate management and policy when it experienced the Black Summer bushfires. These fires followed the hottest and driest year on record (2019) (Bureau of Meteorology 2020), but despite scientific evidence, the Coalition government did not act on the impending environmental crisis. Instead, it ignored warnings.

Unchastened by the bushfires and the calls for urgent climate change action following them, the Coalition approved the Scarborough Gas Project in 2020, a project which is expected to add 800 million to 1.6 billion tons of GHG emissions over the life of the project. The same government went on to support further new coal and gas projects in 2021 and allocated $52.9 million to gas infrastructure projects and a gas-fired power station but did not provide any new support for renewable energy or electric vehicles (Climate Council 2022: 6). Meanwhile, the "plan" that former Prime Minister Morrison prepared for the United Nation's COP26 summit in Glasgow in November 2021 reflected little commitment to sincere climate change action. It was said to contain "a curious mix of arrogance, insularity and lack of ambition" (Schultz 2022: 11), while the chair of the UK Climate Change Committee, Lord Deben, spoke "with (imperial) paternal distress when he described ['the plan'] as 'disappointing' and 'sad'" (Schultz 2022: 11). It was a loose pastiche of half-baked ideas that did not permit Australia to reach net-zero emissions by 2050 (see Climate Council 2022: 28).

Later in 2022, Australians witnessed another environmental crisis in the form of the fourth mass bleaching event of the Great Barrier Reef in seven years. Despite this, the then government submitted a report to UNESCO that downplayed the government's own appallingly inadequate response to the harmful effects of climate change on the Reef. It also blocked the listing of the Reef as "endangered" (Climate Council 2022: 6, 50).

In relation to spending on climate change research and initiatives, the CSIRO budget was cut severely over many years, and of the almost $180 billion Australia spent on COVID recovery efforts, only around 2% will help cut emissions. By contrast, "Germany and France allocated half of their respective COVID recovery spending to climate solutions" (Climate Council 2022: II, 12). It is, therefore, not difficult to see the Coalition years as a period of squandered opportunity in terms of climate action.

Political writer Sean Kelly observed in 2021 that "we [Australians] have all become incredibly skilled at finding ways not to look at what is in front of us" (Kelly 2021: 171). However, if that were true, it changed a little in May 2022. Voters looked at what was in front of them and did not like what they saw. They rejected the Coalition government at the ballot box by way of a swing in favor of the Labor Party and, in particular, the newly emerging "Teals" – a loose group of independent candidates who boldly prioritized issues such as climate change. The new Labor Prime Minister, Albanese, entered office, inheriting a nation reeling from climate-related disasters. Between May 2022 and January 2023, he toured disaster zones in five states (ABC 2023). It is likely that such disasters will continue if, instead of reducing GHG emissions by way of measures such as a tax on fossil fuels and the subsidization of renewable energy, government policy continues to support the sequestering of carbon pollution from present and future fossil fuel projects.

It is against this backdrop, and driven by intense frustration and concern, that the children in the *Sharma* case felt compelled to take legal action. They sacrificed part of their childhood to litigate issues they thought responsible adults should have been addressing but were not. Their thinking aligned with the key message of the 2020 Final Report into the Environment, Protection, and Biodiversity Conservation Act 1999 (EPBCA) (Samuel 2020):

> Australia's natural environment and iconic places are in an overall state of decline and are under increasing threat. The environment is not sufficiently resilient to withstand current, emerging or future threats, including climate change. The environmental trajectory is currently unsustainable.

The *Sharma* Case

The *Sharma* case involved a negligence claim in relation to the extension of a coal mine. Vickery Coal wanted to expand the scope of a pre-existing approval given to its parent company, Whitehaven Coal Pty Ltd. The enlarged approval would permit coal extraction at the site to increase by approximately 33 million tons, and total coal production would go up from 135 million to 168 million tons, producing 100 million tons of CO_2.

The environmental implications of the proposal meant that it was a "controlled action" under the EPBCA, so it could not proceed without the approval of the Commonwealth Environment Minister.

The children sought (1) a declaration that the Minister owed a duty of care not to cause Australian children harm resulting from the extraction of coal and the CO_2 emissions into the Earth's atmosphere that would occur as part of the process, and that the Minister had to exercise her powers under the EPBCA with reasonable care to avoid causing harm, and (2) an injunction to restrain the Minister from an apprehended/alleged breach of the asserted duty of care. They claimed that a positive act by the Minister to approve the mine extension could expose the children to future harm and should be prevented before harm could occur. The children claimed the Minister had to avoid mental and physical injury, economic harm, and property damage that was likely to arise as a result of climate change. It was, therefore, a very extensive claim.

The case dealt with a duty of care in the tort of negligence, but the situation was complex because the purported duty of care had been overlaid or grafted onto a statutory power in the EPBCA. The children reasoned that the private law duty of care operated alongside the Minister's normal statutory obligations. The matter was not, therefore, as straightforward as might be the contest between two private parties in dispute about a simple common-law breach of duty.

The children argued it was reasonably foreseeable that the Minister's positive act of approving the mine could expose them to future harm. In terms of causation and reasonable foreseeability, they reasoned thus: mine approval ▶ increased coal extraction ▶ increased CO_2 emissions ▶ increased global surface temperatures ▶ increased climatic hazards and extreme weather events ▶ consequent exposure to harm (*Sharma* case: 195–197).

Justice Bromberg's judgment in the *Sharma* case made national and international headlines because it found that there was, indeed, a novel duty of care owed by the Minister to Australian children to exercise her powers under ss 130 and 133 of the EPBCA with reasonable care so as not to cause children harm, as a result of emissions from the approval of the Vickery Coal Mine Extension project. His Honor, however, limited the duty of care to health, not property impacts of climate change, and he did not grant an injunction because he found the children did not establish a reasonable apprehension of a breach of a duty of care. Further, he concluded that he could make a reasonable response to the foreseeable harm, by conditioning any approval of the mine extension.

It is telling that Justice Bromberg found the Minister should reasonably foresee the likelihood that the mine would cause future harm to the children. He found (at 286) that the Minister had knowledge of the risk of climate-induced harm and (at 296) that she was in a position of control with respect to that harm. Further, (at 299) he found that the children were vulnerable and reliant on the Minister to protect them.

It was, however, challenging for the Court to reconcile the superimposition of a private law duty (under tort) with a public statutory power. The reconciliation raised issues of incoherence and inconsistency that Justice Bromberg approached by considering the purpose of the statute. He observed that the objects of the EPBCA included the protection of the environment and the economically sustainable use of natural resources. He also noted that Sec. 528 of the Act considered future generations because it stated that ecologically sustainable use had to ensure that "the benefit of the use to the present generation does not diminish the potential to meet the needs and aspirations of future generations." Further, he noted that intergenerational equity, as a principle of ecological sustainable development, was referred to in Sec. 3A of the Act. This allowed him to read protection of human safety as an *implied* relevant mandatory consideration for the purposes of the Minister's decision-making, a reading that became key to his analysis of the private duty of care. He also noted that human safety would "almost always cut across the exercise or performance of a statutory power including a broad discretionary power" (398), adding that "[t]he lives and safety of the Children are not optional considerations but have to be taken into account by the Minister when determining whether to approve or not approve the controlled action" (404).

Justice Bromberg's judgment was bold and exciting. His Honor adopted the scientific findings of experts (such as Professor Steppen), and he was willing to think creatively as to how a common-law duty of care might coexist or be grafted onto a statutory duty and how that duty of care might be expanded to extend to harm from particular environmental decision-making.

His Honor interpreted the law in ways that pushed doctrinal boundaries and simultaneously assisted desperate litigants, but bold decisions are commonly subject to criticism, and the *Sharma* case was no exception. It was open to the criticism of judicial overreach and concerns about the separation of powers, for example. It is perhaps unsurprising that rather than taking the Court's decision into account when making her own decision under the EPBCA, the Minister instead appealed the primary decision to the Full Federal Court.

The *Sharma* Appeal

The three judges of the appellate court were unanimous in finding that no duty of care existed, but they all had different reasons. A detailed analysis of those comprehensive reasons is beyond the scope of this chapter. However, it is worth noting that of the three, Justice Beach's reasons for judgment offered the greatest hope to future climate litigants. Although he found that (1) there was insufficient closeness between the Minister's exercise of statutory power and the likely risk of harm, (2) a duty of care would result in an indeterminate liability, and (3) climate change not being limited to one catastrophic event impacted on the nature and extent of potential liability, he also acknowledged the necessity of taking immediate action to limit GHG emissions (*Sharma* appeal: 750, 752), even if the benefits are not observable for decades.

Further, he commended the trial judge for planting the seed of a novel duty of care, but he observed that it was up to the High Court "to engineer new seed varieties for sustainable duties of care" and to modify existing concepts, such as "sufficient closeness and directedness" and "indeterminacy" to adequately address the accelerating complexity of causal relations, acknowledging that such concepts in their present form may have "a limited shelf life" (*Sharma* appeal: 754).

Impact of *Sharma* Litigation

While climate change activists and members of civil society were disappointed by the *Sharma* appellate decision, the very existence of the *Sharma* litigation may be seen as positive. It was strategically astute for the children to commence proceedings *before* the Minister made her decision to approve or reject the mine extension because it allowed them to effectively run a merits-based case. Had they waited until after the decision was made, they would have been limited to judicial review. By being proactive, they were able to put climate change issues on the court record (notably, the climate change findings at first instance were adopted by the appellate court). The litigation also created opportunities for the judiciary to engage deeply with climate change issues and wrestle with the relevant law, and because of its extensive media coverage, the litigation, although ultimately unsuccessful, also performed an important educational and persuasive role.

Justice Bromberg's primary judgment acts as an exemplar for superior courts on how they might, on different (better) facts, reimagine established legal concepts to more effectively serve ecological integrity. It may be, however, that superior courts choose different legal concepts to refashion or that they modify the duty of care in alternative ways, but Justice Bromberg's bold decision invites superior courts to meet the challenge before them. It serves as a model of thoughtful, creative, and dynamic judicial interpretation and decision-making that responds to a critical problem, and it, along with Justice Beach's reasons for judgment, acts as a signal to the High Court that there is a readiness to see the law develop in ways that more readily reflect community values and help address one of the greatest moral challenges of our time – climate change.

In that sense, the challenge is not unlike the one faced by the High Court in the landmark *Mabo* decision of 1992, recognizing the Native title of Indigenous Australians (*Mabo* v. *Queensland (No 2)* 1992). There, the High Court identified links between the expectations of the international community and "the contemporary values of Australian people" [42] that needed better alignment. Although *Mabo* dealt with a different body of law from *Sharma*, the importance of law reflecting community values is equally significant in both contexts. When the law becomes seriously out of kilter with community values and the legislature demonstrates a lack of willingness to harmonize law and values in a timely fashion, there is an argument that the highest court in the land has a sensitive, considered, but active role to play.

As Justice Brennan stated in *Mabo*:

> The peace and order of Australian society is built on the legal system. It can be modified to bring it into conformity with contemporary notions of justice and human rights, but it cannot be destroyed. . . . [N]o case can command unquestioning adherence if the rule it expresses seriously offends the values of justice and human rights (especially equality before the law) which are aspirations of the contemporary Australian legal system. If a postulated rule of the common law expressed in earlier cases seriously offends those contemporary values, the question arises whether the rule should be maintained and applied. . . . [I]t is necessary to assess whether the particular rule is an essential doctrine of our legal system and whether, if the rule were to be overturned, the disturbance to be apprehended would be disproportionate to the benefit flowing from the overturning [29].

These are issues the High Court may have the opportunity to consider if *Pabai*, a climate change case presently in the pipeline, reaches the High Court.

Pabai is a class action brought by Torres Strait Islanders alleging that the Commonwealth owes a duty of care to the plaintiffs, who will suffer loss and damage as a result of the failures of the Commonwealth government in respect to climate change. Rising sea levels will, for example, mean that they will lose their homes. Presumably, these facts make them more overtly "vulnerable" than the children in *Sharma*, and causation may be easier to establish. If the *Pabai* claim is ultimately successful, the Court may possibly make orders similar to the Dutch case of *Urgenda* (2019) and *Milieudefensie et al. v. Royal Dutch Shell plc* (2021), meaning, the defendants would have to take positive climate change action.

Alternatively, it may be that the "greenwashing" case (not yet at trial) being brought by the Australian Centre for Corporate Responsibility (ACCR) against Santos becomes the seminal Australian climate case. It is "the first case in the world to challenge the veracity of a company's net-zero emission target, and [it] raises important questions about the role of carbon capture and storage (CCS) and blue hydrogen in the transition to zero emissions" (ACCR 2022). ACCR is challenging Santos's claim that natural gas provides a "clean energy" and that it (Santos) has a "credible and clear plan" to achieve "net zero" emissions by 2040 by way of "carbon capture," claims that Santos made in its annual report (ACCR 2022). ACCR, relying on the Corporations Act 2002 Cth (Commonwealth), will argue that Santos is engaging in misleading or deceptive conduct in contravention of both corporate and consumer laws.

Conclusion

The need for government to address climate change remains urgent because successive federal governments squandered opportunities to embark on robust climate action. Those governments repealed legislation that would have led to reduced GHG emissions, failed to act within the spirit of the Paris Agreement, set low emissions targets, cut spending on climate research, and revealed a contempt for collaborative, global climate action. Such inadequate responses contributed to the decision of eight children and a nun to litigate.

Although the bold and creative primary decision in that litigation was overturned on appeal, there are residual benefits. The case put climate change squarely on the judicial agenda, and it acts as a model for more reticent courts to deeply engage with climate research, while the media's coverage of the case performed a valuable role in bringing climate issues to the public in a digestible form.

The litigation represents a jurisprudence of aspiration. It arguably serves to embolden or catalyze others to act, as well as encourages the more creative and novel employment of traditional causes of action. But litigation alone will not be enough. Climate change action will need to be multifaceted. It will be necessary to tap into the opportunities created by the 2022 change of government and the influx of Teal politicians.[3] It will be important to develop amalgamations of the seemingly incompatible and harness the support of unlikely bedfellows (such as conservative rural organizations). It may also be helpful to align with renegade, progressive businesspeople, such as those who opposed the demerger of one of Australia's energy companies. Environmental lawyers may also seek to partner with media and defamation lawyers to challenge the legitimacy of media monopolies, in order to stem climate change disinformation.

Environmental lobbies, media, protest, resistance, and disruption, in concert with science and law, may all be necessary to propel decision-makers into action. Further, as social researcher Huntley explains, it was not reading a report from the IPPC or sitting through a presentation from a climate scientist that caused her to shift from being concerned about climate change to being genuinely alarmed. It was reacting to a "crowd of children holding up signs in the streets, girls who were only a few years older than my eldest daughter. Suddenly it was personal" (Huntley 2020: 4), or as Hochschild puts it, change and understanding come through emotional connection (Hochschild 2016). Perhaps an emotional connection to the children of the *Sharma* case may have been more influential in mobilizing the community around climate

3 Postscript, August 2023. The Australian Senate has referred the first private member's bill, the Climate Change Amendment (Duty of Care and Intergenerational Equity) Bill 2023, of newly elected senator David Pocock to the Environment and Communications Legislation Committee. The bill was devised in consultation with Anjali Sharma.

change issues than the evidence adduced in the case. If so, that is certainly another lesson for the future.

References

ABC (7 January 2023) Television Program.

Australia Institute (2020) *Key Gillard-Era Reform Carbon Price Would Have Saved 72 Million Tonnes of Emissions.* Available at: https://australiainstitute.org.au/post/key-gillard-era-reform-carbon-price-would-have-saved-72-million-tonnes-of-emissions/ (Accessed 11 January 2023).

Australian Centre for Corporate Responsibility [ACCR] (2022) *ACCR Expands Landmark Case Against Santos.* Available at: www.accr.org.au/news/australasian-centre-for-corporate-responsibility-expands-landmark-federal-court-case-against-santos/ (Accessed 15 January 2023).

Australian Government (2009) *Speech, Transcript 16882, Prime Minister Transcripts, Department of Prime Minister and Cabinet.* Available at: https://pmtranscripts.pmc.gov.au/release/transcript-16882 (Accessed 11 January 2023).

Bauman, D., Fortunel, C., & Delhaye, G. (2022) "Tropical tree mortality has increased with rising atmospheric water stress," *Nature.* Available at: https://doi.org.10.1038/s41586-022-04737-7.

Browne, B., Campbell, R., & Cass, D. (2018) *Saved by the Bench*, The Australia Institute. Available at: https://australiainstitute.org.au/report/saved-by-the-bench/ (Accessed 11 January 2023).

Bureau of Meteorology (2020) *Annual Climate Statement 2020: Australia's Fourth-Warmest Year on Record.* Available at: https://media.bom.gov.au/social/blog/2530/annual-climate-statement-2020-australias-fourth-warmest-year-on-record/#:~:text=But%20it%20wasn't%20as,above%20the%201961%E2%80%931990%20average (Accessed 12 January 2023).

Clean Energy Act 2011 (Cth).

Climate Council (2022) *The Lost Years: Counting the Costs of Climate Inaction in Australia.* Available at: www.climatecouncil.org.au/wp-content/uploads/2022/03/FINAL_CC_The-Lost-Decade_Low-Res.pdf (Accessed 12 January 2023).

Commonwealth Scientific and Industrial Research Organisation [CSIRO] (2021) *What Are the Sources of Australia's Green House Gas Emissions?* Available at: www.csiro.au/en/research/environmental-impacts/climate-change/climate-change-qa/sources-of-co2 (Accessed 10 January 2023).

Commonwealth Scientific and Industrial Research Organisation [CSIRO] (2022) *State of the Climate Report.* Available at: www.csiro.au/en/research/environmental-impacts/climate-change/state-of-the-climate (Accessed 10 January 2023).

Environment Protection Biodiversity and Conservation Act 1999 (Cth).

Equity Generation Lawyers (2022) *Sharma Litigants' Statement, "They Will Not Forget Our Names".* Available at: https://equitygenerationlawyers.com/wp/wp-content/uploads/2022/04/220412-Statement-from-Sharma-litigants-on-High-Court-appeal.pdf (Accessed 10 January 2023).

Grudnoff, M. (2020) *The Carbon Pricing Mechanism under the Gillard Government*, Australia Institute. Available at: https://australiainstitute.org.au/wp-content/uploads/2020/12/Carbon-price-10-years-on-web.pdf (Accessed 11 January 2023).

Hochshild, A. (2016) *Strangers in Their Own Land*, New York: The New Press.

Huntley, R. (2020) *How to Talk About Climate Change in A Way That Makes a Difference*, Sydney: Murdoch Books.

Institut Public de Sondage d'Opinion Secteur [IPSOS] (2022) *8 in 10 Australians Are Concerned About Climate Change with a Clear Public Expectation of Government*

Action. Available at: www.ipsos.com/en-au/8-10-australians-are-concerned-about-climate-change (Accessed 9 January 2023).
Intergovernmental Panel on Climate Change [IPCC] (2022) *6th Assessment Report: Impacts, Adaptation and Vulnerability*. Available at: www.ipcc.ch/report/ar6/wg2/ (Accessed 10 January 2023).
Kelly, S. (2021) *The Game*, Carlton: Black Inc. Books.
Longden, T., Quilty, S., Haywood, P., Hunter, A., & Gruen, R. (2020) "Heat-related mortality," *The Lancet Planetary Health*, 4(5), p. e171. Available at: https://doi.org/10.1016/S2542-5196(20)30100-5 (Accessed 1 June 2022).
Mabo v. *Queensland (No 2)* (1992) 175 CLR 1.
Milieudefensie et al. v. *Royal Dutch Shell plc*, District Court of the Hague (26 May 2021) C/09/571932/HA ZA 19–379.
Minister for the Environment v. *Sharma* (2022) FCAFC 35.
Pabai v. Commonwealth of Australia (2022) FCA 836.
Samuel, G. (2020) "Key Message," *Independent Review of EPBCA, 1999*. Available at: https://epbcactreview.environment.gov.au/resources/final-report/key-messages (Accessed 14 January 2023).
School Strike 4 Climate (2023). Available at: www.schoolstrike4climate.com/ (Accessed 9 January 2023).
Schultz, J. (2022) *The Idea of Australia*, Sydney: Allen and Unwin.
Sharma v. Minister for the Environment (2021) FCA 560.
The State of the Netherlands v. *Urgenda Foundation*, The Supreme Court of the Netherlands, 20 September 2019, Case 19/00135.
Vohra, K., Vodonos, A., Schwartz., J, Marais, E., Sulprizio, M., & Mickley, L. (2021) "Global mortality from outdoor fine particle pollution generated by fossil fuel combustion: Results from GEOS-Chem," *Environmental Research*, 195, 110754. Available at: www.sciencedirect.com/science/article/abs/pii/S0013935121000487 (Accessed 10 January 2023).

Section 3

Evolutions in Understanding Ecological Integrity

Chapter 13

Moving From Monocultures to Resilience

Geoffrey Garver and Cristiane Derani

Introduction to the Problem of Monocultures

The word "monoculture" typically brings to mind cultivation of a single plant in agriculture or forestry: endless fields of corn or wheat in the Midwest of the United States, vast olive groves on Mediterranean hillsides, rich tropical forests converted to soy in South America, or even-rowed plantations of palm oil trees in Asia. One might also conjure up the enormous, highly mechanized, and tightly managed industrial hog, cattle, or chicken factories that worldwide process about 50 billion animals for meat consumption annually (Cockshaw 2021). The mind might even wander to considering all kinds of single-purpose industrial processes and industrialization in general. All are valid examples of material monocultures.

Monocultures are not only material, however. More broadly, a *monoculture* is "a culture dominated by a single element: a prevailing culture marked by homogeneity" (Merriam-Webster 2022). This definition includes conceptual monocultures in addition to material ones. Material monocultures involve uniform, single-purpose organization of a material process, whereas conceptual monocultures reflect a universal worldview or other conceptual framing that is put forth as absolute and unquestionable, or superior to, dominant over, or exclusive of all others. Conceptual monocultures include the use of money as a unitary expression of value and state-centered legal systems as a unitary form of law and governance. They also include depiction of humans in conventional economics as rational consumers seeking to maximize their private gain, and Western science as a superior form of knowledge. The sweeping central thesis in *Ecological Thinking: The Politics of Epistemic Location* (Code 2006: 9) is that

> the dominant model of knowledge and epistemology in Anglo-American philosophy produces an epistemological monoculture both in the academy and in everyday life, whose consequences are to suppress and choke out ways of knowing that depart from the stringent dictates of an exaggerated ideal of scientific knowledge making.

DOI: 10.4324/9781003440871-16

Conceptual monocultures also exist in education in the form of siloed university disciplines; in social culture and "happiness" in the form of global consumer brands, the worldwide appeal of Hollywood blockbusters, and globalized consumer phenomena such as Black Friday; of language in the form of internationally dominant languages, such as English, Mandarin, and Spanish; and of morality and spirituality in the form of the world's main monotheistic religions.

These material and conceptual monocultures are interconnected and reinforcing. Industrial agriculture and crop monocultures are tied to a monocultural view of the economy, which in turn is supported by monocultures of law and governance. Thus, interlocking complexes of material and conceptual monocultures can come together in sweeping societal narratives that can endure for decades or centuries, such as the religious narrative of medieval Europe, the scientific narrative of the Enlightenment, and the contemporary narrative of globalization and insistence on economic growth (Michaels 2011; Capra & Mattei 2015). These broad narratives are, in themselves, monocultures, in which an overarching "monoculture of the mind" built around Western ideologies (de Araújo et al. 2020) manifests in material monocultures of production and consumption around food, energy, transport, housing, health, infrastructure, urbanization, and more. This chapter explains how material and conceptual monocultures and the larger narratives they form together undermine socioecological integrity and resilience necessary to the long-term security and stability of social and ecological systems and proposes alternatives that favor diversity and pluralistic approaches.

Common Features of Monocultures

Monocultures, whether material or conceptual, share several features. All monocultures are systems in themselves and typically parts of larger ones as well, and all are interdependent with Earth's social and ecological systems. For example, crop monocultures replace uncultivated ecosystems, and the soils, nutrients, water, and climatic conditions on which they rely or exact impacts all derive from Earth's ecosystems (Sundstrom et al. 2022). The monoculture of money not only operates according to systems dynamics but also has enormous direct and indirect impacts on Earth's social and ecological systems. Thus, the common features of monocultures are most readily seen from a systems perspective.

Because monocultures involve attempts to dominate and control various types of social and ecological systems, it is useful to recall characteristics of "natural" systems that have not been reduced to monocultures. First, natural ecosystems do not behave in linear, gradual, predictable ways but instead are prone to unpredictable episodic change that, at the extreme, can shift the system to a new state (Holling 1996). Second, natural systems are not uniform, and "from the leaf to the individual, the vegetation patch, the landscape, and the planet," their structure and functioning cannot be scaled up or down in linear ways (Holling 1996: 40). Scale matters in terms of system structure and function. Third, ecosystems do not have a stationary state of equilibrium but instead are typically in far-from-equilibrium states and evolve and maintain overall diversity, structure, and function according to a complex interplay of destabilizing and stabilizing forces. Fourth,

> [p]olicies and management that apply fixed rules for achieving constant yields (such as constant carrying capacity of cattle or wildlife or constant yield of fish, wood, or water), independent of scale, lead to systems that gradually lose resilience and suddenly break down in the face of disturbances that previously could be absorbed.
>
> (Holling 1996: 41)

Holling (1996) describes two different kinds of systems resilience, engineering resilience and ecological resilience. *Engineering resilience* describes the ability of a system to resist disturbances or to return quickly to a desired near equilibrium state when it is disturbed (Holling 1996). Systems with engineering resilience have a narrow range of workable or desirable structures and functions, and their chief attributes are uniform and predictable outcomes, as well as efficiency in the achievement of those outcomes. Ironically, however, because it requires protection against disturbances or strategies to quickly recover from them, engineering resilience renders systems increasingly vulnerable to collapse when those protections or response strategies fail. Engineering resilience is the resilience of monocultures. By contrast, *ecological resilience* describes the ability of a system to absorb disturbances while maintaining overall structure and function. Ecological resilience emphasizes far-from-equilibrium conditions and the system's ability to adapt to many stability domains without losing its structure and function (Holling 1996: 41–42). Systems with strong ecological resilience have higher adaptive capacity than those managed for engineering resilience, because they can persist with their overall structure and function intact as they absorb disturbances and shift between a wide range of stability regimes in response to them. In systems with ecological resilience, diversity, unpredictability, flexibility, and redundancy of system functions are key attributes, all of which contribute to lower vulnerability of such systems to collapse due to disturbance. Monocultures typically erode ecological resilience.

The systems features of monocultures are most recognizable in material monocultures, such as monocultural agriculture. Agricultural monocultures involve vast plantations of grains, cotton, palm, timber, or other cultivated commodities, managed to maximize yields, with as much consistency and uniformity in quality as possible. Yet these monocultures are in a constant race against, and therefore are vulnerable to, disturbances that threaten to reduce yields and product quality, such as an evolving array of pests, climatic events, soil depletion, and so on. Protecting against these ever-changing disturbances requires technological innovations, such as new pesticides, fertilizers, GMOs, and irrigation sources. However, agriculture does not have to be purely monocultural, following an engineering resilience approach. Although agriculture usually involves a decrease in biodiversity, agriculture systems that maintain overall biodiversity have existed throughout history. Alternative agricultural paradigms, such as agroecology and regenerative agriculture, that follow principles of ecological resilience taking into account not only the cultivation of food or other commodities but also all the social and ecological systems with which cultivation is associated, are emerging or re-emerging (Sundstrom et al. 2022).

Conceptual monocultures also demonstrate these system commonalities. For example, studies in ecological economics reveal that monocultural economic systems organized around capitalism and insistence on economic growth measured in gross domestic product tend to become uneconomic, in that harmful activities contribute to GDP growth, such as oil spills (for which cleanup costs contribute to GDP) and automobile accidents (for which ambulances, hospitalizations, funeral costs, and so on contribute to GDP), whereas beneficial non-market activities, such as child and elderly care, often do not (Daly 1997). Moreover, growth-insistent economics drives negative ecological impacts, such as global heating, climate change, and biodiversity loss, that create serious long-term economic vulnerabilities (Garver 2021).

In addition to systems-based commonalities, the most potent material and conceptual monocultures in contemporary globalized societies also have common links to colonialism, imperialism, capitalism, and modernity. Patel and Moore (2017: 14–18) show how modern corporations and global capitalism arose out of European colonialism beginning in the 15th century. They highlight Madeira, "the island of wood," which was rapidly deforested after Portuguese sailors discovered it in the 1400s, first for wheat cultivation, and then for the more lucrative production of sugarcane. Italian and Flemish financiers enabled these ventures, with expectation of profit and expansion. Sugar production on the island, using slave labor, and requiring division of labor and specialization (field-workers could not also process the sugarcane, given the demand for quick processing), was an early prototype of monocultural industrial processes – "[t]he plantation was the original factory" (Patel & Moore 2017: 16) – that are replete in the modern capitalist economy. When the island was deforested and no longer suitable for sugar production, the Portuguese owners and their financiers moved production to the Caribbean and South America, leaving Madeira open for production of the fortified wine associated with the island. Over time, this convergence of colonialism and capitalism became normalized, legalized, and globalized in today's capitalist modern economy.

Detailed Examination of Examples of Material and Conceptual Monocultures

Detailed examples help deepen understanding of the preceding overview of different types of material and conceptual monocultures and their common features. Industrial hog farming is a potent example of a material monoculture, and international law and global capitalism illustrate well the features of conceptual monocultures.

Material Monocultures: The Example of Industrial Hog Farming

Industrial hog farming is a common way of producing pork products for global markets. Intensive hog farming is concentrated primarily in North America, Europe, and China, and in Europe and North America, industrial hog farming is done almost to the exclusion of small and medium hog farms (Gilbert et al. 2015). Global pork production increased by a factor of 4.3 from 1960 to 2010, due largely to "intensifying production through increasing animal densities, constructing larger production units, increasing efficiency of the processing infrastructure, using more concentrated feeds, increasing mechanization and the widespread use of pharmaceuticals such as antimicrobials and vaccinations" (Gilbert et al. 2015: 2). In other words, hog farming and pork production have become increasingly monocultural during the "livestock revolution" in recent decades (Gilbert et al. 2015).

Intensification of pork production requires less land per animal and enables increases in food production (Pan & Bonnet 2022), but it carries significant increasing risks to human health and the environment. Large-scale hog production results in concentration of manure, leading to land and water pollution problems (Gilbert et al. 2015; Pan & Bonnet 2022). As well, breeds used in industrial hog production are vulnerable to disease, the pathogens are particularly virulent and can threaten humans, and the massive amounts of antibiotics to control disease contribute to global concerns about antibiotics resistance that affects both livestock and humans (Gilbert et al. 2015; Pan & Bonnet 2022).

In the United States, restructuring of hog production in the past 40 years has resulted in a rapid shift to a fewer number of large operations producing more pork products (USDA 2022). The number of hog farms has decreased by 70% since 1990, but production has increased significantly (USDA 2022). As well,

> [l]arge operations that specialize in a single phase of production have replaced farrow-to-finish operations that performed all phases of production. The use of production contracts has increased. Structural changes have coincided with efficiency gains and lower production costs. Most of the productivity gains are attributable to increases in the scale of production and technological innovation.
>
> (USDA 2022)

True to the monoculture form, large industrial pork producers "[aim] to maximize their efficiency by producing standardized pigs, which [grow] at predictable rates and [produce] predictably uniform meat" (Johnson 2006).

A stunning degree of mechanization and technological specialization exists throughout modern US hog production, yielding "a tower of efficiency-maximizing products, one stacked atop the next, each innovation fixing the problem the last fix created" (Johnson 2006). Moreover, the industry depends on a very narrow range of pig breeds and is dominated by fewer and fewer large parent companies, with a loss of individuality and autonomy for hog farms and farmers, who become beholden to the large parents through strict and demanding contracts (Johnson 2006). Yet the result of all the genetic manipulation, mechanization, stressful living conditions for pigs, and intensification in the industry is that most pork sold *en masse* to consumers – the cheapest "commodity pork" available in supermarkets – has become more acidic and, therefore, less tasty, with a rubbery, dry texture (Johnson 2006). Maintaining quality has become a serious problem, and although technological workarounds have helped stem it, "[e]ach new liberty for pork producers depends on further control, further domination of the pig" (Johnson 2006).

The monoculture of industrial hog production interacts with other material and conceptual monocultures. The feedstock for industrial hog production comes primarily from large monoculture crops, especially corn. More broadly, industrial hog production is part of a monocultural economic system, described in detail later, in which hogs are objectified as a means to make profit, rather than seen as intelligent living beings, and humans are reduced to selfish, greedy consumers. A monocultural legal system supports, in turn, this monocultural economic system. The United States has a weak antitrust regime to prevent consolidation of the industry in a few powerful companies, animal cruelty laws are relatively impotent in the industry, the US Department of Agriculture promotes meat-friendly food diagrams and healthy diet recommendations, trade and investment laws are heavily influenced by and friendly to the industry, the application of environmental laws to concentrated animal feedlot operations (CAFOs) is notoriously weak, and the industry is replete with government subsidies. In some states, "ag-gag" laws criminalize investigative activities, such as undercover photos or videos, that reveal the living conditions of animals in industrial farms (ASPCA 2023).

Conceptual Monocultures

Conceptual monocultures dominate modern legal and economic systems. These monocultures are evident in the current system of international law organized around sovereign states and in the capitalist market system that dominates the global economy. Capra and Mattei (2015) explain how mutually supporting private property owners and sovereign states are locked into a mechanistic paradigm that no longer comports with evolving scientific understandings of how the Earth and its interacting social and ecological systems function. The interdependent dominant conceptual monocultures of law and economics are increasingly detached from their cultural, social, and ecological contexts (Polanyi 1944), enabling dangerous erosion of the social and ecological foundations necessary for a flourishing future for humans and Earth's entire community of life.

Monocultures of International Law

Modern law is constructed in many ways around the abstract concept of the "person," and as a result of the French Revolution and other legal transformations, the equality of legal persons has become entrenched in many modern legal systems. Yet equality in law is often monocultural, in that it tends to dispense with and reject diversity (Césaire 2000; Quijano & Ennis 2000). Although equality remains an important value in law, problematic aspects of homogenous, monocultural treatment of entities recognized in law can be seen particularly in the realm of international law.

The theoretical underpinnings of modern international law had their first concrete expression in the Peace of Westphalia. The formation of equal sovereign European states in the 16th and 17th centuries, culminating in the Peace of Westphalia in 1648, largely determined the basic structure of international law that persists today. Notably, the new sovereign states of Western Europe were also engaged in establishing new relationships with other parts of the world – relations of colonization, occupation of territories, and enslavement of peoples to feed and provide for the new political units that were forming in Europe (Césaire 2000; Rodney 2018).

Modern international law has become a monoculture in that it both confirms the idea and centrality of sovereign states and excludes organizations and subjects that are not configured as such (Wolkmer 2001). Non-state peoples, individuals, and other organizations are largely erased from international law, which is first and foremost about rights, not between people, but between states, understood as territorial, political, economic organizations grounded in the European model (Césaire 2000). Therefore, the condition for attaining the protections and relationships inherent in the rights of a sovereign state is to align with the organization of sovereign power following the Westphalian model. The United States was the first non-European territory to meet these conditions, although it paid the price of a war of independence to gain this status. This early expansion of the Westphalian system beyond Europe demonstrated that the formation of unity in the form of a sovereign state is a process of struggle in which stability is attained through force (Césaire 2000; Fanon 2022), which in the case of the state involves violence concentrated in the hands of the state itself (Locke 2003).

As the Westphalian system of sovereign states has matured, the state as a legal unit has emerged as a complex organization built on a unique historical and cultural narrative. The formation of a state typically involves imposition of societal constraints on any pre-existing cultural diversity and plurality. The unification of international law through the adoption of the sovereign state as a subject, in principle equal to all other sovereign states, has imposed a standardization of political, territorial, and economic organization (da Rocha 2021; Santos 2002). Also, the international system of states effectively conditions relationships such that colonial or independent territories can only join the international community if they take the form of a state. During the 19th century, independence movements not only sought separation from the political and economic power of the European sphere but were also movements to build political unity following the European model, cutting down the tree of colonization but leaving its roots intact. This continued with the decolonization movements during the first half of the 20th century.

Territories that resist capitulation to the international system of Westphalian sovereign states, complete with their neo-colonial structures and interrelationships, are dismissed as uncivilized. Indeed, Article 38 of the Statute of the International Court of Justice (UN 1945) makes clear that the Court will be guided by "the general principles of the law recognized by civilized nations" – in other words, those nations that adapted to the unification of political power in the form of a sovereign state. What emerged in Europe as a way to move past a thousand years of conflict is imposed, monoculturally, on peoples and cultures around the world with diverse histories and lifeways (Wolkmer 2001; Torre Rangel 2006).

No culture has the right to eliminate the conditions of existence. However, the monoculture that has been constructed around state sovereignty is a monolith that tends to destroy life and impede the possibility of alternative relationships for the maintenance or recovery of places contaminated and threatened by this cultural form (Mignolo & Walsh 2018; Césaire 2000). However, the need for an alternative has been recognized within the modern international regime, even though progress has been slow. The United Nations organized the largest meeting of member states ever held, in 1992, in Rio de Janeiro, to build on the 1972 Stockholm Conference on the Human Environment (UN 1972) and bring a new perspective based on humanity "in harmony with nature." For the first time, the United Nations General Assembly recognized in 1992 the existence of different cultures and their value in making life on Earth more harmonious (UN 1992). The Convention on Biological Diversity (CBD) signed in Rio declares that it is necessary to recognize and value cultural diversity and biological diversity, which, unlike monocultures, the CBD recognizes are fundamental for maintaining and improving the conditions for life. Biological diversity, maintained and enhanced by cultural diversity, helps ensure that alternatives for food and medicine exist for human societies, in addition to being the key to the balance of ecosystems, their restoration, and increased resilience (Shiva 2003). Faced with all the imbalances associated with the hegemonic legal monoculture responsible for the emergence of the Anthropocene, it is time for international law to retreat from the monotonous discourse of sovereign states. Discussions on the recognition of legal pluralism (Wolkmer 2001), the foundation of culture in different worldviews, the alternative forms of knowledge production (Walsh 2019; Mignolo 2017), and the ways of being and living are beginning to enter the monocultural system of law and now need to take shape, be expanded, and eventually subdue dangerous, monoculturalized structures.

The Monoculture of Global Capitalism

Beginning with the expansion of colonialism since the 15th century, the global economic system has consolidated a monocultural form of economic exchange (Mignolo 2017; Wolkmer 2001): the market and the exchange of goods for money. Markets have been part of many human cultures throughout history, but state-centered capitalist markets that emerged in Europe in the early modern era began a trend of their continual expansion within an international system organized around sovereign states. Anchored in production systems designed to promote profit in the form of liquid money obtained from market exchanges for goods, which has become a means and an end in itself, this economy was fundamentally capitalist – an economy of and for the accumulation and continual reproduction of capital and profits. Thus, the rise of sovereign states coincided with the rise of an elite class of owners of capital and the means of production, and private property systems work in tandem with state sovereignty to maintain and perpetuate monocultural global capitalism (Capra & Mattei 2015).

The capitalist market economy emerged from the expansion of mercantile exchanges to dominate all economic production, and its expansion was greatly facilitated with the exponential increase in the flow of metals and other materials taken from the "discovered" territories outside of Europe (Patel & Moore 2017; Rodney 2018). The economy as a cultural and historical process of construction of diverse forms of existence, rooted in ecological, social, and cultural context, was increasingly reduced to the dynamics of capitalist market dynamics, and particularly the ceaseless quest for profits and accumulation of wealth (Polanyi 1944). This colonization and monoculturalization of the economy were only possible through the colonization of remote peoples and territories holding minerals, particularly silver and gold. In turn, conquest and colonization of remote people and places allowed for the perpetual expansion of exchanges and also paved the economic, social, and cultural paths for the emergence and expansion of industrial means of production and, ultimately, the Industrial Revolution (Marx 1996). Modernity today is firmly grounded in this process of capitalistic colonialism, which became even more virulent with the Industrial Revolution (Mignolo 2011; Patel & Moore 2017).

Notably, the emergence of the market economy and the Industrial Revolution was neither peaceful nor sudden transformation. The transition toward state-centered capitalism and industrialization in the 17th and 18th centuries involved both internal revolts and much external violence. Violent interventions were necessary to transform relations among people and between people and the natural world into monocultures. Human beings, as well as all other-than-human nature, needed to be desubjectified and desacralized in order to transform them into objects to be exploited in order to enable appropriation and accumulation (Fanon 2022; Césaire 2000), for the enrichment of a distinct social class: the owners of capital and capitalist production processes, who profit most from this economic system. The colonial violence experienced in the early modern era, and the movement of metal ores, timber, and agricultural products, as well as enslaved humans, between the continents, built the economy and opulence of Western Europe and, ultimately, enabled the Industrial Revolution (Rodney 2018).

Violent interventions were not limited to the violent conquests of resource-rich lands in the Americas, Africa, and Asia but also included uprisings within European states. It is not surprising that the opportunities for wealth accumulation that the emerging colonial-capitalist complex opened up gave way to restlessness and, eventually, outright revolution by those held in check by European monarchies. The French Revolution, with its ideals of a free individual and secularism, exemplifies how internal clashes expanded the sectors of society that could pursue capitalist wealth accumulation through adherence to an individualist ideology and the sacralization of money and the liberty to pursue it. The individual devoid of human bonds and commitments is the ideal object for the construction of monocultural industrialization, which sees in "labor power" or in "human capital" a necessary and homogenized element for accumulation.

The set of interhuman and human–nature relations inherent in industrialization and, by extension, the globally dominant capitalist system is monocultural in that it relies on money as a single expression of value, and in that the actors within it are reduced to their singular roles as owners, workers, raw materials, or consumers, ignoring all the complexities that distinguish individual beings or different cultures and ecological contexts. As a monoculture, capitalism becomes highly scalable and transferable, in ways that ignore, obscure, or destroy social, cultural, and ecological diversity and complexity (Tsing 2015). This immense monoculture of values and production is built with the support of law that allows the objectification and standardization of transactions, in which everything is reduced to a price. Legally established market relations are invariably two-sided within a capitalist regime. On one hand, the owners of goods and cash, including corporations when they are recognized as legal or moral persons, are the subjects because they own capital and the means of production. On the other side, everyone and everything else, from potatoes to people to promises, become objectified and available for appropriation or exploitation in order to yield profits and wealth accumulation (Patel & Moore 2017).

This monocultural complex is powerful and enduring because it invades and overtakes the human capacity for reasoning and for critical thinking, at the same time that it intoxicates human senses and desires. In contemporary modernity, human beings are born into a situation where everything is reduced to a price and, immersed in this cave, are largely incapable of imagining a different world. With this false illusion of no options, the capitalist monoculture is irrationally relied on for solutions to even the most acute problems that it causes, such as climate change, the accelerating loss of biodiversity, and other exceedances of safe planetary boundaries (Steffen et al. 2015). The collective incapacity to develop and broadly apply alternative social, economic, legal, and political models that would provide more promising, realistic, and hopeful solutions is perhaps the most forceful demonstration of the monoculture of the mind (Shiva 2003), of the spirit, and consequently, of action that constitutes globalized capitalism.

Increasingly sophisticated communications technologies exemplify the ongoing entrenchment of the monoculture built by the capitalist market economy (Mignolo 2017) and the consequent pricing of goods, values, and ideas. The expansion of sophisticated information technology and the telecommunications and media industries facilitates electronic access to monetary and financial codes and even dispenses with the need to cope with different languages, diversity of currencies, geographical distances, and aesthetic and cultural variants. Goods are for sale on Chinese, Brazilian, or Swedish websites regardless of their origin. An intermingled world of work and raw materials from multiple unnamed places is on sale in any shopping center. The work employed and the nature exploited are erased in the abstraction of the virtual market language on which the global financial and commercial market relies. Modern transactions even increasingly obviate the very existence of currency as a concrete means of exchange and a store of value: They are no longer tied to anything real. In modern, data-driven transactions that occur in the virtual world of the internet, the monoculture of the market reaches the apex of its ghostly transformation, in which everything that is solid melts into air (Marx 2002; Berman 2007).

Ultimately, the monoculture of money and global capitalism reflects an abject understanding and a cultural degeneration of what distinguishes humans from other living beings. If the moment when humans started to bury their dead and take care of the wounded, revealing an awareness of collective survival, is when humans became a distinct species, the moment when all existence is given a monetary price to enable fleeting pleasures devoid of meaningful relationships and reliant on oppression of the other marks the time when humanity destroys millennia of human evolution and is lost.

The question that remains is how to be free of the mental and spiritual monoculture cultivated by the capitalist market society. More and more of human life is transactional, not relational, in this monoculture. Modern global society does not understand another language but instead resigns itself collectively to the false tragic conclusion of Margaret Thatcher and others that "there is no alternative." Thus, in the face of collapse, this trapped society seeks responses to contemporary challenges that are mere reaffirmations of the dominant monoculture, such as green growth, monetary valuation of ecosystem services, the financing of unsustainable energy transitions, inadequate taxes on the use of fossil fuels, and more. The answer lies in limiting, not expanding, the domains in which markets reign, while developing and broadly implementing alternative approaches to economics, politics, and law that are grounded in the kind of life-enhancing diversity and social and ecological embeddedness needed for humanity and the life systems it depends on to flourish into the future.

From Monocultures to Diversity

Unsustainable approaches of modernity are built on mutually reinforcing material and conceptual monocultures grounded in growth-insistent economics, strong protection of private property and capital, and state-centered systems of law and governance, much of which is rooted in the emergence of capitalism during colonial regimes in the early modern era. Modern global society is possible because of the legalization, normalization, and perpetuation of human-created material and conceptual monocultures that continue to cause widespread harm. Modernity is full of harmful, delusional messages, most notably in advertising, that maintain this normalized status quo under an illusion of positive progress in social, economic, and environmental arenas: "You worked hard for it, you deserve it!" or "You are unique and can do or buy anything you want!" or "Your car connects you with nature and makes you YOU!" The convergence of harmful material and conceptual monocultures is unsustainable because they drive urgent and worsening existential crises, such as climate change and biodiversity loss, ultimately threatening global peace and security.

Mainstream approaches to solve the looming social and ecological crises that modern monocultures created are based on those same monocultures, in the form of state-led promotion of technological solutions, with private corporations in a leadership role. Climate engineering and other forms of ecomodernism are the full, fantastical, and horrifying expression of a world committed to monocultures, both material and conceptual. Ecomodernists hubristically liken humans to all powerful gods who must act like gods (Alvarado & Sussberg 2021) and complete the quest for total technological domination of nature (Asafu-Adjaye et al. 2015). The dangers of these deluded believers in the Baconian vision of human domination of nature and of an ecologically unlimited world can hardly be overstated.

Wendell Berry responds that "[t]o recover from our disease of limitlessness, we will have to give up the idea that we have a right to be godlike animals, that we are potentially omniscient and omnipotent, ready to discover 'the secrets of the universe'" (Berry 2008: 41). Noting that, as earthly creatures, humans are bound by Earth's ecological limits, Berry wisely says that the most promising path can be found in the "self-restraints implied in neighborliness, stewardship, thrift, temperance, generosity, care, kindness, friendship, loyalty, and love" (Berry 2008: 38).

At the global scale, ridding the Earth of all monocultures is impossible, but recognizing that they have grown too dominant, and right-sizing them, is essential. A key problem of monocultures is that they impose uniformity by forcefully, and often violently, erasing diversity. The desired alternative is to allow commonalities among diverse local places and sources to emerge from bottom-up approaches, as opposed to oppressive, top-down imposition of universality. Decolonization and undoing colonial legacies are key aspects of this kind of right-sizing, and promoting place-based territories of life (ICCA 2023) and nested institutions of placed-based governance (Ostrom 1990) can help. Answers abound in the degrowth movement (Kallis 2018; Schmelzer et al. 2022) and in higher education proposals for "ecological universities" (Kinchi 2022) as well. Ecological thinking (Code 2006; de Araujo et al. 2020), ecological economics (Erickson 2022), and ecological law and governance (ELGA 2016; Garver 2021) are also sources of a hopeful way forward that provides for flourishing human societies within a flourishing community of all life.

References

Alvarado, D. & Sussberg, J. (Directors) (2021) *We Are As Gods* [Film], Stripe Press.

American Society for the Prevention of Cruelty to Animals [ASPCA] (2023) *What Is Ag-Gag Legislation?* Available at: www.aspca.org/improving-laws-animals/public-policy/what-ag-gag-legislation (Accessed 10 January 2023).

Asafu-Adjaye, J., et al. (2015) *An Ecomodernist Manifesto*. Available at: www.ecomodernism.org/manifesto-english (Accessed 10 January 2023).

Berman, M. (2007) *Tudo que é sólido desmancha no ar*, Companhia de Bolso.

Berry, W. (2008) "Faustian economics: Hell hath no limits," *Harper's Magazine*, May issue, pp. 35–42.

Capra, F. & Mattei, U. (2015) *The Ecology of Law: Toward a Legal System in Tune with Nature and Community*, San Francisco: Berrett Koehler.

Césaire, A. (2000) *Discourse on Colonialism*, Monthly Review Press.

Cockshaw, R. (2021) "The End of Factory Farming: Alternatives to Improve Sustainability, Safety, and Health," *Voices in Bioethics*, 7.

Code, L. (2006) *Ecological Thinking: The Politics of Epistemic Location*, New York: Oxford University Press.

Daly, H. (1997) *Beyond Growth*, Boston: Beacon Press.

da Rocha, J. (2021) "Law, decoloniality and multispecie turn," *Direito e Práxis*, 12(2), pp. 885–914.

de Araujo, C.E., Ribeira de Sá, M.J., & da Conceiçao de Almeida, M. (2020) "Resisting the monoculture of mind: An ode to Indigenous knowledge," *Educaçao em Revista*, 36, p. e231352.

Ecological Law and Governance Association [ELGA] (2016) *Oslo Manifesto for Ecological Law and Governance*. Available at: https://elgaworld.org/oslo-manifesto.
Erickson, J. (2022) *The Progress Illusion: Reclaiming Our Future from the Fairytale of Economics*, Washington, DC: Island Press.
Fanon, F. (2022) *Os Condenados Da Terra*, Zahar.
Garver, G. (2021) *Ecological Law and the Planetary Crisis: A Legal Guide to Harmony on Earth*, New York: Routledge.
Gilbert, M., Conchedda, G., Van Boeckel, T.P., et al. (2015) "Income disparities and the global distribution of intensively farmed chicken and pigs," *PLoS ONE*, 10(7), p. e0133381.
Holling, C.S. (1996) "Engineering resilience versus ecological resilience," in: P. Schulze (ed) *Engineering Within Ecological Constraints*, pp. 40–52, Washington, DC: National Academy Press.
Indigenous and Community Conserved Areas [ICCA] Consortium (2023) *ICCAs-Territories of Life*. Available at: https://toolbox.iccaconsortium.org/meanings-and-more/iccas-territories-of-life/ (Accessed 9 January 2023).
Johnson, N. (2006) "Swine of the times: The making of the modern pig," *Harper's Magazine*, May issue, pp. 47–56.
Kallis, G. (2018) *Degrowth*, Newcastle upon Tyne: Agenda Publishing.
Kinchi, I.M. (2022) "The ecological root metaphor for higher education: Searching for evidence of conceptual emergence within university education strategies," *Education. Sciences*, 12, p. 528.
Locke, J. (2003) *Second Treatise of Government*, The Project Gutenberg. Available at: www.gutenberg.org/files/7370/7370-h/7370-h.htm (Accessed 9 January 2023).
Marx, K. (1996) *Capital (Volume One)*, ed. by F. Engels, Washington, DC: Regnery Publishing.
Marx, K. (2002) *The Communist Manifesto*, London: Penguin Books.
Merriam-Webster (2022) *Merriam-Webster.com Dictionary*. Available at: www.merriam-webster.com/dictionary/monoculture (Accessed 20 October 2022).
Michaels, F.S. (2011) *Monoculture: How One Story Is Changing Everything*, Red Clover Press.
Mignolo, W. (2011) *The Darker Side of Western Modernity: Global Futures, Decolonial Options*, Durham: Duke University Press.
Mignolo, W. (2017) "Colonialidade: O Lado Mais Escuro Da Modernidade," *Revista Brasileira De Ciências Sociais*, 32(94), pp. 1–18.
Mignolo, W. & Walsh, C. (2018) *On Decoloniality: Concepts, Analytics, Praxis*, Durham: Duke University Press.
Ostrom, E. (1990) *Governing the Commons: The Evolution of Institutions for Collective Action*, New York: Cambridge University Press.
Pan, W.K. & Bonnet, G. (2022) "Land use change and health," in: L.M. Hunter, C. Gray, & J. Véron (eds) *International Handbook of Population and Environment*, pp. 301–330, Cham: Springer.
Patel, R. & Moore, J.W. (2017) *The History of the World in Seven Cheap Things: A Guide to Capitalism, Nature, and the Future of the Planet*, Oakland: University of California Press.
Polanyi, K. (1944) *The Great Transformation: The Political and Economic Origins of Our Time*, Boston: Beacon Press.
Quijano, A. & Ennis, M. (2000) "Coloniality of power, eurocentrism, and Latin America," *Nepantla: Views From South*, 1(3), pp. 533–580.
Rodney, W. (2018) *How Europe Underdeveloped Africa*, England: Verso.
Santos, B. (2002) "Hacia una concepción multicultural de los derechos humanos," *El Otro Derecho*, 28, pp. 59–83.

Schmelzer, M., Vetter, A., & Vansintjan, A. (2022) *The Future Is Degrowth: A Guide to a World Beyond Capitalism*, Verso.

Shiva, V. (2003) *Monoculturas da mente: perspectivas da biodiversidade e da biotecnologia*, Brazil: Editora Gaia.

Steffen, W., Richardson, K., Rockström, J., et al. (2015) "Planetary boundaries: Guiding human development on a changing planet," *Science*, 347(6223), pp. 1259885–1–1259885–10.

Sundstrom, S.M., Hodbod, J., & Allen, C.R. (2022) "Resilience of working agricultural landscapes," in: C. Ray, S. Muddu, & S. Sharma (eds) *Food, Energy, and Water Nexus*, Cham: Springer.

Torre Rangel, J. (2006) *El derecho como arma de liberación en América Latina: Sociologia jurídica y uso alternativo del derecho*, CENEJUS, Centro de Estudios Jurídicos y Sociales P. Enrique Gutiérrez, Facultad de Derecho de la Universidad Autónoma de San Luis Potosí.

Tsing, A. (2015) *The Mushroom at the End of the World*, Princeton: Princeton University Press.

United Nations [UN] (1945) *Statute of the International Court of Justice*. Available at: www.icj-cij.org/en/statute (Accessed 7 January 2023).

United Nations [UN] (1972) *Report of the United Nations Conference on the Human Environment*. Available at: https://documents-dds-ny.un.org/doc/UNDOC/GEN/NL7/300/05/IMG/NL730005.pdf?OpenElement (Accessed 7 January 2023).

United Nations [UN] (1992) *Convention on Biological Diversity*. Available at: www.cbd.int/doc/legal/cbd-en.pdf (Accessed 7 January 2023).

US Department of Agriculture [USDA] (2022) *Hog & Pork Sector at a Glance*, USDA Economic Research Service. Available at: www.ers.usda.gov/topics/animal-products/hogs-pork/sector-at-a-glance/#Loc3.

Walsh, C. (2019) "Interculturalidade e decolonialidade do poder: um pensamento e posicionamento "outro" a partir da diferença colonial," *Revista Eletrônica da Faculdade de Direito da Universidade Federal de Pelotas*, 5(1), pp. 6–39.

Wolkmer, A. (2001) *Pluralismo jurídico: fundamentos de uma nova cultura no Direito*, Editora Alfa Omega.

Chapter 14

Sustainability and Recycling

The Scientific Approach

Renata Adami, Salvatore De Pasquale, and Maria Sarno

Introduction

The most important challenges for sustainability include climate change and its negative effects, the waste of resources, poverty, and the pollution of urban centers. In recent years, we have already experienced scorching summers, droughts, floods, and rising sea levels that are strictly related to the increase of the temperature of the Earth and to the consequent modification of the climate in all regions of the planet. Another consequence of these changes is the increase in wildfires, especially problematic in the Mediterranean region, affecting the environment, biodiversity, and the urban and rural populations.

Global environmental problems are mainly the result of human overutilization of natural resources, including fossil fuels, minerals, water, land, and biodiversity. In order to compete in the global economy and trade, nations all over the world use resources in an indiscriminate way, producing large amounts of waste and pollution. Moreover, many resources are not sustainably used, either because of a short lifespan or hyperconsumerism, where they are then sent to the landfill or downcycled, with a decrease in quality and in value.

Sustainable development indicates the guidelines aimed at creating an environment where it is possible to live well today, without compromising what the future world will be, with respect to future generations. The three pillars of sustainability are environmental sustainability, economic sustainability, and social sustainability. These should not be seen and treated as separate entities but as part of a larger set in which each part must interact to achieve the sustainable development fundamental for the proper working of each community. And technology plays a crucial role in the strategies adopted for environmental sustainability. These strategies are related to the use of renewable energies, the application of sustainable mobility, the adoption of plans for a "circular economy," and the correct management of waste.

This last aspect is of great interest for the global community; indeed, *waste* is defined as any substance or object which is discarded or intended/required to be discarded. Therefore, the only way to avoid producing waste is to avoid

DOI: 10.4324/9781003440871-17

getting rid of substances or objects. With this assumption, it is clear that it is impossible to have a zero-waste scenario. The best practice for waste management follows the "4R" rule – REDUCE, REUSE, RECYCLE, RECOVER – that should be considered by consumers when they buy a product or take care of its disposal, and by the manufacturers, who should design products and processes that consider its use and life cycle. This rule can be easily explained by analyzing the words that compose it:

1. REDUCTION is literally to produce less at the source. This is mainly related to the packaging of products that become waste as soon as the product is brought home and consumed. The packaging can be made of a smaller amount of material and of materials that can be easily divided, to be collected in disposal.
2. REUSING means that the product must be used several times. This can be applied to any product or object that, if used repeatedly, has an extension of its life, preventing it from quickly becoming waste. In order to be reused several times, it is necessary that the objects are of good quality. In the case of a broken object, a policy of sustainability suggests that it could be repaired and still used until the end of its life.
3. RECYCLING is the transformation of the material that no longer serves its purpose. Different types of material from waste are selected and separately collected so that they can be processed to produce new materials. Unfortunately, not all items can be recycled in the sense of entering again into the same chain process from which they have been produced.
4. RECOVERY is the valorization of waste to obtain secondary material or energy. Some non-reusable and non-recyclable waste can be modified and used to produce objects completely different from the original ones. Some of these wastes may contain compounds of interest in different fields that can be extracted and used for purpose different than the material of origin. The part that cannot be recovered can be burned to produce energy. Through a controlled combustion process, the calorific value of waste is exploited to obtain energy or heat, producing some other waste (solid, liquid, and gaseous) that must also be properly treated. A good valorization of waste reduces the amount of material that enters the burning process.

In particular, the compounds recovered from waste material are known as secondary raw materials (SRM) and consist of production waste, such as side streams and by-products from manufacturing processes, or materials deriving from recycling processes that can be introduced back into the economic system as new raw materials. SRM can be used as raw materials through simple reuse, recycling, or restoration and can be eventually transformed into more valuable products; for example, oils can be chemically transformed into biodiesel, and cellulose can be modified into nanocellulose that can enter several other production processes.

The guidelines that the European Parliament has expressed to update its legislation on waste treatment have the aim of achieving a zero-carbon economy that is environmentally sustainable, free from toxic substances, and completely circular by 2050. The circular economic system is a model of production and consumption that seeks to minimize waste, enhancing it as a resource and inserting it within a continuous life cycle. In this context, the economic system of a country generates secondary raw materials and then markets them, as is the case for raw materials deriving from extraction activities. This is highlighted and advocated by the European Strategic Plan on the Circular Economy (European Commission 2020). The production and subsequent reuse of SRM has the advantage not only of decreasing the generation of waste but also of improving the security of material supply in the European Union (a demand that is constantly growing), thus reducing the pressures of extracting natural resources.

The European Green Deal (European Commission 2019) aims to drive EU toward a sustainable and fair society. The goal is to reach in 2050 no net emissions of greenhouse gases and improve the resource usage for efficient and competitive economic growth. The decarbonization of the transportation sector is a fundamental part of the plan that requires the use of renewable fuels with lower greenhouse gas production during their production and utilization.

Nanotechnologies and Sustainability

Nanotechnologies deal with materials and devices with nanometer dimensions. Broad applications are developed based on the design, production, and use of structures, devices, and systems at the nanoscale level. Many of these types of technologies can offer advanced solutions to several problems related to sustainability and give a concrete start to the next "industrial revolution" that must consider climate change mitigation as a top priority.

The materials that are produced through nanotechnologies have improved properties compared to the equivalent at macroscale. As a first advantage, nanomaterials are lighter and, at the same time, resistant; moreover, they can be produced using less substance and lower energy, leading to a more efficient utilization of resources. Nanomaterial performances are enhanced because of their high surface area, which increases contact with the environment and allows for unique properties and forces differently than materials of a larger scale. The macroscale rules of chemistry and physics do not apply to their small size, a prerequisite to obtain better performances in different fields and applications.

Nanotechnologies can help keep the parameters of strong sustainability constraints for the planet under control, like climate change, biodiversity loss, nitrogen and phosphorus cycles, stratospheric ozone loss, ocean acidification, freshwater use, overexploitation of land, chemical pollution, and dispersion of gaseous species into the atmosphere. Through nanotechnologies, biodegradable and environmentally friendly materials can be produced. Moreover, manufacturing processes can be improved and become less polluting, a catalytical approach can be used to safeguard and purify the environment and water, and innovative energy sources can be produced, or the energy consumption of traditional technologies can be reduced. These improvements do take a long time, however, which is not compatible with the high speed of energy requirements or the rapid growth of mankind. Therefore, substantial change in environmental policies is strongly needed (Cingolani 2014).

Production of Biodiesel From Secondary Raw Materials

One possible application of nanotechnologies for the recovery of secondary raw materials (SRM) is the production of biodiesel from the oils extracted from the by-products of the food industrial processes. Most fuels used in industry and transportation are derived from oil, which is a nonrenewable fossil fuel. Due to the nonrenewable nature of fossil fuels, the increase of their price, and the need to use less-impactful energy sources in order to reduce emissions of CO_2 (as recommended by the European community in Agenda 2030), the need for alternative resources to produce renewable energy is clear.

In this context, biofuels are an affordable and quick alternative to fossil fuels and can contribute to sustainable mobility in the field of renewable energies. In particular, biodiesel is a renewable fuel that can be produced from biomass or edible oils obtained from the separate collection of waste. It is biodegradable, nontoxic, renewable, and less-polluting than diesel derived from petroleum, since its use reduces emissions of CO_2, reduces particulate emissions by up to 65% compared to traditional fuels, does not contain aromatic hydrocarbons and sulfurous compounds, and has a high flash point (therefore, it is not considered a flammable material).

One of the methods to produce biofuel is chemical transformation through the reaction of transesterification of vegetable oils/animal fats with a short-chain alcohol in the presence of a catalyst in order to obtain alkyl esters of fatty acids and glycerol (Ma & Hanna 1999; Sarno et al. 2021). A *catalyst* is a particular chemical species that accelerates a specific chemical reaction without participating in it; in fact, it remains unchanged.

The composition of biodiesel strongly depends on the characteristics of the raw material, as well as on the process conditions, such as temperature and pressure, the presence of water, the ratio of oil to alcohol, and the type of catalyst. The most used resource to produce this biofuel is the oil obtained from plant products, such as rapeseed and soy (Knothe & Razon 2017). The use of oils of vegetable origin is not, however, the optimal choice because of the competition between food and non-food production and the high production costs, of which only the ones of raw materials amount to about 80–85% (Murugesan et al. 2009).

Scientific research is now aimed at the study of new raw materials to overcome the technological and ecological limits related to the use of these sources. In this regard, the use of oils or waxes from industrial waste can be a valid and economical alternative that further allows to exploit waste as a resource. It is also possible to use cooking oil, that is, used food oil. It is a waste substance to be disposed of, it is produced in large quantities, and it is difficult to recycle because it contains many compounds due to the degradation of the oil when used at high temperatures. Used cooking oil, however, allows us to obtain biodiesel with a high yield, while its recovery also improves the environment (Sarno & Iuliano 2019a).

The most commonly used catalysts for these kinds of reactions are acid or basic and have several limitations in terms of high costs and different disadvantages: They can lead to saponification reactions and can be highly toxic and corrosive, and sometimes there are difficulties in recovery after the reaction. Therefore, an alternative can be the use of biocatalysts, represented by enzymes, which are the main catalysts of biological processes (Iuliano et al. 2020). The enzyme can be anchored on a support to improve the conditions of stability, usually quite precarious, and to facilitate its recovery at the end of the reaction (Narwal & Gupta 2013). The method and support used are important for the success of the immobilization process.

Nanomaterials are very effective in the industrial sector, especially in the field of catalysis: Nanoparticles have a high immobilization capacity thanks to the high surface/volume ratio and offer a lower limitation to diffusion compared to other materials. Therefore, the enzyme is mobile in solution and has high activity. Magnetic nanoparticles can also be used as support, since they are nontoxic, economical, biocompatible, and they allow the recovery of enzymes through the application of an external magnetic field – both a fast and cost-effective approach.

The synthesis of magnetic nanoparticles can be performed using a bottom-up process with a single step (Iuliano et al. 2020), then through an eco-friendly approach can be coated with an agent that promotes the physical immobilization of the enzyme (Polichetti et al. 2020). At the end of the chemical reaction, two liquids are collected; one is composed by the methyl esters that make the biodiesel, and the other is the glycerin that can be used for different applications.

The steps for biodiesel production can be reported, in summary, as follows:

1. Extraction of waxes from matrix
2. Synthesis of nanometric support
3. Immobilization of the enzyme (active phase) on the support
4. Catalytic synthesis of biodiesel
5. Removal and recovery of catalyst

Several sources of secondary raw materials for biodiesel have been successfully used, and depending on the composition of the oil extracted, different kinds of enzymes on different supports and with different functionalization have been developed. Some examples of industrial production side stream from which oils or waxes have been derived include tomatoes, spent coffee grounds, brewer spent grains, grape residue from wine factories, and dairy factory waste scum.

Tomatoes are one of the most widely cultivated vegetable crops in the world, and about 60% of the total waste are tomato seeds, which is usually disposed as livestock feed with low value or otherwise dumped in landfills, causing environmental problems. The oil content in tomato seeds is 20–40% on a dry basis, because of the high content of unsaturated fatty acids, especially linoleic acid; therefore, it is a great source of oil that can be transformed to biodiesel (Sarno & Iuliano 2019c).

Spent coffee grounds contain large amounts of organic components (fatty acids, cellulose, etc.), with a high oil content, depending on coffee types, usually about 11–20% of its dry weight. This can be a great source of oil since globally coffee is the second most commercialized product, after petroleum (Sarno & Iuliano 2018). Brewer spent grains are a side stream in many different countries that consume beer, with a valuable source of oil that can be converted into biofuels (Sarno & Iuliano 2019b).

The wine factory waste consists of grape residue, like grape seeds, from which can be effectively extracted the oil that can then be used as raw material for biodiesel production. Magnetic nanoparticles covered with tartaric acid are synthesized through an "eco-friendly" approach and used for direct physical immobilization of the enzyme lipase. The transesterification reactions produce a biodiesel with a high conversion yield (94%); the enzyme is then recovered through magnetic field and can be reused many times (Sarno & Iuliano 2020b).

Dairy factory waste scum oil is produced by dairy industries that deal with raw milk products, such as butter, yogurt, cheese, ice cream, and other related products, and, due to the transformation processes, produce a great amount of waste per day, leading to many difficulties for disposal (Kavitha et al. 2019). Most of the dairies dispose of this scum in a solid waste disposal site or through incineration (Kelessidis & Stasinakis 2012), which is both economically wasteful and generates pollutants. Dairy scum is a floating solid mass with a low density, formed by a mixture of fats, lipids, proteins, and other compounds. Moreover, they contain a large number of triglycerides (more than 80% on dry bases) that could be recovered as an oil and used as an alternative feedstock for biodiesel production (Sarno & Iuliano 2020a).

It is important to note that oils recovered from the waste of food industries can also find applications in areas other than biodiesel production, due to the properties of some of the contained compounds. The seeds of pomegranate (*Punica granatum* L.), for example, are a by-product of the food industry, and each kilogram of fruit contains about 0.04–0.15 kilograms of seeds, with 12–20 wt% of oil (Kalamara et al. 2015; Durdevic et al. 2017).

Pomegranate seeds have specific compounds that help the protection of the body against inflammation and damage from free radicals. Among other components, they contain punicic acid, which has anticarcinogenic, antioxidant, anti-inflammatory, and antidiabetic effects. Pomegranate seed oil can be transformed into liquid wax esters using immobilized enzymes on a nanometric structure in a solvent-free process. The wax ester produced has very interesting characteristics, comparable to jojoba oils or other natural oils used in cosmetics. Moreover, they show a good antioxidant power and moisturizing properties (Sarno & Iuliano 2020c). It is encouraging for the reduction of waste and the valorization of by-products that all the sources of waxes and oils are promising in the production of biofuels or products with several applications, and that many of these are easily found in the processes already commonly used in different industries.

Production of Nanocellulose From Secondary Raw Materials

Another interesting use of nanotechnology for sustainability is the transformation of waste materials in nanometric compounds that can be used in several applications. Among possible eco-friendly materials, cellulose is increasing in importance as a nanostructured material in two forms at the nanoscale size: nanocrystalline and micro/nanofibrillar. Nanocellulose has excellent mechanical properties, such as high strength and elastic modulus with a high surface area, high dimensional stability and moisture absorption properties, and thermal stability; furthermore, it is widely available, low-cost, and low-density, allowing for the production of lighter products (Dufresne 2013). It is an environmentally friendly material because it is natural, renewable, biodegradable, and biocompatible.

Nanocellulose can also be used in several fields, for example, as a reinforcement filler for polymers, electronic devices, and coating systems, but also as an important component for pharmaceutical drug delivery systems and medical tissue engineering materials. It can also be used in industrial applications for enzyme immobilization (Kalia et al. 2011; Mustafa et al. 2022). Nanocellulose is currently produced using expensive physical, chemical, and enzymatic pretreatment methods with a high energy consumption, but it could be produced in a convenient and environmentally friendly way using processes intended to recover it from different kinds of waste materials (Klemm et al. 2005; Harmsen et al. 2010) and reducing it in size using chemical or physical approaches, such as through milling.

For the first time, nanocellulose has been produced from the waste cellulose material derived from used diapers and incontinence products, applying a chemical approach consisting of a controlled hydrolysis using sulfuric acid solutions (Sarno & Cirillo 2019). Recovered nanocellulose can then be used for the production of absorbent products, pet litter, high-quality paper, chemical building blocks, agricultural nutrients, and energy production via gasification process.

There are countless sources of cellulose and nanocellulose in nature, many of them hidden in waste materials or materials usually not considered for production or simply and carelessly burned in the fields. One example is the Vesuvian pine cone that is considered a waste in the woods at the slopes of a volcano. In recent research, these pine cones have successfully been used as a template to synthesize magnetic nanocellulose with a simple self-assembly method. Vesuvian slopes pine cones have been treated using an alkaline/bleaching process followed by hydrolysis, then, through a simple electrostatic self-assembly approach, nanocellulose was combined with modified magnetic nanoparticles, obtaining a novel magnetic adsorbent. The prepared adsorbent was successfully used for the removal of polluting dye from water but can be used with other applications as well (Iuliano et al. 2022).

Conclusion

Nanotechnologies may offer advanced technological solutions to many of the problems related to sustainability. They can be applied in several processes, improving existing techniques, while also creating new strategies to implement in the production process. Materials produced through nanotechnologies can contribute to a more efficient and sustainable use of resources. The production of nanometric compounds to be used as catalysts in different reactions can contribute to the improvement of sustainable processes. Promising results are already showing in the synthesis of biodiesel using enzymes immobilized on magnetic nanoparticles, with great advantages in the use of waste material as a renewable source of fuel with low environmental impacts.

The use of nanoparticles obtained from waste materials, like nanocellulose from different sources, can reduce the amount of waste and, at the same time, can contribute to the manufacture of composite products with enhanced properties in terms of performances, specific characteristics, and properties of biocompatibility and biodegradability. The development of technologies that make it possible to improve the use of the planet's resources is the key for a more sustainable future of humanity, but it is also fundamental to reduce the differences between the populations of the world and to make the presence of humanity on the planet less invasive.

Reference List

Cingolani, R. (2014) *Il mondo è piccolo come un'arancia, Una discussione semplice sulle nanotecnologie*, ilSaggiatore.

Dufresne, A. (2013) "Nanocellulose: A new ageless bionanomaterial," *Materials Today*, 16(6), pp. 220–227.

Đurđević, S., Milovanović, S., Šavikin, K., Ristić, M., Menković, N., Pljevljakušić, D., Petrović, S., & Bogdanović, A. (2017) "Improvement of supercritical CO2 and

n-hexane extraction of wild growing pomegranate seed oil by microwave pretreatment," *Industrial Crops and Products*, 104, pp. 21–27.
European Commission (2019) *The European Green Deal*, COM 640.
European Commission (2020) *European Commission Action Plan*. Available at: https://environment.ec.europa.eu/strategy/circular-economy-action-plan_en.
Harmsen, P., Huijgen, W., Bermudez, L., & Bakker, R. (2010) *Literature Review of Physical and Chemical Pretreatment Processes for Lignocellulosic Biomass*, Wageningen UR Food & Biobased Research.
Iuliano, M., Ponticorvo, E., Cirillo, C., & Sarno, M. (2022) "A new nanocomposite from vesuvian slope pinecones for Azo-Dyes removal," *Industrial and Engineering Chemistry Research*, 61(5), pp. 1965–1976.
Iuliano, M., Sarno, M., De Pasquale, S., & Ponticorvo, E. (2020) "Candida rugosa lipase for the biodiesel production from renewable sources," *Renewable Energy*, 162, pp. 124–133.
Kalamara, E., Goula, A.M., & Adamopoulos, K.G. (2015) "An integrated process for utilization of pomegranate wastes – seeds," *Innovative Food Science and Emerging Technologies*, 27, pp. 144–153.
Kalia, S., Dufresne, A., Cherian, B.M., Kaith, B.S., Avérous, L., Njuguna, J., & Nassiopoulos, E. (2011) "Cellulose-based bio- and nanocomposites: A review," *International Journal of Polymer Science*, 2011, pp. 837875.
Kavitha, V., Geetha, V., & Jennita Jacqueline, P. (2019) "Production of biodiesel from dairy waste scum using eggshell waste," *Process Safety and Environmental Protection*, 125, pp. 279–287.
Kelessidis, A. & Stasinakis, A.S. (2012) "Comparative study of the methods used for treatment and final disposal of sewage sludge in European Countries," *Waste Management*, 32, pp. 1186–1195.
Klemm, D., Heublein, B., Fink, H.P., & Bohn, A. (2005) "Cellulose: Fascinating biopolymer and sustainable raw material," *Angewandte Chemie-International Edition*, 44(22), pp. 3358–3393.
Knothe, G. & Razon, L.F. (2017) "Biodiesel fuels," *Progress in Energy and Combustion Science*, 58, pp. 36–59.
Ma, F.R. & Hanna, M.A. (1999) "Biodiesel production: A review," *Bioresource Technology*, 70(1), pp. 1–15.
Murugesan, A., Umarani, C., Subramanian, R., & Nedunchezhian, N. (2009) "Bio-diesel as an alternative fuel for diesel engines-a review," *Renewable & Sustainable Energy Review*, 13(3), pp. 653–662.
Mustafa, A.A., Dassenoy, W.F., Sarno, M., & Senatore, A. (2022) "A review on potentials and challenges of nanolubricants as promising lubricants for electric vehicles," *Lubrication Science*, 34(1), pp. 1–29.
Narwal, S.K. & Gupta, R. (2013) "Biodiesel production by transesterification using immobilized lipase," *Biotechnology Letters*, 35(4), pp. 479–490.
Polichetti, M., Modestino, M., Galluzzi, A., Pace, S., Iuliano, M., Ciambelli, P., & Sarno, M. (2020) "Influence of citric acid and oleic acid coating on the dc magnetic properties of Fe3O4 magnetic nanoparticles," *Materials Today: Proceedings*, 20(1), pp. 21–24.
Sarno, M. & Cirillo, C. (2019) "Production of nanocellulose from waste cellulose," *Chemical Engineering Transactions*, 73, pp. 103–108.
Sarno, M. & Iuliano, M. (2018) "Active biocatalyst for biodiesel production from spent coffee ground," *Bioresource Technology*, 266, pp. 431–438.
Sarno, M. & Iuliano, M. (2019a) "Biodiesel production from waste cooking oil," *Green Processing and Synthesis*, 8(1), pp. 828–836.
Sarno, M. & Iuliano, M. (2019b) "Enzymatic production of biodiesel from Brewers' spent grains," in *ECOS 2019-Proceedings of the 32nd International Conference on*

Efficiency, Cost, Optimization, Simulation and Environmental Impact of Energy Systems, Institute of Thermal Technology, vol. 2019, pp. 3827–3839. Institute of Thermal Technology.

Sarno, M. & Iuliano, M. (2019c) "Highly active and stable Fe3O4/Au nanoparticles supporting lipase catalyst for biodiesel production from waste tomato," *Applied Surface Science*, 474, pp. 135–146.

Sarno, M. & Iuliano, M. (2020a) "Biodiesel production from dairy waste scum by using an efficient nano-biocatalyst," *Chemical Engineering Transactions*, 79, pp. 181–186.

Sarno, M. & Iuliano, M. (2020b) "Enzymatic production of biodiesel from grapeseed oil," *Chemical Engineering Transactions*, 80, pp. 301–306.

Sarno, M. & Iuliano, M. (2020c) "G_Fe3O4/Ag supporting Candida rugosa lipase for the 'green' synthesis of pomegranate seed oil derived liquid wax esters," *Applied Surface Science*, 510.

Sarno, M., Iuliano, M., Viscusi, G., Zarli, A., & Ciambelli, P. (2021) "A Nickel/Palladium/Ruthenium-Graphene based nanocatalyst for selective catalytic hydrogenation of vegetable oils," *Industrial Crops and Products*, 170.

Chapter 15

Which Legal Principles for the European Ecological Transition? The Case of the Circular Economy

Massimiliano Montini

Introduction: The European Green Deal and the Ecological Transition

The launch of the European Green Deal in 2019 has opened a new phase in European environmental policy. In the European Commission's view, the European Green Deal should promote economic growth while tackling climate change and environmentally related challenges. The European Green Deal is highly influenced by sustainable development as a key priority goal and operates within the global framework of the UN Agenda 2030, with the view to provide a European contribution to the implementation of the UN SDGs. To this end, the European Green Deal tries to promote at the same time economic, social, and environmental goals while pursuing several interrelated objectives. Among those, a prominent role is played by the carbon neutrality objective, to be achieved by 2050, and by the aim of decoupling economic growth from resource use, which is, however, not assigned a specific deadline. Within such a framework, one of the top priorities of the European Green Deal is the promotion of a green transition, which is sometimes called an ecological transition.

To this respect, the first question which arises is the following one: Can the term *green transition* be equated to the term *ecological transition*? In fact, the term "green transition" seems to call for the need of a general greening of all policies, with the view of tackling climate change and promoting more sustainable patterns of development. Instead, the term "ecological transition" seems to refer to something much more advanced than the mere greening of traditional policies. In fact, such a term seems to evoke the pursuit of a completely new and different approach to sustainable development, whereby the environmental dimension is considered as a floor upon which economic and social development of humans may occur. Under an ecological transition approach, a new holistic process of sustainable development is proposed, whereby a serious attempt is made to achieve a satisfactory balance between the three dimensions of sustainable development, namely, the environmental, social, and economic ones. By doing so, humanity might try to flourish in

DOI: 10.4324/9781003440871-18

harmony with nature while respecting the planetary boundaries constraints. The distinction between a green transition and an ecological transition is therefore not a mere terminological one but rather an ontological one. Such a distinction, however, is not present in the European Green Deal planning documents and legal acts.

Despite this shortcoming, it should be recognized that the European Green Deal was conceived as a genuine effort to gradually revise the dominant model of development in use in the European Union, which tends to put a greater emphasis on economic interests as compared to environmental and social ones. However, the original aims of the European Green Deal were affected by and had to be partially revised due to two main unforeseeable events which happened in the last few years after the publication of the original Green Deal launching document.

The first unforeseeable event was the spread of the COVID-19 pandemic, which started in the first months of 2020. In the aftermath of the initial spreading of the pandemic, the European Commission, as the main governing institution of the European Union, strongly restated that despite the health and economic consequences of the pandemic, the green transition remained a reference objective also in the process of providing an economic relief to member states under pressure because of the pandemic. The European Commission launched a recovery and resilience program named Next Generation EU, which provided to the most-affected member states a huge economic relief, through both grants and loans. The program aims at distributing European funds to member states based on specific recovery and resilience national plans, to be approved by the European Commission and other EU institutions. The sums allocated by the EU to its member states have to be used in a relatively short period of time (up to 2026 at the latest) but will be repaid by the member states, for the quota represented by loans, in a very long period of time, of around 30 years. Such a payment and reimbursement structure, in the plans of the European Commission, should enable a prompt recovery of European member states' economies put under pressure by the COVID-19 pandemic.

The second unforeseeable event was the Ukrainian war, which started with the Russian attack in Ukraine at the beginning of 2022. The Ukrainian war caused a sense of urgency in the European Union and its member states about the need to revise their energy policies, trying to get rid of Russian imports of oil and gas and to substitute them in the short term with similar imports from other suppliers, and in the medium-long term with a greater recourse to the production of renewable energy through plants based in the European territory. This urgency imposed by the Ukrainian war, coupled with the carbon neutrality objective to be achieved by 2050 that was already present in the European Green Deal, prompted the planning of big investments that should accompany the energy transition in the next few years. It is within such a context that the possibility of promoting an ecological transition in Europe should be placed and carefully assessed.

Ecological Transition Equals Energy Transition?

The combined effect of the pandemics and the Ukrainian war prompted the European Union and its member states to consider that the highest priority should be an energy transformation in Europe. This seems to be unavoidable. However, what is controversial is the tendency to equate the need of an "ecological transition" with an "energy transition." The two terms are obviously not equal ones. In fact, the energy transition represents just one dimension of the general transition that is needed, which should include a profound reassessment and reorganization of the current economic development model. Moreover, the energy transition, particularly if driven by an agenda based on a sense of urgency, might not necessarily contribute to the promotion of an ecological transition. This is well exemplified by the immediate European reaction to the Ukrainian war. In fact, the outburst of the Ukrainian war greatly influenced – and not necessarily in the right way – the debate on the reorganization of European energy policies and the definition of the European taxonomy of energy sources, which had started well before the beginning of the war in the framework of the implementation of the European Green Deal.

This is going to have short-, medium-, and long-term effects. The short-term effect of the Ukrainian war has been the tendency to promote a shift from a huge Russian dependence on the supply of oil and gas toward new fossil fuel energy suppliers for European Union's countries and consumers. This has obviously nothing to do with the ecological transition. The medium-term effect of the Ukrainian war has been a huge pressure to revise the initial view related to the definition of a taxonomy of green energy sources that could benefit from the support of public and private funds. The sense of urgency caused by the war affected the redesigning of the European Union's energy supply. It reinforced the view according to which some fossil fuels and some other (not necessarily clean) traditional energy sources should be kept in the basket of the so-called green energy sources. After a long and controversial debate, the European Parliament gave a green light to the inclusion of gas and nuclear energy within the basket of green energy sources. This controversial choice, which might be understandable in purely economic terms, obviously poses many doubts when observed from an ecological point of view.

The long-term effect of the Ukrainian war, coupled with the process of implementation of the European Green Deal, will most probably be an increased need of greater energy independence of the European Union. Therefore, it will most probably prompt a greater promotion of renewable energy production in the European Union. This is certainly a very positive objective. However, a closer observation shows that an increased production of energy from renewable energy sources might not necessarily come without cost for the environment. The major and more urgent problem in this sense is related to the determination of which specific locations, within the European territory, are suitable to host new renewable energy plants (from wind, solar, geothermal, or biomass sources), which sometimes have huge dimensions and impacts.

The aforementioned analysis of the undergoing European energy transition shows, in my opinion, that it is not correct to equate an energy transition to an ecological transition. To promote a truly ecological transition, the necessary fight against climate change should not be the only objective. The ecological transition should instead promote a reorienting of all human activities aimed at finding a better balance between environmental, social, and economic considerations. In the ecological transition, the environmental dimension should become the floor upon which proper economic and social development is based. In other terms, an ecological transition should be underpinned by the protection of the ecological integrity of all ecosystems. This should enable the pursuit of a harmonious development of all species, including not only humans but also all other animal and vegetable species. To promote a proper ecological transition technology and finance may help, but they should not dictate the transition agenda.

For all these reasons, it is not possible to argue that a mere energy transition may substitute for an ecological transition. This also demonstrates that even if a successful energy transition will be achieved in the next few decades and the planned European carbon neutrality by 2050 will be reached, this will not necessarily imply that an ecological transition will be achieved at the same time. In fact, an ecological transition needs to be placed on proper values and principles and coupled with a very clear and detailed planning, with the view to promote truly sustainable patterns of human development. It is for this reason that it is useful to start looking at the relevance of the legal dimension for the promotion of an ecological transition. More specifically, it is necessary to start considering which legal principles may be identified to promote and govern the ecological transition.

In Search of the Legal Principles to Govern the Ecological Transition

Which legal principles may be identified as useful tools to promote and govern the ecological transition? Among the many possible principles which are already well established or are simply emerging in the broad environmental protection field, a few ones may be highlighted as possible references for a successful and meaningful process of ecological transition.

The present analysis will focus on four principles. Two of them are existing environmental law principles that are already present in the European Union legal order, namely, the principle of integration and the principle of sustainable development. Besides, two of them are new, emerging principles in the environmental protection field, namely, the non-regression principle and the *in dubio pro natura* principle.

The Principle of Integration

The principle of integration finds its roots in early international environmental law. In particular, the first reference to the principle may be found in Principle 13 of the 1972 Stockholm Declaration. In such a context, the parties agreed that in order to achieve a more rational management of resources and to improve the environment, states should adopt an integrated and co-ordinated approach to development planning. At that time, the principle of integration was mainly seen as an instrument to be relied upon to make sure that economic development did not run counter to environmental protection concerns for the protection of the human environment. Later, the principle of integration became firmly related to the principle of sustainable development. The close connection between the two principles became evident in the 1992 Rio Declaration, where in Principle 4 it is stated that in order to achieve sustainable development, environmental protection shall constitute an integral part of the development process and cannot be considered in isolation from it.

During the evolution of international environmental law, starting with the 2002 Johannesburg Summit up to the 2012 Rio+20 Conference, the principle of integration slightly changed its focus. More specifically, the principle evolved from being an instrument to integrate environmental consideration into developmental and economic planning and started to be considered as the main tool to promote the interplay and the intersection between the three pillars of sustainable development, namely, the economic, environmental, and social dimensions. Within such a framework, the principle of integration became the main tool for the fulfillment of the three concentric circles representation of the concept of sustainable development. Meanwhile, the principle of integration was recognized in international environmental case law. For instance, the relevance of the principle was implicitly recognized by the International Court of Justice in the 1997 *Gabcikovo–Nagymaros* case, alongside the principle of sustainable development, while the Permanent Court of Arbitration recognized it as a principle of general international law in the 2005 *Iron Rhine* arbitration.

In the context of European Union environmental law, the principle of integration was firstly inserted into the European Community (EC) Treaty in 1987, alongside most of the other main European environmental principles, by means of the so-called Single European Act. At that time, the principle was included among the European environmental principles with a very simple phrasing, according to which "[e]nvironmental protection requirements shall be a component of the Community's other policies." Later, the principle of integration was subject to a few amendments in the pattern of evolution of the EC treaties. Most notably, with the 1999 Amsterdam Treaty, the principle of integration was subject to two very relevant changes. Firstly, it was clearly stated that the duty of integration of environmental concerns into all other European (non-environmental) policies should be pursued with a view to promoting sustainable development. Secondly, the principle was given a greater emphasis through its relocation from the environmental section of the EC Treaty to the section of the European general legal principles placed at the beginning of the treaty. A few years later, however, with the latest amendments brought by the 2007 Treaty of Lisbon, which reorganized the system of the European treaties, the principle of integration was placed in Article 11 of the new Treaty on the Functioning of the European Union (TFEU), with the following wording: "Environmental protection requirements must be integrated into the definition and implementation of the Union's policies and activities, in particular with a view to promoting sustainable development."

As one can see, the wording of the provision dedicated to the principle of integration did not change much as compared to previous versions. However, its relevance decreased, since it lost its special status of a general principle of law, which it had previously enjoyed, and was inserted in the initial part of the Treaty among the provisions of general application. This change gave rise to a situation where the relevance of the principle of integration in the EU legal order diminished in the following years (Jans 2011; Montini 2021). In brief, in the current version of Article 11 TFEU, the principle of integration commands environmental protection requirements to be integrated in the design and implementation of all other policies. However, the practice has shown that the principle of integration has a limited capacity to give environmental concerns a certain weight and to afford them priority over other competing interests. In addition, the principle cannot be successfully used as a tool to challenge European policies in courts and measures that do not adequately consider environmental protection interests (Montini 2018). However, a potentially stronger role of the principle of integration might come from a different reading of its close connection with the principle of sustainable development, as it will be described in the next paragraph (Montini 2021; Sjåfjell 2018).

The Principle of Sustainable Development

The origin of the principle of sustainable development is normally referred to the 1987 Brundtland Report, where it was given the following well-known definition: "Sustainable development is development that meets the needs of the present without compromising the ability of future generations to meet their own needs." From such a wording, two basic constitutive concepts may be enucleated. On the one side, there is the (explicit) concept of the needs, which recognizes in every state the sovereign right to satisfy its own national needs on the basis of its environmental and developmental policies. On the other side, there is the (implicit) concept of limits, which derives from the 1972 Club of Rome Report, "Limits to Growth." According to such a concept, the needs of the present generation can be satisfied only insofar as the respect of equal rights of future generations to satisfy their own needs can be adequately guaranteed. This means, in practice, that the present generation should act in a way that does not negatively impact on the maintenance of the stock of natural resources that they received from previous generations and must hand over to future ones.

In more recent times, the principle of sustainable development has been identified as the leading value upon which the United Nations Agenda 2030 on sustainable development is built. That is the reason the 17 objectives for the international pursuit of sustainable development at a global level up to 2030, approved by the UN General Assembly in 2015, have been named Sustainable Development Goals (SDGs). The choice of the reference to the concept of sustainable development as the reference value is deliberate and intends to mark a stark difference with respect to the previous global objectives, namely, the Millennium Development Goals (MDGs), which acted as UN reference objectives for economic development in the period 2000–2015. Such previous objectives were, in fact, mostly focused on the pursuit of economic development goals, while leaving aside social and environmental considerations. In sum, it may be argued that the pivotal role assigned to the principle of sustainable development in the framework of the SDGs is a great window of opportunity for its global recognition and application. However, it also entails some risks, as its capacity to truly act as a guiding value and reference concept will be concretely tested in the global process of implementation of the SDGs (Montini & Volpe 2017)

In European Union law, the principle of sustainable development was inserted in the original European Community Treaty (ECT) relatively late, by means of the Treaty of Amsterdam of 1999. In the current European Union legal treaty system, there are a few references to the principle in the two main EU treaties, namely, the Treaty on the European Union (TEU) and the Treaty on the Functioning of the European Union (TFEU). The principle is conceived as a guiding objective and a general value upon which EU policies should be based. However, nowhere in the treaties can a specific definition of the principle be found. One of the most interesting references to the principle of sustainable development is contained in the already-mentioned Article 11 TFEU, where the principle of sustainable development is associated with the principle of integration. In such a context, the duty of integrating environmental consideration in the development and implementation of all other EU policies is strictly related to the pursuit of the principle of sustainable development. In fact, the provision at stake prescribes that the duty of integration must be pursued with the view to achieving sustainable development. In such a way, it seems that the duty of integration has become a sort of operational tool which may be used to achieve the overarching objective of sustainable development. This view is confirmed also by the wording of Article 37 of the European Charter of Fundamental Rights, which reads as follows: "A high level of environmental protection and the improvement of the quality of the environment must be integrated into the policies of the Union and ensured in accordance with the principle of sustainable development" (Montini 2021).

The principle of sustainable development might be given a more meaningful role as an instrument to support the ecological transition in general and more specifically the promotion of a circular economy approach. To this effect, a few changes appear to be necessary in the qualification and interpretation of the principle. The first main change regards the general understanding of the principle. Traditionally, the principle is referred to as composed of three concentric and partially overlapping circles which represent the economic, social, and environmental dimensions. Within such a view, sustainable development patterns and solutions should emerge in the tiny common core area where the three circles overlap. This vision has some relevant limits, insofar as it considers the three circles as having all the same importance and dignity. This makes it difficult (if not almost impossible) to have a proper balance of the conflicting interests which exist within the three concentric circles. To support an ecological transition and move from a linear to a circular economy approach, it would be necessary to adopt a new vision for the principle of sustainable development. The concentric circle depiction should be substituted by a pyramidal representation. Within such a new understanding of the principle, the environmental dimension should represent the floor upon which the social and economic dimensions may properly operate. In other terms, according to such a new and different vision, the health and integrity of the ecosystem, which is the core value within the environmental dimension, should be recognized as the necessary and inescapable basis for promoting social and economic patterns of human development which may enable humans to flourish in harmony with nature (Jørgensen 2015).

The second main change for the principle of sustainable development regards the need to rediscover and properly re-evaluate its ecological core (Grober 2012; Bosselmann 2016). A close analysis of the historical origin as well as the logical and ontological nature of the principle of sustainable development has shown that the principle has a fundamental ecological core which derives from it being strictly related to the need to make sure that human development respects the basic conditions upon which natural systems operate. By doing so, socioeconomic development should not occur at the expense of the integrity of the natural environment and should move away from the logic of domination of humans over nature. A new holistic vision should be promoted instead, whereby humans may consider themselves as firmly embedded within the ecosystem where they live. This should create the conditions to promote a human flourishing within the limits and the boundaries imposed by nature and enable patterns of socioeconomic human development that are in a dynamic equilibrium with the health and integrity of ecosystems (Montini 2020).

The Non-Regression Principle

The principle of non-regression has been developed in legal literature, in particular by the French jurist Michel Prieur, as a new environmental law principle that aims at preventing any backtracking of environmental law legislation, both at the international as well as the national level (Prieur & Sozzo 2012). The premise for the development of this new principle was the understanding that the traditional environmental principles are not suited to prevent and avoid attempts at reducing environmental protection, as guaranteed by previous legislation, under the pressure of competing interests, such as economic and social ones. In particular, the principle of non-regression is conceived as an instrument that can guarantee the maintenance of the already-acquired levels of environmental protection that may be contested and attacked in the process of the simplification and derogation of existing environmental protections, which often occurs nowadays in most jurisdictions.

The core element of the principle is the requirement that no subsequent legislative act should be allowed to reduce the level of protection of the environment that has been established by a previous legislative act. The aim of the principle is, therefore, to guarantee that already-acquired level of environmental protection and nature conservation can be maintained over time, despite the possible call for revision made on the basis of other competing interests (Prieur 2018; Prieur & Sozzo 2012).

A specific feature of the principle of environmental non-regression is the close connection with the realm of human rights protection. Such a connection has two distinct relevant features. The first feature concerns the "inspirational" connection between the two contexts where the principle may be applied. In fact, the proponents of the recognition of a general principle of environmental non-regression ground their proposal on the existence of a general principle against backtracking in the level of protection guaranteed by international as well as national human rights legislation. In particular, in the human rights field, there is a general understanding that deliberate retrogressive measures are not permitted, while, quite on the contrary, states should aim at a progressive improvement of the level of human rights protection, both at international as well as at national levels.

The second feature regards the practical connection between environmental non-regression and human rights non-regression. In this sense, the progressive recognition of the right to a healthy environment, guaranteed to any person as a new type of human right, is gradually emerging as a constitutional right in many jurisdictions. Moreover, in 2021, the existence of a new human right to a healthy environment was recognized for the first time by the Human Rights Council (HRC). More precisely, the HRC recognized "the right to a safe, clean, healthy, and sustainable environment as a human right that is important for the enjoyment of human rights" (HRC Resolution 2021).

The principle of environmental non-regression has been initially supported mostly in legal literature, as well as in non-binding soft law legal documents and declarations (Prieur 2018). For instance, in the preparation process for the Rio+ 20 Conference, the European Parliament adopted a resolution in support of a recognition of the principle in the specific environmental protection context. The European Parliament specifically called for "the recognition of the principle of non-regression in the context of environmental protection as well as fundamental rights" (EP 2012). However, because of the strong opposition of many countries, a specific reference to the non-regression principle was dropped from the final declaration "The Future We Want" adopted at Rio+ 20. This notwithstanding, in the aforementioned document, the parties affirmed that "it is critical that we do not backtrack from our commitment to the outcome of the UN conference on environment and development" (UN 2012).

Later, the principle was inserted in other soft law documents, such as, for instance, the 2015 IUCN Draft international covenant on environment and development. In such a context, Article 10 rubricated "non-regression" read as follows:

> Substantive and procedural rules for environmental conservation shall be maintained without regression, and interpreted and applied in favor of ecological integrity, unless compelling reasons of public interest require otherwise. The necessity of any measures of regression shall be revisited and re-examined on a periodic basis in order to restore or enhance pre-existing levels of environmental conservation.
>
> (IUCN 2015)

More recently, the principle of non-regression found a firm and clear recognition in a binding international environmental law treaty. The principle was explicitly referred to in the 2018 Escazú Agreement, namely, the Regional Agreement on Access to Information, Public Participation, and Justice in Environmental Matters in Latin America and the Caribbean. In particular, in such a context, the principle was included in Article 3(c) within the list of environmental principles which should guide the implementation of the agreement.

It is yet too early to affirm that the principle of non-regression is an already-established principle of environmental law (Mitchell & Munro 2023). However, it is rapidly evolving and may well become in the years to come a reference principle to underpin a proper ecological transition, which should have among its fundamental pillars a general prohibition of backtracking of environmental legislation.

The In Dubio Pro Natura *Principle*

The *in dubio pro natura* principle is an emerging principle of environmental law. It calls for the recognition of a certain priority to the protection of the environment and the preservation of natural resources when a doubt exists on the existence of possible risks of a serious and irreversible damage. According to such a principle, when environmental interests are opposed to other conflicting interests and science is not conclusive on providing appropriate elements for balancing the competing positions, the question at stake should be decided in a way that favors environmental protection needs. In other words, the principle calls for a *favor naturae* in the interpretation and application of the law, when a doubt exists about the possibility that a certain activity may pose an urgent threat of a serious and irreversible risk for the environment and natural resources.

In more specific terms, the *in dubio pro natura* principle shares with the precautionary principle a common inspiration that is represented by the general criterion of *favor naturae.* In fact, the two principles have a basic common feature, which consists in the requirement that in case of a serious and irreversible risk for the environment, appropriate actions are taken to protect the environment that may err in caution, even in the absence of a full and conclusive scientific evidence. However, there is a major difference in the scope of application of the two principles. In fact, the precautionary principle may be referred to in cases when full scientific evidence about certain risks for the environment is not available, whilst the *in dubio pro natura* principle may be relied upon when the available science leaves decision-makers and courts with a doubt on the right actions to be undertaken. In other terms, according to the principle under scrutiny, in case of doubt, priority should be given to choices and actions which minimize the risks for the environment.

So far, the *in dubio pro natura* principle has been mainly affirmed and developed in the jurisdictions of Latin American countries. A notable example of its legal recognition is represented by Article 395(4) of the 2008 Constitution of Ecuador. In such context, within a provision dedicated to the protection of nature and the environment, the principle has been recognized in the following terms: "in the event of doubt about the scope of legal provisions for environmental issues, it is the most favorable interpretation of their effective force for the protection of nature that shall prevail" (Carducci et al. 2020).

At the international level, the principle has not yet received proper recognition in binding legal instruments. However, it has been incorporated in Principle 5 of the IUCN "World Declaration on the Environmental Rule of Law," according to which:

> In cases of doubt, all matters before courts, administrative agencies, and other decision-makers shall be resolved in a way most likely to favor the

> protection and conservation of the environment, with preference to be given to alternatives that are least harmful to the environment. Actions shall not be undertaken when their potential adverse impacts on the environment are disproportionate or excessive in relation to the benefits derived therefrom.
>
> (IUCN 2016)

In the European Union legal order, the principle has not yet found a legal recognition. However, it shares similar (and partially common) features with the precautionary principle, one of the fundamental principles of EU environmental law. Therefore, the possibility of an interplay between the two principles deserves closer scrutiny. This might pave the way in the long term for a meaningful role of such a new principle in EU law. This possibility appears to be particularly interesting at a time when the interpretation and implementation of the European Green Deal pose several potential threats for the maintenance of the same level of protection of the environment and the preservation of environmental resources throughout the European territory.

Conclusion: The Ecological Transition in the Case of the (European) Circular Economy

The circular economy objective, together with the climate neutrality goal, is one of the main pillars of the European Green Deal. In this sense, it is a key objective in the efforts to promote an ecological transition. In brief, the main aim of the circular economy is to promote a decoupling of the increase of economic growth from the consumption of finite resources, thus limiting the need of more extraction of new raw materials and reducing all environmental negative consequences associated to such extraction activities. In other words, the circular economy aims at reducing waste, in the sense that all raw materials that are discarded in the production processes, as well as all products that are no longer used, should be reused and recycled, rather than be disposed of as waste.

The circular economy approach includes both an end-of-pipe dimension, which aims at reducing waste by improving reuse and recycling of materials, and an initial phase dimension, which aims at promoting eco-design, in order to reduce the foreseeable impacts of products on the environment. By improving the initial phase of all economic processes, the eco-design practice should make sure that products are designed and manufactured with the aim of reducing the quantity of the raw materials that are needed to manufacture them and that all products can be easily reused or recycled at the end of their life cycle. In this sense, the European circular economy approach appears to be fully in line with the requirements of UN SDG 12, which calls for ensuring sustainable and responsible consumption and production patterns.

This is the reason that the promotion of a true ecological transition needs the pillar of the circular economy. It will not be enough to promote the carbon neutrality objective or aim at an energy transition if these efforts are not coupled with the introduction of a circular economy approach. This should help overcome old linear economic approaches that cause severe environmental externalities, mostly in the form of environmental pollution and extraction of raw materials, as well as negative social consequences on workers and populations. However, it should be underlined that circular economy objectives should be implemented with a broad holistic view that aims at promoting the respect of ecological integrity and at reducing the negative environmental and social effects caused by economic development activities. By so doing, the promotion of a circular economy approach might help promote a true ecological transition of the European society.

Reference List

Bosselmann, K. (2016) *The Principle of Sustainability. Transforming Law and Governance*, 2nd ed., Abingdon: Routledge.

Carducci, M., Bagni, S., Montini, M., Mumta, I., Lorubbio, V., Barreca, A., Di Francesco Maesa, C., Musarò, E., Spinks, L., & Powlesland, P. (2020) *Towards an EU Charter of the Fundamental Rights of Nature. Study*, Brussels: European Economic and Social Committee.

European Parliament [EP] (2012) *Resolution of 29 September 2011 on Developing a Common EU Position Ahead of the United Nations Conference on Sustainable Development (Rio+20)*, P7_TA (2011) 0430.

Grober, U. (2012) *Sustainability. A Cultural History*, Cambridge: Green Books.

Human Rights Council [HRC] (2021) Resolution 48/13, 'The Human Right to a Safe, Clean, Healthy and Sustainable Environment' (Human Rights Council, A/HRC/48/L.23/Rev.1, 8 October 2021).

IUCN (2015) *Draft International Covenant on Environment and Development: Implementing Sustainability*, Principle 10. Available at: www.iucn.org/resources/publication/draft-international-covenant-environment-and-development-implementing.

IUCN (2016) *IUCN World Declaration on the Environmental Rule of Law (2016)*, Principle 5, Available at: www.iucn.org/our-union/commissions/world-commission-environmental-law/our-work/history/foundational-documents-4.

Jans, J. (2011) "Stop the integration principle?" *Fordham International Law Journal*, 33(5), pp. 1533–1547.

Jørgensen, D. (2015) "Rethinking rewilding," *Geoforum*, 65, pp. 482–488.

Mitchell, A. & Munro, J. (2023) "An international law principle of non-regression from environmental protections," *International & Comparative Law Quarterly*, 72(1), pp. 35–71.

Montini, M. (2018) "The principle of integration," in: L. Kramer & E. Orlando (eds) *Principles of Environmental Law*, pp. 139–149, Cheltenham: Edward Elgar Publishing.

Montini, M. (2020) "Designing law for sustainability," in: V. Mauerhofer, et al. (eds) *Sustainability and Law*, pp. 33–48, Cham: Springer.

Montini, M. (2021) "The European Green Deal from an environmental protection perspective: The missing role of the environmental integration principle," in: K.J.

De Graaf, B. Marseille, S. Prechal, R. Widdershoven, & H. Winter (eds) *Grensoverstijgende rechtsbeoefening: Liber amicorum Jan Jans*, pp. 97–104, Paris: Zutphen.
Montini, M. & Volpe, F. (2017) "Sustainable development: Renaissance or sunset boulevard?" in: L. Westra, et al. (eds) *The Role of Integrity in the Governance of the Commons*, pp. 169–178, Cham: Springer.
Prieur, M. (2018) "The principle of non-regression," in: L. Kramer & E. Orlando (eds) *Principles of Environmental Law*, pp. 251–259, Cheltenham: Edward Elgar Publishing.
Prieur, M. & Sozzo, G. (eds) (2012) *La non régression en droit de l'environnement*, Bruxelles.
Sjåfjell, B. (2018) "The environmental integration principle: A necessary step towards policy coherence for sustainability," in: F. Ippolito, M.E. Bartoloni, & M. Condinanzi (eds) *The EU and the Proliferation of Integration Principles under the Lisbon Treaty*, Abingdon: Routledge.
United Nations [UN] (2012) The Future We Want Declaration, § 20.

Chapter 16

The Double-Edged Sword

Reconciling How Developments in Public Health Equate to Increased Life Expectancy and Resource Degradation

Shauna M. Lange

Introduction

In this chapter we explore the concept of the "double-edged sword" as it applies to longer life expectancies and the depletion of natural resources, and through the lens of ecological integrity, justice, and sustainability. There is a direct relationship between population growth, life expectancy, public health advancements, and the degradation or use of natural resources. Natural resources are not limitless, and humanity must recognize that they can and will be exhausted. As global life expectancy has more than doubled since 1900, it is important to take note that the increase in quantity to our lives does not necessarily equate to an increase in quality. Air, soil, and water pollution; food shortages; and overstressed ecosystems pose a real threat to nature and the global community. It is pertinent to evaluate, consider, and attempt to remedy our mistakes and the damages caused before the effects are irreversible. Each day, more and more opportunities for sustainable, ethical engagement present themselves. It is time as a global society to recognize the Earth as a borrowed space – from our children, from our grandchildren – and to act and govern ourselves accordingly.

Throughout history, humankind has been faced with the constant question of priority between improving the quantity and the quality of life. The correlation and the juxtaposition of the two are undeniable. As advancements are made in public health that increase the human lifespan, it is crucial to also consider the impact these added years of existence, and consumption of resources, will have on the ecosystems and global community to which we belong. The Woodcutter, by Ferdinand Hodler in 1910 (National Museum Zurich 2022), is a powerful image to consider in an evaluation of this conundrum. The piece was commissioned to illustrate the advancements of humanity that also directly affect and/or causes man's demise. The image shows a man, average in stature, average in appearance, chopping down a tree with an axe. His clothing appears to be that of a working man, perhaps even a lumberjack, making his living from chopping down trees. The tree itself appears quite barren, perhaps from the consumption of its other resources

DOI: 10.4324/9781003440871-19

(bark, leaves, sap, etc.), and the image itself is also empty, lacking in flora, fauna, or other positive engagement (Sidelnikova 2023).

> [In the] "Woodcutter," the human figure is itself both a man and a tree, which he will have to defeat. A man firmly buried in the ground with his feet. The man whom nature feeds. A man who himself is part of nature. Manpower versus manpower.

We are the soil from which we take. We are the air from which we breathe. We are made of the water from which we drink and survive. With each taking and each benefit received, we deprive humanity in another way. It may not happen instantly, but it does happen and continues to occur, ultimately to the detriment of us all. More than the basic cause and effect, there is almost an unspoken karma.

The phrase "a double-edged sword" has been used across cultures and languages. However, whether double-edged sword in English, *Zweischneid iges Schwer* in German, *spada a doppio taglio* in Italian, *epee a double tranchan* in French, or *espada de doble filo* in Spanish, the message is the same. It is something that is both a blessing and a curse. Perhaps the most impactful example is that as a human society, while having a longer life has benefits to our shared relationships, as individuals, families, and nations, with noted disparities among marginalized peoples and nations, such as access to health care, food, economic stability, the increased consumption of more humans, alongside the increased care that people require as they age, also has detrimental effects on the Earth and its resources, and so detrimental effects on the future of human society as a whole.

Thinking again of the woodcutter, he cuts the tree to give himself and others benefits, and yet in doing so, he destroys so much. Similarly, as Bruno Manser, a Swiss Malaysian rainforest activist stated (Ricq 2020),

> there is, a direct correlation between population growth, life expectancy, and the degradation or use of natural resources, in this case the forest and its resources whether it be the rainforest or otherwise.

As people cut down the trees to build their homes and fuel their fires and feed their families, entire ecosystems are destroyed, and the lives and livelihoods of all those who rely upon them. Critical short-term needs are met, but to what end? And what methodologies for "life," and their underlying principles, such as evolving for survival and adapting society's consumption to change to ensure the continuation of life, are being set to be modeled – or countered – by future generations?

An understanding of the evolution of life expectancy helps us further delve into an understanding of population growth and consumption of resources. Overconsumption is ultimately a crisis of the wealthy caused by the wealthy, of those who "have" versus those who "want," of those who continue to unconsciously or consciously gather more than they need, and with limited regard for those that not only do not have but also do not have because of those who have. Those made poor, made sick by the policies and powers of others, and with a methodology of consumption rooted in discrimination, exploitation, and destruction that has continued from colonialism.

Green Is Gone: A Brief History of Global Life Expectancy and Mortality Rates

Since 1900, the average global life expectancy has more than doubled (Our World in Data n.d.). Advancements in technology, medical care, and food safety have allowed humanity, as a global community, to live longer than ever before. However, it is imperative that the disparity amongst the poorer, underprivileged, and underrepresented people of the world is recognized. Not only do they suffer the wrath – and it is a violence, a "policy violence" – of the lack and degradation of resources, but they also do not even receive the benefits of that harmful consumption, as experienced by the wealthy. As Kathryn Gwiazdon, a legal scholar in ethics and justice, so eloquently and powerfully put it, "[t]hey don't live longer, they don't even get the double-edged sword – they just get the sword" (Gwiazdon 2024). Similar to climate change harms, another crisis of the wealthy, where the wealthy benefit and the poor are harmed, the poor suffer the harms of a consumption that they do not equally partake in. This is why health is a justice issue at its core: The lives of those "without" are not any less deserving of a healthy and full life than the lives of those "with," yet our law and governance systems determine that to be so.

"Green is good" is a commonplace phrase in sustainability and environmental care movements, largely because of the powerful imagery that "green is life." Yet the "green" of life is being manifestly and unjustly replaced by the "gray" of development, and for the health of some, for the financial benefit of some, and at the expense of all life on Earth. Juxtaposed from the "life" of diverse forests, mangroves, wetlands, and desert ecosystems are the monotonous grids of mass agriculture and urban sprawl, the waste that forms mountains and chokes rivers, the concrete that creates roads by covering rivers, the dried and cracked Earth of drought and desertification, and the white bleaching of once a kaleidoscope of coral reefs. This carries powerful imagery. Simple Google Earth searches for "changes in green land space" will give jarring pictures of how areas around the world have lost green space for gray space, or monoculture space. Green is gone, or at least, green is going.

Instead of living in harmony with nature, the ruling governance systems of humanity seem to advance living in disharmony with one another and the planet, rejecting the science and the sacred, as seen through Indigenous cosmologies, of our interdependence with the natural environment and one another, and the injustice at its heart. Cities and urban areas, by their very nature, provide a stark example of this change in habitats and the unequal impacts of public health. Many scholars and policymakers are trying to highlight these particular socioeconomic and discrimination considerations, which also provide an enormous opportunity for cities to be at the center for action toward a more just, sustainable future (see generally Wolch et al. 2014; see generally Local Governments for Sustainability [ICLEI] 2024).

As the science to increase longevity has advanced, so has the science on the interdependence of human life on the environment. But this science has to be adequately relayed. Access to information is a fundamental human right because of the personal and societal change it can affect, and as such, it has a critical role to play in the interconnectedness of ecological integrity and human integrity. For example, environmental degradation due to increased life expectancy and population growth may be better understood by recognizing that resources have a limit and growth is not limitless, a cornerstone of the degrowth movement (discussed later).

Better understanding one's own impact on the environment and others could also lead to change. In the commotion of daily life, it can be easy to forget how everyone leaves a mark, an ecological footprint, a trail of use, before, during, or after every action (see generally Global Footprint Network n.d.). Every aspect of human life, and the consumption, production, and proliferation that follow, has a ripple effect, with repercussions far beyond itself. If society begins to change its thinking to view natural "resources" as necessary foundations of life, and individual actors as global citizens, society would have a better foundation when working toward a more just and sustainable law and governance, where nature is to be respected and is owed responsibilities and local acts have global repercussions.

And with access to information come important conversations on the role of media. Known as the "fourth pillar of democracy," media can move citizen action and political will to protect the environment, to advocate for justice, and to demand affordable, adequate health care. It is imperative to note, however, the increasing concerns regarding the particular impact of social media on public health, whether it be impacts on mental health or the role of policymakers in "silencing science" and disinformation, as is seen with both climate change and COVID-19 (see Sabin Center for Climate Change Law 2024; see generally National Academies of Sciences 2023; Viglione 2020). The general public may have more access to robust science than ever before, with information on how to live longer and healthier lives, and through the protection of the environment, but once again, it is confronted with issues of equity and justice, particularly in countries that do not offer universal health care, privatize the medical and "wellness" sectors, or concentrate harmful industries in already-vulnerable areas (see generally WHO 2023; Hosseinpoor et al. 2023; Natural Resources Defense Council [NRDC] 2023). Generally speaking, though with more robust science and increased access to information, our bodies and minds are better equipped for longer lives – if you can afford it.

Of course, less understandable for the general public may be the direct impacts from biodiversity loss and climate change, or even air and water pollution, on public health, until it reaches a critical point that triggers personal health issues, news cycles, or hopefully, political and legislative action. The result of this general lack of awareness of the interconnectedness of these issues with our lives is that business as usual continues, the harms continue, and the segmented thinking that reinforces hierarchies continues. Political will is meant to follow citizen demands, at least in a democratic society; however, too often, until an issue creates a tangible, personal harm, the demand is not made, the political will is not crystallized, and change does not occur.

One can review statistics and data regarding the changes in life expectancy over the last 100 years from Our World in Data, and the comparisons are dramatic (n.d.). Life expectancy worldwide in 1919 and 2019 are 54.7 years and 78.7 years, respectively, with life expectancy increasing by more than 20 years in many states. Furthermore, an analysis of the last 40 years shows that these numbers continue to increase. The next logical question to ask ourselves is, "How?" What has led to such a dramatic increase in life expectancy?

Better health care and hygiene, healthier lifestyles, sufficient food, improved medical care, and reduced child mortality together mean that humans can now expect to live much longer than our ancestors did from just a few generations ago (Brown 2015). Life expectancy at birth in the EU was about 69 years in 1960, and about 80 years in 2010, which corresponds to a rate of increase in life expectancy of 2.2 years per decade (Brown 2015). And although a longer life is to be celebrated, it must be considered that by continuing "business as usual" in our consumption, the Earth's "resources" will be depleted, and in a manner not able to be compensated for by technological advancements. Even as space exploration becomes a private business venture, not even the wealthiest of our planet will be able to escape a destroyed Earth for a better life elsewhere.

It is important to recognize that with the benefits of living longer, the juxtaposition of quantity versus quality has to also be acknowledged and addressed. The simple fact that a certain number of years are being added to a life does not automatically translate into quality also being added; "years are being added to our lives, life is not being added to our years: the extra years are being added at the very end of our lives and are of poor quality" (Brown 2015). As bodies age, organs and tissues age, and abilities become limited. The end-of-life process is far longer than in prior centuries, and too often longevity of life becomes longevity with suffering. This is not to suggest that there are no benefits to a longer life, but what cannot be ignored is that as life has lengthened, so has the dying process.

Within this analysis, it is important to also understand how changes in causes of death have affected our life expectancy. In 1900, the leading causes of death were pneumonia, tuberculosis, and diarrhea and enteritis (Centers for Disease Control and Prevention n.d.), all tackled by advancements in science and public information. However, in 1997, heart disease and cancer accounted for a vast majority of all deaths (Centers for Disease Control and Prevention n.d.), related in no small part to the degradation of our natural environment, the increased access to low-quality and highly processed foods, and in some countries, the increased costs of health care. So while technological and medical advancements have had a major impact on our ability to sustain life, our manner of death has changed dramatically.

Population Growth and Nature's Degrowth

How does one track the degradation of natural resources as connected to population growth? A groundbreaking Massachusetts Institute of Technology (MIT) study in 1972 titled "Limits to Growth" (LtG) was the first to try to solve the problem based on empirical data and computational analysis (Club of Rome 2023). This study analyzed global resource consumption and production levels based on data from many fields, including population, fertility rates, pollution levels, food production, and industrial output. Today, the "degrowth movement" acknowledges these limits to growth and the necessary response for justice and sustainability (see generally Degrowth 2024; World Economic Forum 2022; Hickel et al. 2022; Garver 2013).

Using the data acquired in this initial study, the scientists were able to outline 12 possible scenarios for the future, based on different forecasts of human behavior. Incessant economic growth eventually overshoots the world's natural resources, rendering further economic growth impossible (Club of Rome 2023). In other words, resources are limited, the modern-day global economy is tied to the presence and availability of those resources, and it is on track to deplete those resources.

One of the most famous scenarios is known as "business as usual" (Club of Rome 2023). This scenario illustrates how, if the global community continued on its current path of growth and use and development, the world's economy would begin to lose ground in the 2040s, and shortly thereafter, the world's population, food availability, and other resources would decrease. Areas that are already water-stressed would face extreme shortages as significant decreases in wheat production occurred. Intermittent scarcities would occur from the overuse of coal resources by Asian power plants, leading to coal mine closures and shipping and transportation bottlenecks. Maize is believed to be the product with the strongest international price rise, citing 20% by 2020, 80% by 2030, and 100% by 2050 (Club of Rome 2023). Harvest shocks would be common, and energy costs would increase on a global scale. These global systems shocks to all critical infrastructures – transportation, energy, and agriculture – would cause global insecurity, threats to democratic governance, and great loss of life.

Some of those predictions are already coming true. Natural resource degradation is exacerbating social tensions that lead to conflict. Climate change is exacerbating the vulnerability risks associated with natural resource degradation (Nguyen et al. 2023), with "[n]atural resource degradation increas[ing] susceptibility and exposure to climate shocks and further strain[ing] the adaptive capacity of resource-dependent communities" (World Bank 2021).

From the World Bank's 2021 report "The Natural Resource Degradation and Vulnerability Nexus: An Evaluation of the World Bank's Support for Sustainable and Inclusive Natural Resource Management" (2009–19):

> [O]ne-third of all land and one-fifth of all forest cover are severely degraded (UNCCD 2017. Groundwater, which accounts for half of drinking water and 43 percent of irrigation water, is being depleted at an alarming rate (citing Smith et al. 2016). The fraction of fish stocks at biologically sustainable levels fell from 90 percent in 1974 to 67 percent in 2015 (citing FAO 2018b). The growth of the world's population, the intensity of urbanization, and the changes in land use affect water availability to the extent that its reserves would be significantly unable to fulfil the increasing demands or become non-exploitable due to pollution and contamination The above reasons may cause conflicts among different users (urban use, irrigation, industry, and other uses) for accessing the needed water.

Soil is one of the most undervalued natural resources, yet the preservation of terrestrial life on the planet and its economy depends on soil conditions at both the local and global levels (see Ginzky 2019; Heuser 2021). Soil is considered to be a renewable natural resource and is formed at a prolonged low rate (World Bank 2021), which means that when our development and land use damages the soil, it is not instantly reparable, and development of new suitable soils takes extended periods of time.

Urbanization not only utilizes major amounts of soil but also damages and inhibits its ability to repair itself with natural processes. Urbanization through land use changes leads to the fragmentation of ecosystems and poses a significant threat to wildlife. As neighborhoods, cities, and general development sprawl, animals, plants, wildlife ecosystems, and the habitats that give them life generally become torn apart. At the same time, urbanization increases non-food species dominance in specific areas and alters the existing balances developed over time and through adaptation between edible species. In the United States alone, more than 10% of edible species are considered endangered. A species that is endangered is not regenerating at a speed rapid enough to maintain its numbers, so eventually the ability to regenerate will cease. The fragmentation of ecosystems and the expansion of non-food species may be the main factors to such an endangerment of the edible species, as they affect, respectively, 85% and 49% of these species (World Bank 2021).

Diminished Ecological Integrity, Diminished Human Integrity, and Ways Forward

Several studies in medicine and epidemiology have proven that environmental quality is a critical factor affecting health and morbidity: Air and water pollution, depletion of natural resources, soil deterioration, and the like are all capable of increasing human mortality (thus reducing longevity) (Mariani et al. 2009). It goes without saying that polluted air, water, and soil and a lack of resources do not make for a desirable, or livable, neighborhood. Each organism has an impact on the next, and the importance of being able to sustainably coexist cannot be overstated.

Life expectancy has a positive correlation with environmental quality, so while humans may live longer, they do not necessarily live better, more fulfilling, or healthier lives (Mariani et al. 2009). Lives may be prolonged by several years, but if those years are spent suffering due to illness caused by pollution or a lack of necessary resources, it is difficult to see how the benefit would outweigh the cost. In "Life Expectancy and the Environment," the authors describe the environmental performance indicator (Mariani et al. 2009):

> This synthetic indicator takes into account both "environmental health," as defined by child mortality, indoor air pollution, drinking water, adequate sanitation, and urban particulates, and "ecosystem vitality," which includes factors like air quality, water and productive natural resources, biodiversity, and sustainable energy.

This information shows the indisputable correlation between longer life expectancy and CO_2 emissions, rising seas, growing wastelands, deforestation, famine, war, and ultimately, lower life expectancy, especially in poorer or marginalized communities. The article further references that 40 countries experienced a decrease in longevity between 1990 and 2001. Like the woodcutter that opened this chapter, we cause our own dilemma in efforts to ensure our own survival.

Every day, human pressure on the Earth causes growing health risks from large-scale environmental changes. Easter Island, for example, showed a closed system in which insufficient environmental care and excessive pressure on natural resources led to the downfall of the local population. Failing to take into account which resources are irreplaceable or slower to regenerate will lead to the loss of fundamental needs that society cannot live without. But what can be done? How can real, meaningful change be facilitated?

Ecological integrity "refers to the ability of environmental life-support systems to sustain themselves in the face of human-induced impacts" (Sieswerda et al. 2001). In order to live sustainably with nature, we must ethically engage with our world and with one another in a way that does not threaten life on Earth or cause further harm to those who have been marginalized by the governance systems of the powerful. We must advance ecological integrity in law and policy that includes a plan for the future, to preserve and protect those foundations of life for future generations to come.

An ecological system has integrity when it supports and maintains a balanced, integrated, adaptive biological system having the full range of living elements (genes, species, and assemblages) and processes (mutation, demography, biotic interactions, nutrient, energy dynamics, and metapopulation processes) expected in the natural habitat of a region (Loucks et al. 1997). In order to support and maintain these elements and processes, an ecosystem must be able to sustain the life within it and simultaneously sustain itself. Therein lies the challenge. There are many opportunities to increase sustainability, encourage ethical engagement, and motivate change. To do so, ways to counter and combat the destruction caused by humanity's mere existence must be explored.

As with most of humanity's greatest challenges, and related to access to information earlier, there must be education of the crises and the root causes of the crises. Enabling the public to better understand the interconnectedness of all life, humanity's utter dependence on the natural world, and the direct links between human health and the environment is the foundation for democratic governance that then supports action that protects the environment for public health, for future generations, and for the quality of life, not just the quantity of years. With knowledge also comes accountability – political accountability – to respond to and better regulate the actors that are destroying the environment and degrading human life. Education and the will of the people will continue to be limited if political systems continue to be corrupted by corporate powers.

Throughout this chapter has been the power of powerful imagery and the corresponding data, and just as images of injustice can motivate, so too can images of healing: the evidence of a healing Earth and a healing people. The very correlation of reducing consumption and improving quality of life can be quantified when resources begin to recuperate. Just, sustainable growth will allow for just, sustainable growth. The problem will not be quantity versus quality of life but a more basic level of a shared, sustained existence.

Financial resources to support local, state, and international action on public health and the environment will also help advance ecological integrity and human health, as well as hold major corporations and industries accountable for their actions. This type of undertaking will require a real commitment from governments and corporate bodies alike to predict, monitor, and take action to modify development and growth as needed so as not to overutilize the integrity of the resources upon which all life depends, and to also make the clear connections between actions and harms. Strengthen standing in courts, strengthen science, and create specific agencies for zoning and land use, development commissions, population impact, water usage, soil degradation; advance the principle of non-regression and the continuous environmental impact assessment to facilitate responsibility, accountability, and public awareness. This movement will have to go beyond non-binding, "forward-looking" statements and pretty but impotent words about "corporate social responsibility" and move toward binding corporate social obligations that complement existing state responsibilities to prevent harms and address unequally shared benefits and harms. Some courts are already starting to recognize the human rights obligations of private actors, and this needs to be supported and reinforced (e.g., *Milieudefensie et al. v. Royal Dutch Shell* 2019).

Ultimately, a more adequate system of accountability based upon a more simply stated structure, and with a more coherent and accurate worldview, that is, the integrity of the natural environment is inseparable from the integrity of humans and humanity, will need to be advanced. Minus political and corporate corruption, one relatively easy path would be to tax polluting industries in such a way that it encourages a drastic change in methodology or use, with direct implications for not just public health but investments in more just, sustainable technology. Seeking out environmental fiscal reform such as that considered in the US Clean Air Act (CAA) (Eisenberg et al. 2014), a comprehensive federal law to regulate air emissions, gives a springboard to analyze possible ways of acquiring not only stronger regulations but also a method of funding for further awareness and proactivity campaigns. The CAA established standards of quality to protect public health and welfare and limit the amount of hazardous air pollutants that were released by vehicles and other mobile or stationary sources (Eisenberg et al. 2014). There must be a penalty for the unequal and unjust exploitation and destruction of the natural environment, at the ultimate expense of all. This is a critical unfairness – the very thing that the rule of law is meant to address. And perhaps we can use these failures – in the private and public spheres – as part of the remedy forward, where corporations and governments who fail to meet their obligations could be required to fund a level of publicity or public education about not only the degradation that occurred but also the ways it can be avoided in the future.

In conclusion, let us return to the forest. The woodcutter can chop the tree to build his home, to fuel his fire, to create his tools to extend his life, but when the trees cannot regenerate at the rate of his destruction, neither shall his life nor the life of his children continue with any quality worth having. A life without life, until there is no life at all. We must find ways to act and govern ourselves according to a simple truth: There is no human integrity without ecological integrity.

Reference List

Brown, G.C. (2015) *Living Too Long: The Current Focus of Medical Research on Increasing the Quantity, Rather Than the Quality, of Life Is Damaging Our Health and Harming the Economy*, EMBO Reports. Available at: www.ncbi.nlm.nih.gov/pmc/articles/PMC4328740/.

Centers for Disease Control and Prevention (n.d.) *Achievements in Public Health, 1900–1999: Control of Infectious Diseases*. Available at: www.cdc.gov/mmwr/

preview/mmwrhtml/mm4829a1.htm.In%201900%2C%20the%20three%20leading,than%205%20years%20(1).
Club of Rome (2023) *The Limits to Growth +50*. Available at: www.clubofrome.org/ltg50/.
Degrowth (2024). Available at: https://degrowth.info/degrowth.
Eisenberg, S., et al. (2014) *A State Tax Approach to Regulating Greenhouse Gases under the Clean Air Act*, The Climate and Energy Economics Project, Brookings. Available at: www.brookings.edu/wp-content/uploads/2016/06/state_approach_regulating_ghgs_morris.pdf.
Garver, G. (2013) "The rule of ecological law: The legal complement to degrowth economics," *Sustainability*, 5(1), pp. 316–337. Available at: https://doi.org/10.3390/su5010316.
Ginzky, H., et al. (2019) *International Yearbook of Soil Law and Policy 2019*, Cham: Springer.
Global Footprint Network (n.d.). Available at: www.footprintnetwork.org/our-work/ecological-footprint/.
Gwiazdon, K. (2024) Personal communications with author, available upon request.
Heuser, I. (2021) "The legal dimension of soil protection and sustainable use of soils," in: V. Mauerhofer (ed) *The Role of Law in Governing Sustainability*, Milton Park: Routledge.
Hickel, J., et al. (2022) "Degrowth can work – here's how science can help," *Nature*. Available at: www.nature.com/articles/d41586-022-04412-x.
Hosseinpoor, A.R., et al. (2023) "WHO releases health inequality data repository," *The Lancet*, 401(10388), pp. 1565–1566. Available at: https://doi.org/10.1016/S0140-6736(23)00859-0.
Local Governments for Sustainability [ICLEI] (2024). Available at: https://iclei.org/.
Loucks, O., Karr, J.R., Crabbe, P., Westra, L., Rees, W.E., & Soskolne, C. (1997) *Definition of Ecological Integrity*, Presented at the Global Ecological Integrity Projects Workshop, Social Sciences and Humanities Research Council of Canada, Cortona, Italy.
Mariani, F., Perez Barahona, A., & Raffin, N. (2009) "Life expectancy and the environment," *SSRN Electronic Journal* [Preprint]. Available at: https://doi.org/10.2139/ssrn.1506316.
Milieudefensie et al. v. Royal Dutch Shell 2019. Available at: https://climatecasechart.com/non-us-case/milieudefensie-et-al-v-royal-dutch-shell-plc/.
National Academies of Sciences (2023) *Silencing Scientists and Health Workers during the Pandemic*, Webinar Series. Available at: www.nationalacademies.org/our-work/silencing-scientists-and-health-workers-during-the-pandemic-webinar-series.
National Museum Zurich (2022) *In the Forest: A Cultural History at the National Museum Zurich*. Available at: www.landesmuseum.ch/en/exhibitions/temporary/2022/in-the-forest/in-the-forest.
Natural Resources Defense Council [NRDC] (2023) *The Environmental Justice Movement*. Available at: www.nrdc.org/stories/environmental-justice-movement.
Nguyen, T.T., Grote, U., Neubacher, F., Rahut, D.B., Hung Do, M., & Paudel, G.P. (2023) "Security risks from climate change and environmental degradation: Implications for sustainable land use transformation in the Global South," *Current Opinion in Environmental Sustainability*. Available at: www.sciencedirect.com/science/article/pii/S1877343523000696#:~:text=Climate%20change%20exacerbates%20environmental%20degradation,increased%20competition%20for%20natural%20resources.
Our World in Data (n.d.). Available at: https://ourworldindata.org/.

Ricq, I. (2020) "The Day Bruno Manser Disappeared," *The Ecologist*. Available at: https://theecologist.org/2020/feb/10/day-bruno-manser-disappeared.

Sabin Center for Climate Change Law (2024) *Silencing Science Tracker*, Columbia Law School. Available at: https://climate.law.columbia.edu/Silencing-Science-Tracker.

Sidelnikova, A. (2023) *Woodcutter, 1910, 101×131 cm by Ferdinand Hodler: Description of the Artwork*. Available at: https://arthive.com/ferdinandhodler/works/524879~Woodcutter.

Sieswerda, L.E., Soskolne, C., Newman, C., Schopflocher, D., & Smoyer, K.E. (2001) "Toward measuring the impact of ecological disintegrity on human health," *Epidemiology*, 12(1), pp. 28–32. Available at: http://doi.org/10.1097/00001648-200101000-00006.

Viglione, G. (2020) "Four ways Trump has meddled in pandemic science – and why it matters," *Nature*. Available at: www.nature.com/articles/d41586-020-03035-4.

Wolch, J., Byrne, J. & Newell, J. (2014) "Urban green space, public health, and environmental justice: The challenge of making cities 'just green enough,'" *Landscape and Urban Planning*, 125, pp. 234–244. Available at: https://doi.org/10.1016/j.landurbplan.2014.01.017.

World Bank (2021) *The Natural Resource Degradation and Vulnerability Nexus: An Evaluation of the World Bank's Support for Sustainable and Inclusive Natural Resource Management (2009–19)*, Independent Evaluation Group, Washington, DC: World Bank.

World Economic Forum (2022) *Degrowth – What's Behind the Economic Theory and Why Does It Matter Right Now?* Available at: www.weforum.org/agenda/2022/06/what-is-degrowth-economics-climate-change/.

World Health Organization (2023) *Health Inequality Monitor*. Available at: www.who.int/data/inequality-monitor/data.

Chapter 17

Healthy Environment

Elaboration of a Clearly Defined Notion and Forms of Protection by European Administrative Courts

Emanuele Fratto Rosi Grippaudo

Introduction

A recent article on the environmental administrative law of ancient Rome argues that Roman society considered the protection of certain natural resources – such as air, water, or soil – directly related to human safety and health, as a general interest known as *salubritas*, which was the subject of continuing legal protection by the Roman legal system (Ruiz-Pino 2017). Jurists, praetors, and even emperors were very sensitive to the issue of environmental health, recalling the contribution to the fight undertaken by Labeo against the pollution of water heritage (Di Porto 2015: 13) and the attention given by Augustus to the shaded areas in restructuring the outdoor courts (Traversa 2022). More than 2,000 years later, the theme of environmental health is still the focus of scholarship in legal doctrine, which often exalts in its innovative character.

In order to understand why, the answer must be found in the choice, made by doctrinal sources, courts, and legislators, to place public interest alongside a different means of protection of environmental salubrity: the individual right to a healthy environment. Through this right, any person who is endangered by polluted natural elements, such as air, water, or terrain, may ask a court to limit the activity making the environment unhealthy, thus directly and indirectly protecting environmental integrity. This right has not only allowed courts to better protect the health of individuals from anthropogenic pollution but is also proving to be a useful weapon within the framework of climate litigation. Indeed, as shown by recent research, whenever the right to a healthy environment is invoked in a climate litigation case, the success rate increases (Savaresi & Vilchez Moragues 2021).

Although the right to a healthy environment has recently been classified as a human right by the United Nations General Assembly (UNGA), this right is not universally recognized at a national level. Some legislators, with the support of their national legal doctrine, have raised doubts as to its content and the positive obligations that might result from its recognition by the state. It is well understood that the same legal doctrine that supports the recognition

DOI: 10.4324/9781003440871-20

of the right to a healthy environment supposes that this right has always been characterized by a vague and almost "slippery" content (Adinolfi 2018). The Inter-American Court of Human Rights (IACHR), for example, which recognizes the right to a healthy environment as a "universal value" and a "fundamental right for the existence of humankind," asserts that the right to a healthy environment "protects the components of the environment, . . . not only because of its benefits or effects for humanity." Therefore, according to the IACHR, the right to a healthy environment is a human right which has nature, and not exclusively humanity, as its direct beneficiary.[1] To nations rooted in ancient Greek and Roman jurisprudential principles, this could be considered an unconventional concept of human rights that may be more difficult to accept and adopt. However, the progressive worsening of environmental conditions, and particularly its impacts on human health and well-being, has affected the work of administrative courts, which have strengthened the protection of environmental health and delivered some groundbreaking rulings.

This chapter attempts to address these issues with reference to European countries with a civil law legal system. It begins with an investigation into the meaning of a healthy environment and then turns to the public interest in environmental health, examining judgments from Italian administrative courts. The chapter then examines the role of the right to a healthy environment in the international and European legal context, as well as in select European countries that follow a civil law legal system, with particular attention to the Federal Administrative Court of Switzerland. Then, examining the approaches of French administrative courts, it looks at how and against whom this right can be enforced.

This chapter aims at reinforcing the idea that the protection of environmental health is a state prerogative, and that the right to a healthy environment can be an excellent means of complementing this protection, especially with existential crises such as climate change. Moreover, this chapter argues that the right to a healthy environment can be considered as a standalone fundamental right because of its implications for the right to health. Finally, this chapter asserts that the right to a healthy environment can be a source of positive obligations for the state.

The Concept of a Healthy Environment and the Means of Protection Provided by Law

The concept of environmental health is composed of two elements: (1) the notion of environment, which, even in etymological terms, considers nature

1 See *IACHR Case of the Indigenous communities of the Lhaka Honhat (our land) association v. Argentina*, judgment of 6 February 2020, at www.corteidh.or.cr/docs/casos/articulos/seriec_400_ing.pdf, last accessed January 2024.

in a biunivocal relationship with humankind, leaving the concept of environment legally void in the absence of this relationship,[2] and (2) the notion of health, which is connected to the human being and suggests that every limit imposed by law on human activity which has effects on the environment should be functional to present or future human life and not to the protection of the ecosystem itself. *Environmental health* has been defined in many ways, and legal doctrine contributes some of these definitions.[3] The World Health Organization, in a meeting in Sophia, Bulgaria, in 1993, crafted a very broad definition of *environmental health* which "comprises those aspects of human health, including quality of life, that are determined by physical, chemical, biological, social, and psychosocial factors in the environment" (citation). And according to the European Charter on Environment and Health, environmental health "comprises those aspects of human health and disease that are determined by factors in the environment."

These definitions allow us to make some important considerations. First, the term "environment" has a wider meaning than simply the natural (i.e., non-human) environment and instead encompasses any context in which human beings live. Second, the concept of "healthy environment" is markedly anthropocentric, since, unlike environmental integrity, its purpose is not the protection of nature but of human health. Third, from this it follows that, ironically, the two concepts of healthy environment and environmental integrity could be in contrast, since the natural environment could be extremely unhealthy for humankind. For example, marshes are rich ecosystems that have a decisive role in combating climate change but can be extremely dangerous for human health, because in them live mosquitoes responsible for the spread of malaria. The reclamation of a marsh, then, is the result of an administrative procedure in which the adjudicator strikes a balance between public interest in the integrity of the environment and the health of the environment, establishing that the latter must prevail. For example, in Italy, the reclamation of the Pontine marshes eradicated malaria from the affected areas, thus saving thousands of lives but destroying an entire ecosystem.

The protection of environmental health is carried out by two separate means. The first belongs to public law and understands that public interest is in environmental health, while the second pertains to private law and incorporates the right to a healthy environment. These are two sides of the same coin, as evidenced by the Italian and Spanish Constitutions,[4] where health is safeguarded both as a public interest and as a fundamental right of

2 The word *environment* has its etymological roots in the Old French *environ*, meaning "surroundings," suggesting that humans are meant to be posited at the center of the ecological matters.

3 See the definitions listed in Frumkin (2005).

4 See Article 32 of the Italian Constitution and Article 43 of the Spanish Constitution.

the individual. These public and private interests could come into conflict, most recently seen during the COVID-19 pandemic, when legislators imposed quarantine periods on subjects infected with the virus in the name of public interest to health protection.[5]

Healthy Environment as a Public Interest: Italian Administrative Courts and the *Consiglio di Stato*

According to administrative law doctrine, a *public interest* is a general interest enshrined in a law (Corso 2020: 31), policy, or regulation that is in force and represents both the purpose, the limit, and the justification of administrative action (Lombard et al. 2019). The protection of human health, including environmental health and environmental integrity, is a public interest, as well as the protection of the landscape, economic development, urban planning regulation, conservation of cultural heritage, public budget balance, etc. In a democratic society, a public interest is worth more than a private interest because it includes a greater number of private interests (López Calera 2010). The interest in exercising a private economic undertaking is a typical example of private interest. Therefore, a public administration can only adopt an administrative measure requested by the individual – for example, an environmental permit – if the private interest is not in contrast with the public interest (Scoca 2017: 205). An administrative measure aimed at pursuing an interest other than the public interest indicated in the law or that it is the result of an unreasonable weighting of public and private interests is invalid and may be challenged by its addressee before an administrative court.[6]

The Italian administrative justice system consists of regional courts of first instance, which are called TAR for *Tribunale amministrativo regionale*, and a court of appeal based in Rome, which is called the Council of State (*Consiglio di Stato*). On numerous occasions, the Italian administrative courts have intervened to protect the public interest in environmental health. This has been done, for example, through the annulment of hunting regulations allowing the use of lead bullets,[7] by confirming the legality of a plan aimed at containing airport noise and reducing daily flights,[8] by affirming the legality of an infrastructure project striving for reducing road transport,[9] and

5 This topic has been delved into in Grandi (2021).

6 On this issue, see Sánchez Morón (2021).

7 See ruling no. 1206 made by the second chamber of the *TAR* for the Liguria region, on 28 July 2014, at www.giustizia-amministrativa.it, last accessed January 2024.

8 See ruling no. 3239 made by the third chamber of the *TAR* for the Lazio region, on 17 March 2021, at www.giustizia-amministrativa.it, last accessed January 2024.

9 See ruling no. 10849 made by the first chamber of the *TAR* for the Lazio region, on 18 August 2015, at www.giustizia-amministrativa.it, last accessed January 2024.

by confirming an administrative measure to limit aldehyde emissions in the ecosystem.[10]

The Italian administrative courts also played an important role in the infamous "Ilva case," concerning a coal-fired steel plant based on the Gulf of Taranto in the Puglia region. Civil and criminal courts established a causal link between environmental exposure to inhalable substances produced by the steel plant and the development of tumors and cardio-circulatory pathologies in persons living nearby.[11] Overturning the first instance ruling of the TAR for the Puglia region,[12] the Council of State[13] declared the illegitimacy of an order of closure of the plants adopted by the mayor of Taranto, stating that the extension of the environmental authorization, granted by the Ministry of Environment to the steel plant, is an expression of a reasonable and delicate weighting between the public interest in the protection of environmental health and that in the preservation of work in the steel industry. As these cases show, environmental health is not only the object of a public interest but also of an individual right.

Healthy Environment as a Human Right

In Resolution No. 76/300 of 28 July 2022, the UNGA recognized the right to a healthy environment as a human right.[14] This Resolution follows Resolution No. 48/13 of the Human Rights Council (HRC) of 8 October 2021 and represents the culmination of a process that started in 2011, when the HRC requested the Office of the High Commissioner for Human Rights to prepare a study on human rights and the environment (Cima 2022). Resolution No. 76/300 is very important for two reasons. First, because it considers the right to a healthy environment as an autonomous human right (Köböl-Benda 2022). Second, because, notwithstanding its soft law nature, it could be used as a template for later binding treaties, as evidence for emerging customary norms, or codified in domestic constitutions (Günther & Ekardt 2022). However, this Resolution does not identify the right-holder or the obligations arising from the protection of this right, thus leaving to the interpreter the onus of determining its normative content. In fact, practice has already

10 See ruling no. 2344 made by the fourth chamber of the *Consiglio di Stato*, on 30 March 2022, at www.giustizia-amministrativa.it, last accessed January 2024.

11 See the recent ruling no. 1, made by the first *corte d'assise* of the criminal court of Taranto, on 28 November 2022, at www.dejure.it, last accessed January 2024.

12 See the ruling no. 249, made by the first chamber of the *TAR* for the Puglia region, on 13 February 2021, at www.giustizia-amministrativa.it, last accessed January 2024.

13 See the ruling no. 4802, made by the fourth chamber of the *Consiglio di Stato*, on 23 June 2021, at www.giustizia-amministrativa.it, last accessed January 2024.

14 See the text of the Resolution at https://documents-dds-ny.un.org/doc/UNDOC/GEN/N22/442/77/PDF/N2244277.pdf?OpenElement, last accessed January 2024.

shown that the content of the right to a healthy environment is necessarily variable and adjustable according to the context to which it applies (Perruso 2021).

In Europe, the most important interpreter of human rights is the European Court of Human Rights (ECtHR), which, until now, has always denied the inclusion of a right to nature preservation among the rights and freedoms guaranteed by the European Convention on Human Rights (ECHR) (Dogaru 2014). However, the ECtHR has stated that although there is no explicit right in the ECHR to a clean, healthy, and sustainable environment, where an individual is directly and seriously affected by severe environmental harm, an issue may arise under Article 2[15] or Article 8[16] of the Convention, concerning, respectively, the right to life and the right to respect for private and family life. The violation occurs when the adverse effects of the pollution have attained a necessary minimum level, taking into account intensity, duration, physical and mental effects, and a causal link between the alteration of the environment and the enjoyment of the rights protected by the ECHR. This way, the ECtHR provides for indirect protection of the right to a healthy environment, by sanctioning only environmental violations that simultaneously result in an infringement of other human rights already recognized in the Convention (Celentano 2022).

Since the late 1990s, the Parliamentary Assembly of the Council of Europe (PACE) has debated whether an additional protocol "concerning the rights of the individual to a healthy and viable environment" should be supplementing the ECHR in order to incorporate into the Convention a human right which, to date, appears to be only a *construction prétorienne* (Martin & Maljean Dubois 2011). In 2010, the Committee of Ministers rejected the proposal to draw up an additional protocol on the right to a healthy environment made the year before by PACE.[17] On 29 September 2021, PACE adopted Resolution No. 2396,[18] in which it invited member states to open a debate

15 See, for example, the case *Öneryildiz v. Turkey*, app. no. 48939/99 (ECtHR Grand Chamber 30 November 2004) concerning a violation of Article 2 of the Convention on account of the lack of appropriate steps to prevent the accidental death of nine people in a methane explosion that occurred in a waste collection site, at https://hudoc.echr.coe.int/eng#{%22itemid%22:[%22001-67614%22]}, last accessed January 2024.

16 See, for example, the recent case *Kapa and others v. Poland*, app. nos. 75031/13, 75282/13, 75286/13, and 75292/13 (ECtHR 14 October 2021), regarding a case of acoustical pollution, at https://hudoc.echr.coe.int/eng#{%22itemid%22:[%22001-212138%22]}, last accessed January 2024.

17 See the text of the Committee of Ministers recommendation at https://search.coe.int/cm/Pages/result_details.aspx?ObjectID=09000016805b480c#globalcontainer, last accessed January 2024.

18 See the text of this Resolution at https://pace.coe.int/pdf/658d3f594762736ba3c0f378798b2c9529cf4be34aa45a8c38616ecd18fa80c0/resolution%202396.pdf, last accessed January 2024.

on the adoption of an additional protocol to the convention which explicitly recognizes "the right to a safe, clean, healthy and sustainable environment" as a human right. According to the Resolution, this recognition would have two consequences. First, it would allow "a violation of the right to a healthy environment to be found irrespective of whether another right had been breached and would therefore raise the profile of this right." Second, it would be "an incentive for stronger domestic environmental laws and a more protection-focused approach by the Court." The absence in the ECHR of a right to a healthy environment has left European states free to choose whether to introduce this right into their legal system.

Healthy Environment as a Fundamental Right in European Countries With a Civil Law Legal System: Swiss Legislators and the *Tribunal Administratif Fédéral*

The right to a healthy environment finds widespread recognition in the legal systems of 150 states, with express provisions that recognize this right at the constitutional level or at the level of ordinary legislation (Boyd 2018). Where a provision explicitly recognizing this right is lacking, the constitutional courts or courts of last instance have nevertheless considered this right as an essential, albeit implicit, element of the right to life or the right to health (Pauciulo 2022). Focusing on European states with a civil law legal system, some of them, such as Portugal, Spain, France, Belgium, and Romania, enshrine in their Constitutions the right to a healthy environment.[19] Other countries, such as Italy and Switzerland, enshrine in their Constitutions the protection of the environment, but not the right to a healthy environment.[20]

The Italian case is unique. As stated in the 2019 report of the United Nations Special Rapporteur on the issue of human rights,[21] Italy recognizes a

19 See the following provisions: Article 66 of the Portuguese Constitution; Article 45 of the Spanish Constitution; Article 1 of the Charter for the Environment of the French Constitution; Article 23 of the Belgian Constitution; Article 35 of the Romanian Constitution.

20 With reference to the Italian Constitution, see the following provisions: Article 9(3), providing for the protection of the environment, biodiversity, ecosystems, also in the interest of future generations; Article 41(2), stating that private economic enterprise must be carried out in respect of environmental values; Article 41(3), providing that programs and checks determined by the law to direct and coordinate public and private enterprise activity must be oriented not only toward social but also environmental purposes. With reference to the Swiss Constitution, see the following provisions: Article 2(4), stating that the Swiss Confederation is committed to the long-term preservation of natural resources; Article 74(1), giving legislative competence to the Confederation on the protection of the population and its natural environment against damage or nuisance; Article 73, enunciating the principle of sustainable development; Article 74(2), enunciating the polluters pays principle.

21 See *Right to a healthy environment: good practices. Report of the Special Rapporteur on the issue of human rights obligations relating to the enjoyment of a safe, clean, healthy*

constitutional value to the right to a healthy environment only in an indirect way. The Supreme Court of Cassation[22] has acknowledged the existence of a fundamental right to a healthy environment since the 1970s, through the combined interpretation of Articles 2 and 32 of the Constitution. Article 32 protects the right to health in a way that originally was interpreted restrictively as the right of the individual human being to life and physical or psychological integrity. Through time, this protection started to be interpreted in a different manner, characterized by the enrichment of the provision of Article 2 of the Constitution, which guarantees the inviolable rights of the person, both as an individual and in social groups where human personality is expressed, such as the family home, workplace, etc. Consequently, the Supreme Court of Cassation asserted that the right to health had to be understood also as the right of every individual to live and develop their personality in a natural or artificial context, in which nothing imperils their physical and mental health. This interpretation constitutes the right to a healthy environment in the Italian legal system, which can be understood as the social side of the right to health and protects the environment as a setting in which the human being lives, establishes relations, and develops their own potential. In light of this, the right to a healthy environment is, as some authors contend, "too much unbalanced in favor to the right to health" (Fracchia 2022) and, if its object does not get a wider content, risks becoming a superfetation (Alpa 1999). This state of affairs has not changed with the 2022 constitutional reform, which introduced the fundamental principles of environmental protection in the Constitution. In approving the constitutional reform, the Parliament deliberately chose not to consider two draft bills[23] that included the duty of the Republic to protect the health of the environment and climate stability.

The Swiss case has considerable clarity and consistency and can be examined from the points of view of federal legislators, the Federal Supreme Court, and legal doctrine. During the constitutional reform of 1970 and 1973, the legislators expressly decided not to include the right to a healthy environment in the Federal Constitution. This choice was due to the concern of not being able to define this right precisely enough, with doubts as to its possible vertical and horizontal effects (Chautard 2020). On 13 May 2000, the Federal Supreme Court

---

and sustainable environment, 30 December 2019, A/HRC/43/53, at https://undocs.org/en/A/HRC/43/53, last accessed January 2024.

22 See ruling no. 5172 made by the *sezioni unite* ("joined chambers") of the *Suprema Corte di cassazione* on 6 October 1979, in *Foro it.*, no. 1, 1979, p. 2302.

23 The first draft constitutional reform bill is no. 2914/2021; see it at http://documenti.camera.it/leg18/pdl/pdf/leg.18.pdl.camera.2914.18PDL0134610.pdf, last accessed January 2024; the second draft constitutional reform bill is no. 3181/2021, see it at http://documenti.camera.it/leg18/pdl/pdf/leg.18.pdl.camera.3181.18PDL0149100.pdf, last accessed January 2024. On this issue, see Grasso (2022).

stated clearly that there is no right to healthy environment in the Swiss legal system.[24] This Court accepts that the Confederation is obliged to take protective measures in order to ensure the efficient implementation of certain fundamental rights, but these do not include the right to a healthy environment. Swiss legal doctrine has, on more than one occasion, reaffirmed that Swiss federal law does not recognize the right to a healthy environment (Martenet 2007), explaining that the protection of the environment cannot give rise to individual rights, which may be exercised against the state or between individuals, in absence of a precise legal framework. This framework should specify the conditions under which individuals or legal persons may claim the defense of interests for themselves or in favor of nature.[25] At a cantonal level, the Constitution of the Canton of Geneva, adopted on 14 October 2012, enshrines the fundamental right to a healthy environment, conferring on Geneva the status of pioneer in this field in Switzerland.[26] However, both doctrine and the courts still have not established how this provision should be interpreted (Magistro 2017: 2).

The topic of health damages resulting from an unhealthy environment due to climate changes has been addressed in an important ruling made by the Federal Administrative Court (FAC), *Verein KlimaSeniorinnen Schweiz et al. v. Federal Department of Environment, Transport, Energy, and Communications (DETEC)*.[27] In this case, an association of elderly ladies (Association) submitted a petition to the Federal Department of the Environment calling for action to reduce anthropogenic CO_2 emissions. The Association complained about the inadequacy of the measures implemented by Switzerland in compliance with national and international rules on climate change and the consequent risk that would result for the health of the elderly. They claimed that Article 10 of the Federal Constitution (right to life) and Articles 2 (right to life) and 8 (right to respect for private and family life) of the ECHR are a source of positive obligations of the state aimed at protecting the health of individuals from an excessive rise in temperature. The applicants argued that there is a direct link between themselves and the damage caused by insufficient or omitted emission reductions; therefore, they have the status of "victim" within the meaning of Article 34 of the ECHR. In addition to other claims, the applicants requested the adoption of more stringent regulatory acts on emission reductions.

The Federal Department of the Environment rejected the application, and the Association challenged the rejection measure before the FAC. The Court

24 See the rulings of the Swiss Federal tribunal 126 II 300 consid. 5c, JdT 2001 I 674.

25 Authors supporting this theory are listed in Favre (2021).

26 Article 19 of the Constitution of Canton Geneva states that "[e]veryone has the right to live in a healthy environment." See the text of this Constitution at www.fedlex.admin.ch/eli/cc/2013/1846_fga/fr, last accessed January 2024.

27 See the ruling no. A-2992/2017, made by the first chamber of the *Tribunal Administratif Fédéral* on 27 November 2018, at www.klimaseniorinnen.ch/wp-content/uploads/2019/02/Judgment-FAC-2018-11-28-KlimaSeniorinnen-English.pdf, last accessed January 2024.

ultimately rejected the appeal, stating that the applicants had not been affected in the rights referred to in Article 10 of the Federal Constitution and in Articles 2 and 8 of the ECHR with a "sufficient intensity" to be able to say Article 34 of the ECHR had been violated. In addition, it stated that the Association was not entitled to act because it was not prejudiced in a particular way with respect to the rest of the population, and therefore, the action was *actio popularis* and inadmissible. The Federal Supreme Court confirmed the judgment of the FAC, specifying that the Association had to pursue the environmental objectives required in the appeal through political means made available by the Swiss Confederation. The judgment issued by the Federal Supreme Court has been appealed to the ECHR. On 26 April 2022, the Chamber to which the case had been allocated relinquished jurisdiction in favor of the Grand Chamber that has undertaken a public hearing on 23 March 2023 and will deliver the final judgment in the next few months.

While awaiting the ECtHR's decision on the *KlimaSeniorinnen* case, on 20 May 2022, the Committee on Legal Affairs of the National Council of Switzerland, one of the two branches of the Swiss Parliament, rejected the proposal for a constitutional reform to enshrine in the Federal Constitution the fundamental right to a healthy environment.[28] The reason this Committee gave for the rejection was that the concept of healthy environment is "not sufficiently defined" to be enshrined as a fundamental right. Afterward, on 12 December 2022, the National Council of Switzerland decided not to pursue the initiative of incorporating this right any further. Curiously, this decision was made after Switzerland played a key role in the process of the UNGA resolution no. 76/300, bringing countries with different positions on this issue closer together.

The Identity and the Enforcement of the Right to a Healthy Environment: French Administrative Courts and the *Conseil d'Etat*

Assuming that the healthy environment can be the object of a subjective right, in the shape of a human and a fundamental right, determines the emergence of three additional issues. First, according to eminent doctrine (Cranston 1973: 36), a *human right* is defined as

> a universal moral right, something which all men everywhere at all times ought to have, something of which no one may be deprived without a grave affront to justice, something which is owing to every human being simply because he is human.

28 See the documents relating to this matter at www.parlament.ch/it/ratsbetrieb/suche-curia-vista/geschaeft?AffairId=20210440.

Therefore, there is no doubt that the owner of a human right, and the beneficiary of its protection by the state, is a human being, not a legal person or an indefinite entity, such as, for example, nature. Second, as Swiss legal doctrine has pointed out (Flückiger 2014), it would bring no added value if the right to a healthy environment

> were to confine itself to grouping under a common umbrella the environmental dimension of conventional substantive and procedural human rights. If the fundamental right to a healthy environment is not to be an empty shell, it should therefore go beyond the protection conferred by traditional human rights in order to effectively preserve biodiversity, the climate for present and future generations, landscapes, soil fertility, natural resources or the genetic heritage of animals, plants, and other organisms.

So if the right to a healthy environment is to be recognized as a standalone fundamental right, this right must express a different content or, at least, a *quid pluris* as compared to other fundamental rights, especially as compared to the right to health. This *quid pluris* can be identified in the following way. A human right may have a similar core as another human right, but, at the same time, provide for additional negative or positive obligations for the state that recognizes it.

See, for example, the human right to seek asylum in Article 14 of the Universal Declaration of Human Rights (UDHR) and the right to migrate in Article 13 of the UDHR. Both protect the possibility for an individual to leave the state of origin to enter and settle in a state of destination. However, the right to seek asylum provides for a number of additional obligations for the state hosting the migrant, such as the negative obligation of non-refoulement, which includes rejection at the frontier, or the positive obligation of developing appropriate asylum procedures. For this reason, some authors believe that "the right to leave any country and the right to seek asylum are two sides of the same coin in the refugee context" (Edwards 2005), while others say that "the benefit of the right to leave has been equally restated for another specific – and much more significant – group of persons: refugees and stateless persons" (Chetail 2019: 83). Just as the right to seek asylum is intended to protect only that category of migrants that the law qualifies as refugees, the right to a healthy environment is intended to protect individuals only from that category of health damage resulting from an unhealthy environment. This parallel could go further by saying that just as the right to seek asylum indirectly protects the physical integrity of those who exercise it, the right to a healthy environment indirectly protects the integrity of the natural environment.

Third, it is necessary to identify which positive and negative obligations may arise for a state from the recognition of the right to a healthy environment as a fundamental right and whether the failure to comply with them is enforceable. An answer to this question may come from the French experience. In France, the right to live in a balanced and healthy environment is enshrined in Article 1 of the Charter for the Environment, an addendum to the French Constitution, since 2004. The Charter for the Environment combines the right with a number of specific duties for citizens and public authorities, including the duty of environmental surveillance. At first, the right to a healthy environment could only be invoked against public authorities and administrative authorities, but the Constitutional Court, in its decision of 8 April 2011, extended its application to all persons, expanding the scope of the environmental surveillance obligation.[29] Through this ruling, the Constitutional Court made the right to live in a healthy and balanced environment enforceable, since the breach of the surveillance obligation by a private person or a public authority implies their liability before a civil or an administrative court (Crottet 2012; Bary 2013).

A significant breach of the obligation of environmental surveillance was raised in the case of *Notre Affaire à Tous and Others v. France*, also known as the *Affaire du siècle* case, decided by the Administrative Court of Paris.[30] On this occasion, the court did not consider the link between the state's obligation to combat climate change and the protection of the right to a healthy environment. However, it found that the state had failed to meet the greenhouse gas reduction targets it had set via decree and established the existence of a causal link between this fault and the worsening of climate change. Then, it declared the state responsible for the ecological damage caused by the failure to meet the greenhouse gas emission reduction targets and ordered it to take immediate and concrete actions to comply with its commitments on cutting carbon emissions and repairing the damages caused by its inaction by 31 December 2022.

Much greater contributions have been made by the French Council of State, which, in a recent groundbreaking decision,[31] established that

> the right of everyone to live in a balanced and respectful environment, as proclaimed by Article 1 of the Charter of the Environment, is a fundamental freedom within the meaning of Article L. 521–2 of the Code of Administrative Justice.

29 See ruling no. 2011–116 QPC made by the *Conseil constitutionnel* on 8 April 2011, at www.conseil-constitutionnel.fr/decision/2011/2011116QPC.htm, last accessed January 2024.

30 See the ruling no. 1901967, 1904968, 1904972, 1904976/4-1 made by the *Tribunal Administratif* of Paris on 14 October 2021, at http://climatecasechart.com/wp-content/uploads/sites/16/non-us-case-documents/2021/20211021_NA_decision.pdf, last accessed January 2024.

31 See the ruling no. 451129 made by the *Conseil d'Etat* on 20 September 2022, at www.legifrance.gouv.fr/ceta/id/CETATEXT000046316542?init=true&page=1&query=&searchField=ALL&tab_selection=cetat, last accessed January 2024.

This Article regulates an urgent application and allows any person who considers that the administration – local and regional authorities, state services, public institutions – is causing serious and manifestly illegal damage in relation to a fundamental freedom to refer the matter to the "urgent applications judge." This decision represents a real step forward in the protection of environmental health, considering that the judge is required to make a ruling within 48 hours and has the authority to order "all measures required" to safeguard this fundamental freedom. As for the practical consequences of this decision, it is still too early to pronounce.

Conclusions

This chapter has shown how contemporary legislators, doctrine, and jurisprudence have managed to bring attention to ancient concepts of state responsibility and environmental health, giving it a much more ecological meaning than it had in the past. The individual right to a healthy environment, which now flanks the ancient public interest in environmental health, has proved to be an effective pole of attraction of numerous procedural tools present in national legal systems, as well as a useful weapon in climate litigation. Doubts about the content of this human right can be overcome by referring to the concepts of environmental health developed in the international context, while its difference with the right to health could be resolved in terms of analogy by applying considerations already carried out with reference to other human rights. The consequences which may emerge in terms of the positive obligations arising from the recognition of this right are evident in the judgments of the French administrative courts, which, together with the administrative courts of other European states, are playing a key role in protecting environmental health and, more generally, in combating climate change.

Reference List

Adinolfi, G. (2018) "The right to a healthy environment: Delineating the content (and contours) of a slippery notion," in: F.Z. Giustiniani, E. Sommario, F. Casolari, & G. Bartolini (eds) *Routledge Handbook of Human Rights and Disasters*, pp. 211–227, London: Routledge.

Alpa, G. (1999) "Il diritto soggettivo all'ambiente salubre: 'nuovo diritto' o espediente tecnico?" in: S. Grassi, M. Cecchetti, & A. Andronio (eds) *Ambiente e diritto*, pp. 431 et seq., Firenze: Olschki.

Bary, M. (2013) "Le droit a un environnement sain en droit français," *JURIS: Revista do Departamento de Ciências Jurídicas*, pp. 17–39.

Boyd, D.R. (2018) "Catalyst for change: Evaluating forty years of experience in implementing the right to a healthy environment," in: J.H. Knox & R. Pejan (eds) *The Human Right to a Healthy Environment*, pp. 17–41, Cambridge: Cambridge University Press.

Celentano, F.E. (2022) *La giurisprudenza della Corte europea dei diritti dell'uomo in materia di ambiente salubre tra bilanciamento d'interessi e obblighi degli stati: il caso Kapa e altri c. Polonia*, Ordine internazionale e diritti umani, pp. 399–413.

Chautard, M. (2020) "Le droit fondamental à un environnement sain, de 2000 à 2020," *Quid?*, 1, pp. 10–13.
Chetail, V. (2019) *International Migration Law*, Oxford: Oxford University Press.
Cima, E. (2022) "The right to a healthy environment: Reconceptualizing human rights in the face of climate change," *RECIEL*, 31(1), pp. 38–49.
Corso, G. (2020) *Manuale di diritto amministrativo*, Torino: Giappichelli.
Cranston, M. (1973) *What Are Human Rights?* New York: Taplinger Publishing.
Crottet, B. (2012) "Le Conseil constitutionnel et l'obligation de vigilance environnementale," *Revue française de droit constitutionnel*, 2, pp. 239–267.
Di Porto, A. (2015) *Salubritas e forme di tutela in età romana: Il ruolo del civis*, Torino: Giappichelli.
Dogaru, L. (2014) "Preserving the right to a healthy environment: European jurisprudence," *Procedia – Social and Behavioral Sciences*, 141, pp. 1346–1352.
Edwards, A. (2005) "Human rights, refugees, and the right 'to enjoy' Asylum," *International Journal of Refugee Law*, 17, pp. 293–330.
Favre, A.C. (2021) "Art. 74," in: V. Martenet & J. Dubey (eds) *Constitution fédérale, Commentaires*, Vol. I, pp. 1643–1658, Basel: Schulthess.
Flückiger, A. (2014) "Droits de l'homme et environnement," in: M. Hottelier & M. Hertig Randall (eds) *Introduction aux droits de l'homme*, pp. 606–620, Geneva: Schulthess.
Fracchia, F. (2022) "L'ambiente nell'art. 9 della Costituzione: un approccio in 'negative,'" *Il diritto dell'economia*, 107(1), pp. 15–30.
Frumkin, H. (2005) "Introduction," in: H. Frumkin (ed) *Environmental Health*, pp. XXXI–XXXII, New York: Wiley.
Grandi, F. (2021) "L'art. 32 nella pandemia: sbilanciamento di un diritto o "recrudescenza" di un dovere," *Costituzionalismo.it*, 1, pp. 82–136.
Grasso, G. (2022) *La révision de la Constitution italienne sur la protection de l'environnement: qu'est-ce qui pourra changer pour les questions climatiques?* Italian papers on federalism, pp. 46–60.
Günther, P. & Ekardt, F. (2022) "Human rights and large-scale carbon dioxide removal: Potential limits to BECCS and DACCS deployment," *Land*, 11, p. 2153.
Köböl-Benda, V. (2022) "The current status of the right to the environment in the global international law," *Opolskie Studia Administracyjno-Prawne*, 20(1), pp. 117–143.
Lombard, M., Dumont, G., & Sirinelli, J. (2019) *Droit administratif*, Paris: Dalloz.
López Calera, N. (2010) "El interés público: entre la ideología y el derecho," *Anales de la Cátedra Francisco Suárez*, 44, pp. 123–148.
Magistro, F. (2017) *Le droit à un environnement sain revisité: étude de droit suisse, international et comparé*, Genève: Schulthess éd. romandes.
Martenet, V. (2007) "Un droit fondamental à un air sain?" *Umweltrecht in der Praxis*, pp. 922–950.
Martin, J.C. & Maljean Dubois, S. (2011) "La Cour européenne des droits de l'homme et le droit à un environnement sain," *Prévention des risques et responsabilité environnementale*, pp. 37–53.
Pauciulo, D. (2022) "Il diritto umano a un ambiente salubre nella risoluzione 76/300 dell'Assemblea generale delle Nazioni Unite," *Rivista di Diritto Internazionale*, 4, p. 1118 seq.
Perruso, C. (2021) *L'affirmation d'un droit à un environnement propre, sain et durable universel*, Notes sur la Résolution 48/13 du Conseil des droits de l'Homme des Nations Unies, Actualités Droits-Libertés.
Ruiz-Pino, S. (2017) "Algunos precedentes históricos de protección o defensa de los recursos naturales y de la salubritas en Roma," *Hacia un derecho administrativo medioambiental romano, Revista Digital de Derecho Administrativo*, 17, pp. 91–109.

Sánchez Morón, M. (2021) *Derecho Administrativo, Parte general*, Madrid: Tecnos.
Savaresi, A. & Vilchez Moragues, P. (2021) "The right to a healthy environment and climate litigation: A mutually supportive relation?" *Social Science Research Network*, pp. 1–19. Available at: https://ssrn.com/abstract=3829114.
Scoca, F.G. (2017) *L'interesse legittimo: storia e teoria*, Torino: Giappichelli.
Traversa, L. (2022) "Le trasformazioni dei tribunali di età augustea fra salubrità e politica," *Griseldaonline*, 21(2), pp. 13–21.

Chapter 18

The Place of Public Policy Management Councils and Policy Conferences in Alleviating the Health Challenges in the Niger Delta

Nsikan-Abasi Odong

Introduction

Nigeria operates a mono-economy which is solely dependent on the oil resources of the Niger Delta. As of 2022, oil and gas resources from the Niger Delta accounted for 95% of Nigeria's total revenue and also contributed 80% to Nigeria's Gross Domestic Product (Obiam & Amadi 2022). Despite this strategic role, the health-related metrics in the Niger Delta is low when compared to the national average. While life expectancy in the rest of the country is 54.7 years (United Nations Development Programme 2022), it is 40 years in the Niger Delta (Orisakwe 2021: 3). Also, while infant mortality in the rest of the country is 75.7 per 1,000 live births (United Nations Development Programme 2022), it is 400 out of 1,000 (4 per 10) live births in the Niger Delta (Unrepresented Nations and Peoples Organization 2017).

Some of the reasons for these are, first, that the provision of health-care services in the Niger Delta is abysmal (Francis et al. 2011). Second, the infrastructure and critical medical supplies upon which life systems and functional health-care services are dependent are equally lacking. For instance, the "supply of water and electricity [in the Niger Delta] is undependable" (Francis et al. 2011: 40). On critical medical supplies, the story is no different, as "health centers [in the Niger Delta] are even fewer. . . . The most basic supplies . . . and medicines in health care centers, are simply not available" (Francis et al. 2011: 40). Third, these health challenges are exacerbated by the environmental degradation of the Niger Delta as a result of the extraction of crude oil in the region. According to the United Nations Environment Programme (UNEP) report, Environmental Assessment of Ogoniland (United Nations Environment Programme 2011: 10):

> The Ogoni community is exposed to petroleum hydrocarbons in outdoor air and drinking water, sometimes at elevated concentrations. They are also exposed through dermal contacts from contaminated soil, sediments, and surface water. Since average life expectancy in Nigeria is less than 50 years, it is a fair assumption that most members of the current Ogoniland community have lived with chronic oil pollution throughout their lives.

DOI: 10.4324/9781003440871-21

The nexus between environmental degradation and health challenges in the Niger Delta is further underscored by the finding that there are "serious threats to human health from contaminated drinking water to concerns over the viability and productivity of ecosystems . . . [and] that pollution has perhaps gone further and penetrated deeper than many may have previously supposed" (United Nations Environment Programme 2011: 10). The UNEP report also notes that in some part of the Niger Delta (United Nations Environment Programme 2011: 10),

> [people] are drinking water from wells that are contaminated with benzene, a known carcinogen, at levels over 900 times above the World Health Organization (WHO) guideline. . . . Hydrocarbon contamination was found in water taken from 28 wells at 10 communities adjacent to contaminated sites. At seven wells the samples are at least 1,000 times higher than the Nigerian drinking water standard of 3 µg/l. Local communities are aware of the pollution and its dangers but state that they continue to use the water for drinking, bathing, washing, and cooking as they have no alternative. Benzene was detected in all air samples at concentrations ranging from 0.155 to 48.2 µg/m3. Approximately 10 per cent of detected benzene concentrations in Ogoniland were higher than the concentrations WHO and the United States Environmental Protection Agency (USEPA) report as corresponding to a 1 in 10,000 cancer risk.

This chapter focuses on the lack of health-care services in the Niger Delta, the lack of avenues to public participation of the Niger Delta in the health sector, and how the public policy management councils (PPMCs) and policy conferences (PCs) could contribute to changing the narrative. With Brazil as a case study, it will examine the state of health-care delivery in Brazil prior to and after the adoption of PPMCs/PCs as possible guidance for the successful creation/application of PPMCs/PCs in the Niger Delta. It will explore the infrastructure behind the PPMCs/PCs, including the constitutional right to health, the purpose and structure of PPMCs/PCs, the constitutional right to public participation, the principles of subsidiarity, the enabling of legislation to operationalize PPMCs/PCs, and the role of public will to participate in health-care delivery.

Constitutional Right to Health

Health care is a constitutional right in Brazil. Article 196 of the Brazilian Constitution (1988) provides that health

> is the right of all and the duty of the National Government and shall be guaranteed by social and economic policies aimed at reducing the risk of illness and other maladies and by universal and equal access to all activities and services for its promotion, protection and recovery.

In contrast, Section 17(3)(c) (contained in Chapter II) of the Nigerian Constitution (1999) provides that "the health, safety, and welfare of all persons in employment are safeguarded and not endangered or abused." There are at least two implications arising from this section. First, the section is not justiciable because Section 6(6) of the Constitution provides that

> [t]he judicial powers vested in accordance with the foregoing provisions of this section . . . (c) shall not, except as otherwise provided by this Constitution, extend to any issue or question as to whether any act or omission by any authority or person or as to whether any law or any judicial decision is in conformity with the Fundamental Objectives and Directive Principles of State Policy set out in Chapter II of this Constitution.

Second, Section 17 of the Nigerian Constitution has been interpreted to guarantee health care only for the working class. With most of the people in the Niger Delta unemployed (United Nations Development Programme 2006: 25), they have not been able to rely on this provision to access health care. Comparatively, the unequivocal constitutional right to health in Brazil seems to put Brazilians in a much stronger position to hold the government accountable to the provision of better-quality health care for all people. Therefore, what may be needed in Nigeria for the benefit of all who live in the Niger Delta would be an outright constitutional guarantee of the right to health. Additional considerations would be needed, but it would be an important foundation for the people of the Niger Delta to demand improved health-care delivery.

The Purpose and Structure of Public Policy Management Councils and Policy Conferences

Public policy management councils are decision-making platforms consisting of government representatives and civil society members with shared responsibilities, and the failure to carry out agreed-upon decisions usually results in the suspension of federal funding (Avritzer 2012: 117–118). Usually, PPMCs' duties would include the development and approval of policies, as well as oversight powers over the implementation processes (Wampler 2015: 135). In general, the composition of PPMCs – of government and civil society representatives – is to achieve parity of stakeholder voices (Barth 2006: 257). In practice, the federal government ties the transfer of funds to states and municipalities to the existence of PPMCs. In Brazil, because of the impact of these participatory mechanisms, there has been an explosion in the number of PPMCs from about 25,859 in 1999 (Barth 2006: 257) to about 65,000 a decade later (Wampler 2015: 133). There are at least 488 PPMCs per municipality, and with 5,506 municipalities, that means an estimated 2,686,928 PPMCs in Brazil (Barth 2006: 257).

On the other hand, policy conferences are feedstocks to the PPMCs. They are broad-based participatory summits convened by the government to supply the PPMCs with fresh perspectives on policies, recommend broad policy directions to municipalities, and give the citizens opportunity to vote on directive policies and policy benchmarks for relevant PPMCs and government policy departments (Martelli et al. 2019: 2). One unique feature of the PCs is that they utilize a bottom-up approach to policy deliberations, as they usually

> take place in a staggered way, starting at the municipal level and feeding upwards, through representative delegates, to the state and national levels. . . . They are, therefore, "effective mechanisms for channeling social demands and assessing the situation of public services, especially at the municipal level."
>
> (Martelli et al. 2019: 2)

In contrast, Nigeria seems to operate a top-down health-care system, where the people have no opportunity to contribute to health-care delivery (Francis et al. 2011: 37). This top-down approach seems to create a dissonance between policy formulation and implementation in the health sector in Nigeria and the needs of the ultimate beneficiaries, in this case the people of the Niger Delta. Nigeria should embrace public participation in its health-care sector as this would enrich policy formulation through alternative views and improve policy implementation by guaranteeing the buy-in of the public.

Constitutional Right to Public Participation and Public Policy Management Councils/Policy Conferences

Public policy management councils have been described as the "most widely adopted participatory governance institutions in Brazil, as permitted under Brazil's 1988 Constitution" (Wampler 2015: 133). According to Jessica Jerome (2015: 60–61):

> The 1988 Brazilian constitution stipulated that the new health care system be governed according to democratic criteria and include the participation of civil society in its decision-making processes. In order to enact these principles, *conselhos de saúde* (health councils) were to be created at the federal, state, and municipal levels to help plan and supervise health care actions. The councils were to be composed of service providers, government officials, and – most provocatively – local residents, who were to make up a full 50 percent of council membership.

Both the PPMCs and policy conferences draw strength from the public participatory provision of the Brazilian Constitution on public health (Art. 198). Jutta Barth argues that citizen participation in governance in Brazil

> is clearly established by the 1988 Constitution, which defines explicitly in Article 198 . . . that social sectors, such as health . . . have to establish participatory systems and mechanisms in order to formulate public policies and manage the allocation of resources.
>
> (Barth 2006: 257)

This is not the case in Nigeria, as there is no equivalent provision in the Nigerian Constitution where the citizens have the right to public participation in the health sector. In comparison with Brazil, the closest constitutional provision on public participation in Nigeria is broadly construed and not specific to health care. Section 14(2)c) (contained in Chapter II) of the Nigerian Constitution provides that "the participation by the people in their government shall be ensured in accordance with the provisions of this Constitution." Secondly, and as already discussed, Chapter II of the Nigerian Constitution is not directly enforceable. Therefore, it would be recommended that Nigeria amend its Constitution to enshrine the right to public participation in its enforceable part of the Constitution. This would lead to the adoption of public participatory models in health-care delivery and enable the participation of the Niger Delta in the health-care sector.

The Principle of Subsidiarity

The Brazilian Constitution decentralizes the practice of public policy management councils and policy conferences. Article 198 of the Constitution establishes a unified health-care system for the country, but the system is highly decentralized, with each unit of government having a single management in place, as well as a mandate for community participation in public health. The Nigerian Constitution mirrors the Brazilian Constitution by decentralizing the responsibility over health care. It vests the local government areas (LGAs) – the closest tier of government to the people – with the responsibility over health care in Nigeria (Section 2(c) of the 4th Schedule). However, comparatively, Nigeria does not have most of the conditions that are present in Brazil.

Indeed, Nigerians are eager to participate in the health-care system but lack the enabling constitutional/legislative support to do so. For instance, before the enactment of the National Health Act of 2014 (NHA), Nigeria did not have federal legislation on health care (Dan-Azumi 2018: 70). The United Kingdom's Department for International Development (DFID) – now Foreign, Commonwealth, and Development Office – an international donor agency and development partner set up by the United Kingdom to administer foreign development aids, was interested in health sector reform in Nigeria and therefore facilitated the setting up of a change agent program (CAP) made up of people with interest in Nigeria's health-care sector, styled "change agents" (Dan-Azumi 2018: 73). The change agents "critically-reviewed the Nigerian health system . . . [and identified the need for] a legal framework for the provision and management of healthcare in Nigeria," and this resulted in the conception and draft of the National Health Bill (Obi 2014). The change agents held strategic meetings with the various ministries and agencies of government to secure their support and ownership of the bill (Dan-Azumi 2018: 81).

> [T]he Federal Ministry of Health (FMOH) came to depend on the coalition to organise policy dialogues and sometimes provide technical support. . . . The coalition worked to integrate government actors into the movement, helping the state to own the process. . . . It was able to do this horizontally among members, which created a sense of ownership among health professional associations and religious bodies, and vertically between the legislature, executive, and the public.

Civil society organizations (CSOs) also played critical roles. They funded a public workshop on the bill that was organized by the Senate Committee on Health (Obi 2014) and developed a "productive alliance" with the Senate on its passage (Dan-Azumi 2018: 74). There was opposition, especially from faith-based groups, based on deeply rooted religious and cultural beliefs which disagreed with the provisions on the harvesting and exploitation of human embryos and sperm (Dan-Azumi 2018: 74). The Health Sector Reform Coalition (HSRC) was then formed to coalesce and unite the interests of all the various groups toward the passage of the NHA. The HSRC "held several high-level consultations across the country to flesh out the core issues of the bill that need addressing and secure the buy-in of stakeholders so as to avoid some of the hurdles previously faced" (Dan-Azumi 2018: 75). The HSRC also reached out and convinced opinion leaders (national and international) to throw their weight behind its passage (Dan-Azumi 2018: 75).

The bill was passed by the National Assembly (NASS), but President Goodluck Jonathan refused to assent to the bill (Obi 2014), and so it was reintroduced in 2012 (Champions for Change). At the 2013 public hearing conducted by the NASS, the CSOs organized press conferences, produced research and analysis reports, prepared policy briefs, provided technical support, and convened strategic meetings with members of the NASS, including sending twice-daily text messages to over 5,000 people, including legislators, to encourage them to pass the bill, and placed adverts on national dailies, conducted "unrelenting advocacy" to persons of influence and international bodies, such as the DFID and the World Bank, to support the Bill, and advocated for and promoted the interests of Nigerians to be fully considered and included in the legislation (Dan-Azumi 2018: 76–77). The CSOs also used public events and traditional and social media platforms, like Twitter (now X), with the hashtag #nationalhealthbill, to galvanize citizen support for the bill (Dan-Azumi 2018: 81). Pressure was also placed on the NASS when CSOs organized a demonstration by the Market Women Association on 18 May 2011 (Obi 2014). Conscious of the refusal of presidential assent on the previous bill, the CSOs also enlisted the support of international partners and donor agencies, especially the WHO, World Bank, USAID, and DFID, to impress on President Jonathan to assent to the bill (Dan-Azumi 2018: 79). These strategies seemed to work, because the NASS passed the bill into law in 2014, and the following October, President Jonathan signed the bill into law and it became the NHA.

With the enactment of the NHA, the CSOs are not relenting and have adapted their aims slightly with the times. For instance, in partnership with the federal Ministry of Health, they have created a special unit to deliberate on how to operationalize and implement the NHA (Dan-Azumi 2018: 80). Currently, they are mounting "a robust advocacy to ensure compliance and proper implementation [of the NHA]" (Dan-Azumi 2018: 79). The public's role from the conception of the bill to its adoption show that, just as in Brazil, Nigerians are ready to participate in health-care delivery. However, participatory platforms like the PPMs/PCs are lacking.

Enabling Legislation to Operationalize Public Policy Management Councils and Policy Conferences

Apart from the broad constitutional provisions, public policy management councils usually have enabling laws which determine their composition and guide their operations. In Brazil, it is Law No. 8,080, 1990. The law grants the PPMCs "veto power over the plans and accounts of the health secretariat. If the council [PPMCs] rejects the plan and budget that the health secretariat is required to present annually, the Health Ministry does not transfer funds" (Coelho et al. 2005: 176).

Specifically, Article 7 of the law re-enacts the principles of community participation and decentralizes health-care policies and services. To further ensure public participation in the health sector, Article 12 of the law establishes the National Intersectoral Commissions, made up of the representatives of health-related public institutions and civil society, to fashion out public health policies. This suggests that apart from creating an avenue for public participation in policy formulation, PPMCs and PCs have the capacity to reduce official influence in the governance process, and ultimately, these "improve the effectiveness and efficiency of public policies" (Barth 2006: 253).

The NHA in Nigeria has similar provisions to Brazilian Law No. 8,080. Section 1 of the NHA creates a national health-care system to cater to the health needs of Nigerians. Although Section 1(2) of the NHA resembles Brazil's in the decentralization of health-care delivery, in principle, by including the LGAs, wards, and villages, as well as private health care, alternative health care, and traditional health care in the national health system, certain issues critical to effective health-care delivery are still centralized. For instance, Section 2(1) of the NHA mandates the federal Ministry of Health to develop and implement the national health policy without any input from the other arms of government or the public. Furthermore, the federal Ministry of Health is vested with the responsibility to prepare the annual budget for the health sector, without public input (Section 2(2)). Also, the NHA creates the Basic Health Care Provision Fund (Section 11(1)), but it is the National Primary Health Care Development Agency that administers the fund, without public input (Section 11(7)).

Nigeria should embrace decentralization in the health-care sector to improve service delivery. For instance, through Section 4 of the NHA, the federal minister of health determines those who enjoy free health-care services (Section 3(1)). The World Bank puts Nigeria's population at about 213 million (World Bank 2021). It could be overwhelming and ineffective for a single federal minister to be responsible for the identification and management of the health-care needs of such a large number of persons. It would be more effective if the LGAs, being the closest tier of government to these people, and the PPMCs operating at the community levels oversaw these duties, as done in Brazil. The NHA should be amended to reflect this.

The Public Health Impacts of Public Policy Management Councils and Policy Conferences

While the Unified Health System (SUS) is the public health-care scheme provided for in the Brazilian Constitution (Art. 198), public policy management councils and policy conferences regulate the funding needs and decide how the SUS delivers health-care services to the people (Souza 2017: 494). Prior to the adoption of the SUS in Brazil, the health insurance scheme created in 1960 was the main policy framework to provide public health care, but it only covered employees in the public sector, leading to the proliferation of private health care for the rest of the people, such that by 1989, 80% of hospitals in Brazil were privately owned (McGregor et al. 2017: 2). The health insurance scheme left a vast majority of Brazilians who were not in the public service at the mercy of politicians and charitable organizations (Jerome 2015).

Since its establishment in 1990, the SUS has recorded reasonable progress, including expanding health-care coverage for many people (Castro et al. 2019: 345). It has also distributed health-care centers, as the SUS now "boasts a wide network of clinics, health posts, and programs that grant universal access to an array of care free of cost, at the point of service" (McGregor et al. 2017: 2). The programs developed by the SUS and PPMCs/PCs include

> health networks for mental health services, emergency care, and specialised outpatient services. The community health workers programme was

> established . . . to service the poorest areas in northeastern Brazil, followed by the family health programme . . . which provide acute services, health promotion, disease prevention, chronic disease management, and maternal and child services.
>
> (Castro et al. 2019: 348)

Research shows the clear impact of the reforms, as there are visible improvements in the health metric between 1990, when the 1988 Constitutional reforms would have started to take root, and 2015 (Castro et al. 2019). However, there are challenges that would reverse these gains if not addressed. For example, lack of funding could affect sustained investments in health-care services. The 1988 Constitution stipulates that the SUS shall be funded from the security budgets of all tiers of government (Art. 198). However, former President Michel Temer had determined to cut down on public sector spending and pushed through a constitutional amendment bill (Bill PEC 241) to freeze federal health funding at the 2017 levels for 20 years, beginning from 2018 (Castro et al. 2019: 347), with the amendment becoming effective in 2016 (Castro et al. 2019: 347). While public spending has been frozen, the health-care needs of Brazilians keep rising. For instance, from 1989 to 2014, "real health expenditure per capita increased by 149%" (Castro et al. 2019: 347). Insufficient funding has affected the quality of health-care services delivered to the people (Castro et al. 2019: 348), with dissatisfaction for the poor quality of health-care services leading to mass demonstrations in 2013 (McGregor et al. 2017: 1). As of 2017, about 77% of the population expressed dissatisfaction with the quality of health-care services that were being dispensed in Brazil (McGregor et al. 2017: 2). Insufficient funding has also affected the provision of specialist care (Castro et al. 2019: 348).

The funding shortfall has resulted in the gradual return and embrace of an unaffordable privatized health-care system in Brazil, which the SUS was designed to phase out (McGregor et al. 2017: 2). By 2003, 32 million Brazilians had private health-care plans, and the numbers went up to 49 million by 2012 (a 53% increase in that period), meaning, that as at 2012, more than 25% of the population had private health-care plans (McGregor et al. 2017: 2). Comparatively, while investment in public health care was frozen, private sector investment was rising. For instance, "[b]etween 2006 and 2012, the total quantity of privately managed and maintained equipment in the SUS nationwide increased by approximately 88%, while publicly managed equipment grew by only about 51%" (McGregor et al. 2017: 2). Seeing that health care is a constitutional right in Brazil, the government will have to balance the opportunity cost of providing the services with the risk of litigation by enforcers of such rights. Already, from 2008 to 2015,

> federal government expenditure for claims related to medicines rose from 70 million to 1 billion Brazilian reais . . . in some regions, individuals with

> low incomes, who use litigation as an instrument to improve access to care, are the majority of litigants.
>
> (Castro et al. 2019: 348)

Conclusion

The chapter focuses on the lack of health-care services in the Niger Delta, the concomitant lack of avenues to public participation of the Niger Delta in the delivery of health-care services, and how public policy management councils and policy conferences could contribute to changing the narrative. As shown in the chapter, Brazil was in a similar position as the Niger Delta, and the introduction of the PPMCs/PCs in Brazil was an attempt to reposition health-care delivery to provide better-quality services to its people. Despite the continued challenges noted earlier, particularly related to enforcement, funding, and privatization, health care in Brazil has improved due to these innovations (Souza 2017: 494).

> These achievements derive principally from two innovations in the Constitution of 1988: decentralization of health care management and social participation. Decentralization, by making municipal governments legally responsible for the implementation of health activities, stimulated introduction of new funds into the health system – from local treasuries. Social participation has been essential for guiding use of resources in accord with plans approved by the Health Councils in each town. Health Councils, comprising equal numbers of representatives of the public administration and from civil society, monitor implementation of the plans.

Unlike that of Brazil, the Nigerian Constitution does not provide for public participation in the health sector. It only stipulates in Section 17(3)(d) that the federal government "shall direct its policy towards ensuring that . . . there are adequate medical and health care facilities for all persons," and moreover, the section is contained in Chapter II of the Constitution, which is not justiciable. The implication is that people in the Niger Delta are not able to participate in policy formulation and implementation in the health sector and cannot also hold the government accountable for the failure to provide health-care services for the people, including for health problems caused or made worse by environmental harms from the oil and gas sector. It is recommended that Nigeria adopt mechanisms like the PPMCs/PCs for the benefit of its residents, including the Niger Delta.

Reference List

Avritzer, L. (2012) "The different designs of public participation in Brazil: Deliberation, power sharing and public ratification," *Critical Policy Studies*, 6(2), pp. 113–127.

Barth, J. (2006) "Public policy management councils in Brazil: How far does institutionalised participation reach?" *Public Administration and Development*, 26(3), pp. 253–263.

Bill PEC 241 (2016). Available at: www.planalto.gov.br/ccivil_03/Projetos/PEC/2016/msg329-junho2016.htm.

Castro, M., et al. (2019) "Brazil's unified health system: The first 30 years and prospects for the future," *The Lancet (British edition)*, 394(10195), pp. 345–356.

Champions for Change (2012) *Nigeria's National Health Bill Policy Summary*. Available at: https://riseuptogether.org/wp-content/uploads/2018/05/C4C_Health_Care_Briefing_Document_12-30-14_FINAL.pdf.

Coelho, V., Pozzoni, B., & Cifuentes, M. (2005) "Participation and public policies in Brazil," in: J. Gastil & P. Levine (eds) *The Deliberative Democracy Handbook: Strategies for Effective Civic Engagement in the Twenty-First Century*, San Francisco: Jossey-Bass.

The Constitution of the Federal Republic of Brazil (1988).

The Constitution of the Federal Republic of Nigeria (1999).

Dan-Azumi, J. (2018) "CSOs advocacy and the legislative process in Nigeria: A case study of the national health act, 2014," *Journal of Law, Policy and Globalization*, 74(69), p. 70.

Francis, P., LaPin, D., & Rossiasco, P. (2011) *Securing Development and Peace in the Niger Delta a Social and Conflict Analysis for Change*, The Woodrow Wilson International Center for Scholars.

Jerome, J. (2015) *A Right to Health: Medicine, Marginality, and Health Care Reform in Northeastern Brazil*, Austin: University of Texas Press.

Law No. 8,080 of September 19, 1990. Available at: www.planalto.gov.br/ccivil_03/leis/l8080.htm.

Martelli, C., Almeida, C., & Lüchmann, L. (2019) "The meanings of representation and political inclusion in the conferences of public policies in Brazil," *Journal of the Brazilian Political Science Association*, 13(1), pp. 1–23.

McGregor, A., et al. (2017) "Do elections matter for private-sector healthcare management in Brazil? An analysis of municipal health policy," *BMC Health Services Research*, 17(1), pp. 1–13.

National Health Act, Law No. 8 of 2014. Available at: www.publichealth.com.ng/wp-content/uploads/2017/10/National-Health-Act-2014-Explanatory-Memoranda.pdf.

Obi, F. (2014) "The National Health Bill: After Ten Years in the Making Is an End in Sight?" *Nigeria Health Watch*. Available at: https://nigeriahealthwatch.com/the-national-health-bill-after-ten-years-in-the-making-is-an-end-in-sight/.

Obiam, S. & Amadi, O. (2022) "The Nigerian state and development in the Niger Delta region," *World Journal of Advanced Research and Reviews*, 14(1), pp. 125–133.

Orisakwe, O. (2021) "Crude oil and public health issues in Niger Delta, Nigeria: Much Ado About the Inevitable," *Environmental Research*, 194, pp. 1–9.

Souza, L. (2017) "The right to health in Brazil: A constitutional guarantee threatened by fiscal austerity," *Journal of Public Health Policy*, 38(4), pp. 493–502.

United Nations Development Programme (2006) *Niger Delta Human Development Report*, Abuja: UNDP, online. Available at: http://bebor.org/wp-content/uploads/2012/09/UNDP-Niger-Delta-Human-Development-Report.pdf.

United Nations Development Programme (2022) *Human Development Reports*. Available at: https://hdr.undp.org/en/countries/profiles/NGA.

United Nations Environment Programme. (2011) *Environmental Assessment of Ogoniland*, 38.

Unrepresented Nations & Peoples Organization (2017) *Ogoni: Infant Mortality Increases as Niger Delta remains Polluted*. Available at: https://unpo.org/article/20516.
Wampler, B. (2015) *Activating Democracy in Brazil: Popular Participation, Social Justice, and Interlocking Institutions*, South Bend: University of Notre Dame Press.
The World Bank (2021) *Population, total – Nigeria*. Available at: https://data.worldbank.org/indicator/SP.POP.TOTL?locations=NG&view=map.

Further Reading

Ako, R. & Olawuyi, D. (2017) "Environmental Justice in Nigeria Divergent tales, paradoxes and future prospects," in: R. Holifield, J. Chakraborty, & G. Walker (eds) *The Routledge Handbook on Environmental Justice*, pp. 567–577, London: Routledge.
Etemire, U. (2015) *Law and Practice on Public Participation in Environmental Matters: The Nigerian Example in Transnational Comparative Perspective*, London: Routledge.
Onalu, C., et al. (2021) "Mortality experiences for women in riverine areas of the niger delta and utilization of maternal health services," *Health & Social Work*, 46(1), pp. 59–67.

Section 4

Ecological Integrity for Peace

Chapter 19

Teaching Peace and Public Health Using Indigenous Law

The Educational and Pedagogical Challenges

Kathleen Mahoney

Introduction

Indigenous law has existed for thousands of years, sustaining Indigenous populations with rules, principles, and customs to govern everyday living and supporting relationships required to survive. Teaching concepts such as peace and public health through Indigenous law requires explanations of its basic principles, its sources, and its use in solving problems. A major barrier preventing Indigenous law from taking its place in the Canadian legal system is the fact that the lawyers and judges that operate within it are ill-prepared by their legal education to comprehend or reconcile the relationship between colonial legal systems and Indigenous systems of law. If it is offered in law school, Indigenous law is usually in the form of optional courses, even though it is vital to crafting appropriate solutions to all of the most important problems Canada faces today, not the least of which are peace and public health, and their relationship with Canada and the Indigenous peoples of the country.

In post-colonial societies, traditional legal approaches to injuries and harms motivated by systemic discrimination and cultural genocide fall short of achieving justice in the eyes of those harmed. This chapter argues that to achieve a better system of justice, legal processes and remedies must be rethought to allow for Indigenous perspectives and theories of law to inform law schools, the bar, the judiciary, and continuing professional education programs, so that the first peoples can enjoy their rights under their own legal rules and processes in the same way as the settler societies enjoy theirs.

I have taught two courses on Indigenous law at the University of Calgary Law School in Canada over a 6-year period. One course is on Indigenous Law and Litigation, the other, Understanding Indigenous Law. Prior to teaching these two courses, I taught numerous mainstream law courses, including torts, human rights, international human rights, international humanitarian law, feminist legal theory, and tort legal theory in the JD curriculum over a 35-year period. All these courses are based largely on British and Canadian common law or Euro-centric law and attempt to deal with conflict resolution,

DOI: 10.4324/9781003440871-23

human rights, personal injury, antiracism, and equality, or one might say, peace and public health, writ large.

The Indigenous law courses were added to the law school curriculum in 2015 because of the findings and "calls to action" of the Truth and Reconciliation Commission that dealt with the 150 years of cultural genocide that was perpetrated by Canada on Indigenous peoples. The key "call to action" relevant to Canadian law schools reads as follows (TRC Canada 2015a: no. 28):

> We call upon law schools in Canada to require all law students to take a course in Aboriginal people and the law, which includes the history and legacy of residential schools, the *United Nations Declaration on the Rights of Indigenous Peoples*, Treaties and Aboriginal rights, Indigenous law, and Aboriginal-Crown relations. This will require skills-based training in intercultural competency, conflict resolution, human rights, and antiracism.

In the following pages, I describe the methods and content I have developed to teach both Indigenous Law and Litigation and Understanding Indigenous Law. Both courses present challenges of acceptance by students because of deeply embedded ignorance and bias with respect to the relevance and even the existence of Indigenous law in the legal academy, the judiciary, and at the bar. These realities create special challenges for the law teacher not encountered in other courses, except perhaps for critical theoretical courses on feminism and race (see Alexander-Floyd 2010).

Understanding the Indigenous Law Course

The overarching goal of the Understanding Indigenous Law course is to demonstrate the value of Indigenous law in contemporary society and how Indigenous legal solutions can lead to better outcomes for Indigenous peoples than those available under the mainstream laws. As their teacher, I want to ensure that my students do not merely see Indigenous law as simply an academic "boutique" experience. I want them to learn that Indigenous law exists, that it has the potential for real and transformative impacts on society, and that it can enhance the legal knowledge and practice of lawyers, judges, and law professors, adding to the sum of justice in the world. Whether students take the course out of curiosity or merely to fulfill their formal legal theory paper requirement, it requires them to rethink principles they thought were immutable. Ideally, they will discover creative ways in which Indigenous legal principles can be used to find solutions that are impossible to achieve under the mainstream legal system.

My methods and organizational strategies reflect the goals, but as a non-Indigenous professor, I find teaching Indigenous law and legal theory presents numerous pedagogical and credibility challenges. One of the threshold challenges is whether a non-Indigenous person should be teaching

Indigenous law at all, and whether she is able to teach Indigenous law and theory in culturally appropriate and credible ways. Another threshold question is, How does one organize the courses? Should they be organized in traditional ways, under conventional common law frameworks, or according to Indigenous frameworks (Borrows 2016)? Alternatively, is it best to take a pluralistic approach blending different legal systems, comparing and contrasting outcomes and solutions? The great diversity of legal traditions among First Nations, Métis, and Inuit communities raises a further challenge, namely: Can Indigenous law be organized or explained by one overarching theory, or do the students need to be exposed to theories from many Indigenous traditions to grasp underpinnings, explanations, and traditions of Indigenous law?

My approach is a pluralistic one in both pedagogy and content. I have tried to mitigate my lack of Indigenous heritage by including respected elders from different First Nation tribes as guest lecturers, as well as inviting Indigenous political leaders, lawyers, judges, and grassroots activists to participate in classroom discussions. I use video materials showing discussions with knowledge keepers, as well as assign journal articles and case critiques written by experts in Indigenous law and theory. Students go on field trips to local First Nations, meeting face-to-face with elders and Indigenous scholars who guide them through their community's history and cultural practices.[1]

The introductory content of the course uses these methods to familiarize law students with the history of Indigenous peoples in Canada and their relationship with Canada's laws. Traditional mainstream legal and historical educational materials suggest that Indigenous peoples were lawless prior to European contact. This false idea is challenged through a variety of materials and content provided by the museum visits, lectures, and interactive sessions with experts and Indigenous elders and storytellers. Students learn that Indigenous peoples never gave up their laws voluntarily and that they have been forced to live in a foreign legal system that was imposed on them. From this starting point, students begin to understand the significance of that imposition on Indigenous peoples' ways of living, their customs, languages, and cultures, learning how they have been systematically and deliberately destroyed. They begin to realize that colonial laws took or attempted to take away Indigenous sense of spirituality, governance, order, and identity and replace them with European norms to eliminate what was perceived as the "Indian problem." With these understandings, combined with the knowledge that prior to colonization, complex legal systems existed for millennia in Indigenous communities, and continue to exist, students begin to shift their thinking and perspectives to do the necessary intellectual work to move past

1 For example, the students visited the Tsuu'tina cultural museum, which borders on Calgary's city limits.

stereotypes and assumptions and contemplate different approaches to law and justice.

The next phase of the course requires students to watch three video lectures and conversations with Indigenous scholars who have pioneered Indigenous law teaching and research into the existence of Indigenous law.[2] They address sources of Indigenous law, gender, and sexuality in Indigenous law and participate in a conversation about "tough questions" regarding Indigenous law. Students discuss these preliminary issues and debate them with each other. Revealing Indigenous law principles and their sources enables students to see that just like mainstream law, Indigenous law has sources, principles, and challenges that require debate and discussion in the contemporary context. The difference here is that students begin to understand that Indigenous laws have the unique capacity and objective to heal and restore relationships, rather than emphasizing retribution, deterrence, and punishment.

We then follow up by reading and analyzing civil and criminal judgments compared to Indigenous stories and incidents that have dealt with similar wrongdoing. Students compare the processes and substantive principles used in both and evaluate them in terms of consequences. An example of this is the case of *Thomas v. Norris* (1992). The case was heard in the British Columbia Supreme Court, and at its central issue is whether the group rights of an Indigenous community to conduct ceremonies can be used to supersede the common-law rights of an individual Indigenous person against personal violations of bodily security. The facts and reasons for judgment provide an excellent opportunity to examine the legal language and devices used in judicial reasoning and their limitations with respect to ensuring that Indigenous peoples receive the full benefit of their rights.[3]

The facts of the case were that the plaintiff, a member of a First Nations community, was "grabbed" from a friend's house and taken to the band's longhouse by the defendants to address his unacceptable violent behavior toward his wife. He was held at the longhouse for four days, where he underwent an initiation ceremony for spirit dancing which, in mainstream legal language, included assault, battery, and unlawful imprisonment. From the Indigenous perspective, however, the ceremony was a customary way to correct unacceptable behavior and to raise the offender to a higher spiritual awareness. The plaintiff brought a case in Court for damages for assault, battery, and unlawful imprisonment. The Court held that the law gave no protection to spirit dancing against unwilling participants. Individual rights to

2 These videos were made by the Indigenous Law Research Unit of the University of Victoria Faculty of Law for the Video on Demand project (2015a, b, c). They include *Indigenous Law: An Introduction*; *Indigenous Law, Gender, and Sexuality*; and *Indigenous Law: Tough Questions*.

3 For a discussion of the case and its implications, see *Individual vs Collective Rights: Aboriginal People and the Significance of Thomas v. Norris* (Isaac 1992).

bodily security trumped any collective right that may exist for the practice of the spirit dance. The language used by the Court in its discussion of "rights" ignored the relevance of Indigenous laws with respect to "grabbing," and there was little, if any, discussion on how the two systems of law could be harmonized.[4] The decision demonstrates how the language, analysis, and legal devices used in judicial reasoning make it very difficult for Indigenous rights to be respected or acknowledged in Canadian courtrooms.

Other cases the students study include those where judges have struggled with sentencing decisions for Indigenous offenders. It is widely understood that one of the most obvious failings of the criminal justice system in Canada and other colonized countries is the gross overrepresentation of Indigenous men, women, and youth in the prison systems (Rudin 2019). The Royal Commission on Aboriginal Peoples called the overrepresentation "injustice personified" (1995: 28). Students learn that once Aboriginal persons are arrested, prosecuted, and convicted, they are more likely to be sentenced to prison than non-Aboriginal people. This overrepresentation continues to grow. In Canada in 1995–1996, Aboriginal people made up 16% of all those sentenced to custody. By 2011–2012, that number had grown to 28% of all admissions to sentenced custody, even though Aboriginal people make up only 4% of the Canadian adult population (Statistics Canada 1995–1996). The overincarceration of women is even more disproportionate: in 2011–2012, 43% of admissions of women to sentenced custody were Aboriginal (Statistics Canada 2011–2012a). Aboriginal girls make up 49% of the youth admitted to custody, and Aboriginal boys are 36% of those admitted to custody (Statistics Canada 2011–2012b).

To encourage students to think deeply about this issue, they are assigned to study the case of *R. v. Moses* (1992), where the trial judge spoke very critically about the failings of the structure and principles of the justice system as it applies to Indigenous people. Instead of using conventional sentencing methods to sentence an Indigenous offender convicted of theft and illegal possession of a weapon, the judge opted for a sentencing circle – a uniquely Indigenous mode of resolving disputes, where all the participants meet as equals and speak openly with the objective of healing the injured party as well as the accused person and the community.[5] Mutual respect and community responsibility guide the process. Unlike the adversarial processes of the

4 The reality was that the elders and members of the band who "grabbed" the plaintiff were required under Indigenous law to seek permission of the plaintiff's family and the advice of elders before "grabbing" the offending individual, neither of which was done by the defendants. Had the judge acknowledged this violation of Indigenous law, he could have effectively adjudicated in favor of individual rights under both Indigenous and Canadian laws without diminishing the importance of collective rights to the spirit dance.

5 For a discussion of the case, see *R. v. Moses and Sentencing Circles: A Case Comment* (Benevides 1994).

conventional courtroom, the participants in a criminal sentencing circle are those who will be most affected by the sentence but are normally excluded at this crucial stage of the criminal process: victims, offenders, the police officers involved, and members and officials of the accused's First Nation community. Each person seated in the circle has an equal opportunity to speak about the needs of the offender and of the victim, constraints on the community's participation in the rehabilitation of the offender, or any other concerns the participants might have.

Certain members of the judiciary use sentencing circles as a novel way to replace sentencing required by the formal justice system. Because it is considered "novel," it has achieved some success but is not widely used or understood. To function, it first requires judges and legal counsel to have knowledge of the process. Second, the Crown and defense counsel must co-operate in the process of establishing the membership of the circle. Thirdly, they must obtain consent and trust from both the accused person and the victim, and finally, the community, and to a lesser extent, victims and offenders, must be recognized as equal partners with the justice system in the resolution of criminal disputes.

The *Moses* case provides the students with lessons on many levels. They not only debate the efficacy, legitimacy, and practicality of sentencing circles but also come to understand the inability of the criminal law to deal effectively with Indigenous offenders. In addition, the case also invites students to think deeply about the ethics of using mainstream law to deal with Indigenous issues in the criminal context, where most violations occur because of alcohol addictions, which themselves are most often related to poverty and dislocation associated with colonization. By studying sentencing circles, students learn that the method can resolve many disputes, not just criminal ones. Its flexibility lends itself to resolving family law matters and child welfare issues, as well as for use in litigation, mediation, conflict resolution, negotiations, or general problem-solving.

Indigenous Law and Litigation Course

For the Indigenous Law and Litigation course, I start with the report of the Truth and Reconciliation Commission (TRC Canada 2015b) and the calls to action (TRC Canada 2015a) to provide a context for the centerpiece of the course, which is the Indian Residential School Settlement Agreement (IRSSA 2006). It is with the pedagogical goals of demonstrating the efficacy of Indigenous legal principles compared to mainstream law in mind that I chose the IRSSA as the focus of the course. The settlement agreement, the largest of its kind in Canadian history and unique in the world, is one of the best examples of how the use of Indigenous legal principles can result in positive outcomes that are fair and just, not just for Indigenous peoples, but for the country as a whole.

The agreement, which was judicially approved of by all nine jurisdictions in Canada, is also a dramatic example of the shortcomings of the Euro-Christian Canadian legal structure in remedying the range of harms caused by mass violations of human rights. The students learn how the agreement was formulated, negotiated, and implemented primarily using Indigenous legal principles and Indigenous law, and how the legal team of the Assembly of First Nations (AFN)[6] strategized to lead the negotiations using Indigenous legal processes. As the chief negotiator for the Assembly of First Nations, I give the students a firsthand account of the steps that were taken during the negotiations, as well as the preparations leading up to the negotiations, consisting of consultations with Indigenous elders, leaders, and groups of survivors.

Before the students study the settlement agreement in detail, they study the history of Indian residential schools and the impact they had on the lives of the survivors, their families, and communities. They learn that the government of Canada implemented a policy starting in the late 1800s and ending in 1996 under which it compelled Indigenous children to leave their homes and attend Indian residential schools (IRS) that were supervised by Canada and run by various churches (The Davin Report 1879; see also TRC Canada 2015b: 56–57). The 130 schools were located in all the provinces and territories of Canada, except New Brunswick and Prince Edward Island. The government's assimilationist policy was specifically designed to assimilate Aboriginal people into European ways of life by forcing them to abandon their language, culture, and Indigenous ways of life and adopt the language, culture, and religions of non-Indigenous Canadians (Scott 1920; see also TRC Canada 2015b). Deliberate, often brutal, strategies were adopted to destroy family and community bonds. While attending the boarding schools, children were denied any meaningful contact with their parents, sometimes for their entire childhoods. About one child in three was abused physically, sexually, and emotionally, and the damage they suffered adversely affected generations of Indigenous peoples.[7] Children were subjected to medical experiments, forced labor without pay, and inferior nutrition and education (TRC Canada 2015b: 91).

The Truth and Reconciliation Commission's research into deaths in residential schools found that some 3,201 deaths could be documented (TRC Canada 2015b: 92). The Commission points out that the number could be much higher but cannot be proven because of the government's policy of destroying

6 The Assembly of First Nations is a national political body representing 634 First Nations across Canada.

7 Of the 105,000 claimants, 38,257 made claims for abuse. See Indigenous and Northern Affairs Canada, *Statistics on the Implementation of the Indian Residential School Settlement Agreement*, available at: www.aadnc-aandc.gc.ca/eng/1315320539682/1315320692192.

health records of the residential schools (TRC Canada 2015b: 91).[8] Many of the children who died in the schools were buried on school sites, sometimes in unmarked graves (TRC Canada 2015b). Often, no notice was given to their families, and their bodies were never returned to their families and communities.

The students come to understand that the government knew what was happening in the schools. The undisputed evidence is that Dr. Peter Henderson Bryce, Chief Medical Officer for the federal government and persistent advocate for the health of Indigenous children living in residential schools, spoke out as early as 1907 about the unacceptable conditions and deaths of children in the schools, saying the churches and the federal government had the means to save many lives but failed to take adequate action (The Bryce Report 1907). In his report, Dr. Bryce states the following:

> It suffices for us to know . . . that of a total of 1,537 pupils reported upon nearly 25 per cent are dead, of one school with an absolutely accurate statement, 69 per cent of ex-pupils are dead, and that everywhere the almost invariable cause of death given is tuberculosis.
>
> (The Bryce Report 1907: 18)

Further evidence from Dr. Bryce's inspections suggested that the number of student deaths over time were much higher, when taking into account that many children died shortly after leaving the schools (Milloy 1999). At the same time this was occurring, the forced assimilation policy continued to be implemented with the view that the government "could not kill the Indian but it could kill the Indian in the child" (see TRC Canada 2015b).[9]

The harms that were committed against at least 150,000 Indigenous children, their families, and their communities lasted over a period of 150 years. Their effects, in turn, caused hundreds of thousands of Indigenous people over generations to become impoverished and illiterate, with limited occupation opportunities and lost income,[10] addictions, psychological disorders, physical injuries and deformities, sexual dysfunction, and numerous other problems.[11] These harms continue to this day.

8 Between 1936 and 1944, 200,000 Indian Affairs files were destroyed.

9 The term has been attributed to Duncan Campbell Scott, but more accurately, it originated in the US military.

10 The settlement agreement provides for loss of opportunity described as one of chronic inability to obtain employment, chronic inability to retain employment, periodic inability to obtain or retain employment, inability to undertake or complete education or training resulting in underemployment and/or unemployment, or diminished work capacity. Claims can alternatively be made for actual income loss. For the settlement agreement, see IRSSA 2006.

11 Some of the harms listed as compensable in the compensation model are loss of self-esteem, pregnancy, forced abortions, forced adoptions, psychotic disorganization, PTSD, self- injury,

Following the lectures and readings on the history, students are taken on a field trip to the Tsuut'ina First Nation to have a guided tour of the nation's cultural museum. They meet elders and Indigenous scholars and watch a video, *Stolen Children: Residential School Survivors speak out*, featuring the firsthand stories of residential school survivors, as well as intergenerational survivors (CBC 2015). In the video, survivors speak about the sexual, physical, and psychological abuse inflicted in the schools, as well as the consequential harms to themselves and harms to their immediate families, harms to the community, and intergenerational harms to their descendants. Hearing the testimonies of the survivors firsthand in the video, as well as from the elders in attendance, students come to understand that the residential school harms were both collective, such as loss of language and culture, loss of family and community life, spiritual harms, loss of identity, and continuing intergenerational dysfunction, as well as profound individual personal injuries in cases of sexual, psychological, and physical abuse.

An exercise assigned at the conclusion of the field trip requires the students to write a 1,500-word personal letter to any one of the residential school survivors in the video.[12] Students are instructed to write their letters from their perspective as a law student, sharing with the recipient their reflections on what they were told about the residential schools. The letters reveal that many of the students were shocked to hear about the gross violations of human rights, torts, and crimes that were committed in the residential schools. Most of the students relate in the letters that their education has been radically incomplete. From kindergarten to post-graduate studies, they say the story of residential schools was never told to them. This exercise forces the students to react in a personal way, not only to the victims of the abuse, but also to an education system which systematically denied them crucial historical information.

Once they understand the extent of the harms inflicted on the individual students as well as the harms suffered by Indigenous communities, cultures, and intergenerational survivors, the next assignment is for the students to write a legal opinion on what the Canadian laws of tort or human rights would provide to survivors seeking reparations. Students find that legal obstructions such as Crown immunity, limitation periods, remoteness, causation, duty of care, vicarious liability, and standards of care stand in the way of a comprehensive response to the wide array of harms suffered. In many cases, no remedies would be available at all. They then compare the results of

sexual dysfunction, inability to form or retain relationships, eating disorders, severe anxiety, guilt or self-blame, lack of trust in others, addictions, nightmares, aggression, hypervigilance, anger, retaliatory rage, and humiliation.

12 The letters were not sent to the survivors. The exercise was to have student confront their own feelings, stereotypes, and assumptions. They submit the letters for their class participation mark.

their research to the terms of the settlement agreement that was achieved by using Indigenous law principles. Compensation for loss of language and culture; intergenerational compensation; healing funds; memorialization funds; individual compensation for physical, sexual, and psychological abuse; a truth and reconciliation commission; a research center and archive; and a series of apologies – all were provided through the agreement.

This exercise not only teaches students the differences in the two legal regimes in addressing the harms suffered but also brings them to a critical understanding of the shortcomings of mainstream law and its consequences, including its perpetuation of systemic discrimination of the colonized group by the very laws that are designed to compensate victims fairly. Most of the students expressed frustration in the knowledge that, for the most part, the harms suffered by the survivors cannot be remedied under Canadian law, yet the Indigenous solutions to remedy the harms are complete from both collective and individual perspectives.

They also conclude through this exercise that their law school education has been incomplete. They say their legal training must recognize and include Indigenous law as part of Canada's legal system. A further benefit of this exercise is that it shatters many racist stereotypes and provides students with deeper understandings of how and why so many Indigenous peoples experience poverty, unemployment, underemployment, poor health, disproportionate numbers of suicides, and social dysfunction.

Conclusion

When governments, like the Canadian government, have the stated policy objective of reconciliation to correct the colonial wrongs of the past (Harper 2008), the procedures and principles required to achieve it require solutions far beyond those rooted in traditional legal methods and principles inherited from Canada's colonial masters. Instead, it is the principles of Indigenous feminist theory, Indigenous legal theory, and Indigenous legal traditions that recognize the wide range of harms colonialism causes and that can empower the victims to articulate what they want, justify individual and collective reparations, and lay the groundwork for new relationships with the colonial institutions that operationalized exclusionary and discriminatory systems of governance.

The challenge is for law teachers, judges, and lawyers to embrace the existence of Indigenous law, to include it in the curricula of their courses, their legal practice, or their judgements, and to engender understanding of its practical usage to resolve many of the problems that the mainstream justice system has proven itself incapable of resolving in a fair and just manner. I believe the courses outlined earlier go some distance in meeting those objectives.

Reference List

Alexander- Floyd, N. (2010) "Critical race black feminism: A jurisprudence of resistance and the transformation of the academy," *Signs: Journal of Women in Culture and Society*, 35(4).

Benevides, H.J. (1994) "*R. v. Moses* and sentencing circles: A case comment," *Dalhousie Journal of Legal Studies*, 3, p. 134. Available at: https://commentary.canlii.org/w/canlii/1994CanLIIDocs13#!fragment//BQCwhgziBcwMYgK4DsD-WszIQewE4BUBTADwBdoByCgSgBpltTCIBFRQ3AT0otokLC4EbDtyp8BQk-AGU8pAELcASgFEAMioBqAQQByAYRW1SYAEbRS2ONWpA.

Borrows, J. (2016) "Heros, tricksters, monsters and caretakers: Indigenous law and legal education," *McGill Law Journal*, 61(4), p. 795. Available at: https://lawjournal.mcgill.ca/article/heroes-tricksters-monsters-and-caretakers-indigenous-law-and-legal-education/.

Bryce, P.H. [The Bryce Report] (1907) *Report on the Indian Schools of Manitoba and the Northwest Territories*, Department of Indian Affairs, Ottawa: Government Printing Bureau. Available at: http://peel.library.ualberta.ca/bibliography/3024.html.

Canadian Broadcasting Corporation [CBC] (2015) *Stolen Children: Residential School Survivors Speak Out*. Available at: www.youtube.com/watch?v=vdR9HcmiXLA.

Davin, N.F. [The Davin Report] (1879) *Report on Industrial Schools for Indians and Half-Breeds*, Canada Depart of the Interior. Available at: https://dev.nctr.ca/wp-content/uploads/2021/01/Davin-Report.pdf.

Harper, S. (2008) *Statement of Apology to Former Students of Residential Schools, Prime Minister Stephen Harper*. Available at: www.rcaanc-cirnac.gc.ca/eng/1100100015644/1571589171655.

Indian Residential School Settlement Agreement [IRSSA] (2006) *Official Court Notice*. Available at: www.residentialschoolsettlement.ca/settlement.html.

Indigenous Law Research Unit at the University of Victoria Faculty of Law (2015a) *Indigenous Law: An Introduction*, ed. by K. Todd [video]. Available at: www.youtube.com/watch?v=7uNgq7raxk4.

Indigenous Law Research Unit at the University of Victoria Faculty of Law (2015b) *Indigenous Law, Gender, and Sexuality*, ed. by K. Todd [video]. Available at: www.youtube.com/watch?v=m9Lsda4lDoo.a.

Indigenous Law Research Unit at the University of Victoria Faculty of Law (2015c) *Indigenous Law: Tough Questions*, ed. by K. Todd [video]. Available at: www.youtube.com/watch?v=Uuzd0hN7QI8.

Isaac, T. (1992) "Individual vs collective rights: Aboriginal people and the significance of Thomas v Norris," *CanLli Docs*, 134.

Milloy, J. (1999) *A National Crime: The Canadian Government and the Residential School System, 1879 to1986*, Winnipeg: University of Manitoba Press.

R. v. Moses (1992), *11 C.R.(4th) 357* (Y.T. Terr. Ct.).

Royal Commission on Aboriginal Peoples (1995) *Bridging the Cultural Divide*, Ottawa: Canada Communications Group.

Rudin, J. (2019) *Aboriginal Peoples and the Criminal Justice System*, Ipperwash Inquiry. Available at: www.archives.gov.on.ca/en/e_records/ipperwash/policy_part/research/pdf/Rudin.pdf.

Scott, D.C. (1920) *Evidence of D.C. Scott to the Special Committee of the House of Commons Investigating the Indian Act Amendments of 1920*, Library and Archives Canada, RG 10, vol. 6810, file 470-2-3, vol. 7 (L-2) and (N3).

Statistics Canada (1995–1996) *Adult Correctional Services in Canada, 1995–1996*. Available at: www.statcan.gc.ca/pub/85-002-x/85-002-x1997004-eng.pdf.

Statistics Canada (2011–2012a) *Adult Correctional Services in Canada, 2011–2012.* Available at: www.statcan.gc.ca/pub/85-002-x/2014001/article/11918-eng.htm#a5.
Statistics Canada (2011–2012b) *Youth Correctional Services in Canada, 2011–2012.* Available at: www.statcan.gc.ca/pub/85-002-x/2014001/article/11917-eng.htm#a5.
Thomas v. Norris (1992) 2 C.N.L.R. 139 (B.C.S.C.).
Truth and Reconciliation Commission of Canada [TRC Canada] (2015a) *Calls to Action.* Available at: https://ehprnh2mwo3.exactdn.com/wp-content/uploads/2021/01/Calls_to_Action_English2.pdf.
Truth and Reconciliation Commission of Canada [TRC Canada] (2015b) The Survivors Speak: A Report of the Truth and Reconciliation Commission of Canada. Available at: https://publications.gc.ca/collections/collection_2015/trc/IR4-5-2015-eng.pdf.

Chapter 20

Indigenous Peacebuilding and Environmental Restorative Justice

Giada Giacomini

Introduction

This chapter explores the interrelation between Indigenous peacebuilding and environmental restorative justice. This research field has yet to be meaningfully developed, and the present chapter aims at setting a first objective regarding the theoretical interrelation between these two areas of study. It will do so by introducing the concept of environmental restorative justice (ERJ) and proposes an interpretation of this approach to reparation that interlinks with Indigenous peacebuilding and peace ecology. It will also consider the conceptualization of the environment in Indigenous peacebuilding techniques, highlighting how crucial the environmental dimension is in conflict management and resolution, arguing that Indigenous peacebuilding in its environmental dimension can be considered a form of ERJ. ERJ can, indeed, be considered both as an environmental agenda that can contribute to restorative justice while also describing how restorative justice can be used in the context of environmental harm. Finally, the chapter will analyze a case study represented by the law of victim reparations in Colombia as a form of Indigenous environmental restorative justice in the context of a post-conflict situation. This law sets a crucial recognition of the environment as a victim in the context of conflict and harm; it is inspired by Indigenous cosmovision and their sacred conception of the territory. ERJ and peace ecology can thus be considered a component of ecological integrity: Through environmental restoration and recognition of the territory as a victim, the environment is transformed into a subject of peace processes and peacebuilding.

Environmental Restorative Justice

Environmental restorative justice is a term that encapsulates both an aim (obtaining justice) and a process (restoration of relationships) within the environmental dimension of a conflict. So it refers both to the restoration of environmental conditions to the pre-conflict situation, with an emphasis on reparations, and to the healing of the relationship between human communities

DOI: 10.4324/9781003440871-24

and the harmed environment, with a guarantee of non-reiteration. Another important conceptualization of this approach can be found under the term "Earth restorative justice," which deals with restoring the broken human–nature relationship and finding ways to include ecological relationships in human rights, legislation, and the democratic processes (Bosselmann 2022).

Before explaining the conceptual underpinnings of ERJ, it is necessary to clarify the meaning of *restorative justice* and why it is important to add an environmental dimension to it. Restorative justice is a complex concept that refers to a multitude of processes. According to Boyes-Watson (2019), there are at least four areas of activism that have proved to be fundamental in the contemporary restorative justice moment: justice reform in the criminal legal system; reforms in child welfare, education, and social services; movements taking place in a post-conflict situation and asking for peacebuilding in a democratic society; and the movements to establish political autonomy of Indigenous groups in the aftermath of colonization. The last demand of these is fundamental if we are to build more just societies in the contemporary context, and it interrelates with the current struggles of Indigenous peoples for self-determination and the emergence of a rights-based approach to their issues in national and international law.[1]

Restorative justice, broadly intended, is an approach to addressing harm and the risk of harm or conflict through the engagement of all parties affected, with the objective being to reach a common understanding of the wrongdoing, apologize, and try to repair the wrongdoing or achieve justice. A restorative approach to justice aims to rethink the conceptualization of how justice can be achieved. Rather than focusing on a mere punitive approach to the guilty, it seeks to focus on harms, needs, and obligations toward the victim. Restorative justice is more preoccupied with the re-empowerment of those affected by the wrongdoing and the re-establishment of the balance within a community (Zehr 2008). Examples of this type of restorative justice approach can be found in the institution of various reconciliation and truth commissions in post-conflict or genocidal situations (e.g., Peru, Canada, Rwanda, South Africa).

EJR moves from this restorative approach to include the environment both as a victim of harm, therefore in need of reparations, and as an important factor to include in the justice, truth, or peacebuilding process. EJR thus operates in two directions: the first concerns processes of reparation of environmental harm, while the second focuses on the role of the environment in building peaceful societies and on the interconnection between humankind and nature. The second approach is therefore more tightly linked to environmental justice and to the building of environmental democracies. However, as argued in this chapter, ERJ can also be inserted into holistic discourses of

1 For an account of this, please refer to Barelli (2016).

Indigenous environmental peacebuilding and in the recognition of the environment as a victim per se of conflict, war, and polluting practices. This would enable the epistemic coexistence of multiple sources of knowledge, Indigenous and Western, into the definition and application of justice approaches to environmental harm.

Regarding the first point, reparation of environmental harm, there is incontrovertible evidence that climate change, biotic impoverishment, pollution, and failure of environmental regulation are at the root of the current environmental crisis. Adopting an ERJ approach in this sense means applying the principle and functioning of restorative justice into the environmental regulatory realm. According to Forsyth et al. (2021), there are several characteristics and processes that can be applied to the ERJ framework. First, ERJ is directed toward the healing of harms caused to nature, so a more-than-human approach. Second, the participation of victims and offenders is required in the reconciliation process, and this should include the agency of more-than-humans. This point will be revisited later in the chapter with regard to Colombian law. Third, ERJ processes are based on storytelling and dialogue. And fourth, they are based upon precise harms, identifiable victims, and offenders that need to take responsibility for the harm. Finally, ERJ aims at establishing accountability for those who have caused the harm.

This approach has been used in courts. For example, New Zealand has shown one of the more consistent uses of ERJ for environmental crime. There is a total of at least 49 restorative justice proceedings under the Resource Management Act 1991 (NZ). Conferences occur within the broader proceeding context as an intermediate phase, and then the matter is returned to court for the decision, which takes into consideration the outcomes of the ERJ conference. In particular, the regional prosecutorial authority Environment Canterbury utilizes "alternative environmental justice" program conferencing as an alternative to prosecution (Hamilton 2021).

In New South Wales, Australia, the Land and Environment Court has also applied restorative justice in its proceedings. In the case *Garret v. Williams* (2007), the defendant, Craig Williams of Pinnacle Mines Pty. Ltd., was charged with several offenses under the National Parks and Wildlife Act 1974 (NSW).[2] The mining operation involved the construction of certain infrastructure, during which several Indigenous artifacts were destroyed, and damage was caused to the Aboriginal protected area. The ERJ proceedings involved a conference in 2006 in Broken Hill, where the defendant narrated the events that took place and the victims spoke of how they were affected in order to ascertain the possibilities for repairing harm and avoiding future offenses (White 2014; Hamilton 2014; Preston 2011; Al-Alosi & Hamilton 2019).

2 *Garrett v. Williams* [2007] NSWLEC 96.

Concerning the second point, EJR can also be linked to a discourse on reparative environmental justice. Environmental justice as a doctrine deals with issues of distribution of environmental resources (including ensuring the right to a healthy environment), participation in environmental policymaking and the decision-making processes, and contesting the structural causes of environmental inequalities (Schlosberg 2007). Reparative environmental justice adds a dimension to reparative and restorative justice, insofar as it wishes to apply environmental justice to ecological destruction and restoration, climate change, and animal abuse, through a non-anthropocentric point of view (see generally Almassi 2020). In this sense, Indigenous worldviews, peacebuilding processes, and peace ecology are a crucial constitutive part of ERJ approaches.

Indigenous Peacebuilding and Peace Ecology

Indigenous peacebuilding, Mac Ginty argues, has sparked the interest of the international community, especially in recent years. Such approaches to peacebuilding were especially "trendy" with NGOs, bilateral donors, and academics because they are considered local and sustainable (Mac Ginty 2011: 47). Because of its local and localized nature, Indigenous peacebuilding is often presented as the alternative to "liberal peace," or a peace imposed by external forces following Westernized standards and procedures (Mac Ginty 2008). Some important examples of Indigenous peacebuilding processes are found in the post-genocide Gacaca tribunals of Rwanda, or the Pokot community conflict resolution mechanism in Kenya (Mac Ginty 2008). At the international level, the Manila Declaration on Conflict Resolution, Peace Building, Sustainable Development, and Indigenous Peoples, which was adopted in Manila, Philippines, in December 2000 for the International Conference on Conflict Resolution, wished to create a concrete response to a peacebuilding initiative.[3]

At the United Nations level, the 15th session of the United Nations Permanent Forum on Indigenous Issues, which took place at United Nations Headquarters in New York from 9 to 20 May 2016, was devoted to the theme "Indigenous peoples: conflict, peace, and resolution." To prevent conflict, the Forum had urged consensus-building based on the principles enshrined in the UN Declaration on the Rights of Indigenous Peoples. Relevant articles in this sense are Article 30, which affirms that military activities shall not take place in the lands or territories of Indigenous peoples (unless justified by a relevant public interest or otherwise freely agreed with or requested by the Indigenous peoples concerned); Article 32, which establishes the right to free, prior, and informed consent of Indigenous peoples and asks states to provide

3 The text of the Declaration is available at: www.twn.my/title/manila.htm.

effective mechanisms for just and fair redress for any such activities, and to take appropriate measures to mitigate adverse environmental, economic, social, cultural, or spiritual impacts; Article 40, which states that Indigenous peoples have the right to access to, and prompt decision through, just and fair procedures for the resolution of conflicts and disputes with states or other parties, as well as to effective remedies for all infringements of their individual and collective rights.

The rationale for dealing with Indigenous peacebuilding and environmental restorative justice in the same chapter is that the former focuses, among other characteristics, on processes of restoration of the human/environment balance. Sustainability is an increasingly important factor in peacemaking, as it deals with the utilization of local human personnel in the reconstruction of a post-conflict society and not with dependence on externally provided aid. It also enables communities to rely on their natural resources and the environment for the rebuilding of their societies. The promotion of such practices in post-conflict situations supports the restoration of ecological balance and the reconstruction of the territory (see also Vayda 1976).

Some characteristics of Indigenous peacebuilding are common to restorative justice: the emphasis on truth-telling; dealing with relational, affective, and psycological rehabilitation of the victim and the offender; and a process that aims more at restoring the balance within a community rather than punishing the offender. For what concerns the environmental dimension, Indigenous peacebuilding offers a different perspective on peace and justice compared to that of the Western tradition: "indigenous traditions are the repository of vast experience and deep insight on achieving balance and harmony" (Alfred 2009: 21). The environmental justice movements of Indigenous peoples have been quite vocal over the years about the current planetary crisis and the genocidal approach that is being undertaken against Mother Earth.[4] These strong declarations are connected to issues of recognition of territorial rights, legal and cultural pluralism, and respect for Indigenous distinctiveness within settler states. In fact, Indigenous peoples have not only been historically disposessed of their ancestral lands but are also being systematically targeted by extractivism, pollution practices, and environmental violence (Whyte 2016).

4 Examples of Indigenous demands for the planet are found in the Indigenous Peoples' Global Summit on Climate Change, The Anchorage Declaration, 24 April 2009, at https://unfccc.int/resource/docs/2009/smsn/ngo/168.pdf; the Kari-Oca II Declaration Indigenous Peoples Global Conference on Rio+20 and Mother Earth, 18 June 2012 at www.scribd.com/document/98585247/Kari-Oca-II-Declaration#; the Mandaluyong Declaration of the Global Conference on Indigenous Women, Climate Change and REDD Plus 18–19 November 2010 at www.asianindigenouswomen.org/index.php/climate-change-biodiversity-and-traditional-knowledge/climate-change/64-mandaluyong.-declaration-of-the-global-conference-on-indigenous-women-climate-change-and-redd-plus/file, last access January 2023.

Respect for the natural world and the establishment of a relational ontology between the human and the other-than-human communities are deeply embedded in Indigenous legal systems. For example, in the book *Kayanerenkó:wa: The Great Law of Peace*, Kayanesenh Paul Williams provides an account of how the five nations – Mohawk, Oneida, Onondaga, Cayuga, and Seneca – established a lasting system of peace. In such a system, respect and care for the natural world have an essential part in maintaing peaceful relationships among community members (Williams 2018). Another example of this approach to peace and ecological balance is found in the Onjisay Aki International Climate Calls to Action, a declaration adopted by consensus at the Onjisay aki-Our Changing Earth-International Climate Summit held in Sagkeeng First Nation in Manitoba, Canada, in 2017.[5] The call to action states that

> [t]he traditional wisdom of Indigenous Peoples, rooted in *laws of peaceful conduct*, and a love for and spiritual relationship with the land, air, water, fire, and plant, animal and human and celestial worlds, has allowed them to live sustainably within diverse homelands for millennia.
>
> (McGregor 2021, emphasis added)

The concept of peace ecology is pertinent to such discourses as it can be considered "the worldview underpinning environmental peace-making" (Kyrou 2007). From the ontological point of view, peace ecology considers how all life on Earth is interconnected and how proper ecological conditions enable the maintenance of peace and peaceful relations. It also considers how environmental destruction is being used as a method to target and destroy entire ethnic groups by cutting their vital resources, such as the case of Native Americans. From the practical point of view, peace ecology adopts a positive conceptualization of peace, stressing the role of human responsibility toward environmental protection. Therefore, Indigenous cosmovisions constitute a crucial value in the application of a peace ecology approach. The realization of a peace ecology approach is based on a multiagency epistemic model that considers alternatives to the capitalistic, extractivist model of development. Indigenous knowledge, alongside deep ecology, eco-feminism, and eco-centric law, is one such way.[6]

For example, Alberto Gomes describes the lifeways and practices of the Orang Asli (Aborigines) from Malaysia that inspire an ecological approach to peace. He argues there are at least three aspects of Orang Asli cultural distinctiveness that foster an ecological approach to peace. The first relates to a democratic and inclusive aspect of their society. Second, they maintain an

5 The Call is available at: https://onjisay-aki.org/onjisay-aki-international-climate-calls-action.

6 For a general account on peace ecology, see Amster (2015).

egalitarian approach to access to natural resources and combat inequalities. Third, their system is based on collective property ownership that prevents environmental resources from being controlled by an individual or group. The Orang Asli people are also featured among the most nonviolent and peaceful societies in the world, which count 22 out of 25 Indigenous peoples (Gomes 2018).

While peace ecology is a fairly recent literary phenomenon, from such premises it is evident how it is deeply interlinked with Indigenous knowledge and ERJ. An example of such an approach to environmental peacebuilding is found in the Victims and Land Restitution Law of Colombia, which was drafted in consultation with Indigenous peoples.

The Victims and Land Restitution Law in Colombia: Toward a New Notion of Victimhood for the Environment?

Colombia has undergone a process of violent conflict since the 1960s. Such complex and long-lasting hostilities featured the confrontation of the government of Colombia, far-right paramilitary groups, and far-left guerrilla groups, such as the Revolutionary Armed Forces of Colombia (FARC), the National Liberation Army (ELN), and the Popular Liberation Army (EPL), fighting over their influence in the nation. In 2012, the Havana Peace Talks were first initiated between the government of President Juan Manuel Santos and the FARC. The negotiations resulted in an agreement reached in 2016.

The lists of atrocious crimes committed during this long and complex conflict are innumerable, and they include the forced displacement of millions of people, the participation of young people in guerrilla warfare, the use of land mines, and the murder of tens of thousands of civilians. The conflict, however, has not only created human victims; the environment and natural resources in Colombia have also been a victim of this conflict: deforestation, mining, coca plantations, and the mercury pollution of waters for gold mining are just some examples of how the environment has been depleted, in some cases permanently (Pereira et al. 2021). While the Colombian case is very complex and it cannot be discussed here in a thorough and fair way, this chapter will highlight how the participation of Indigenous peoples in the peacebuilding process has led to the legal recognition of the territory as a victim of armed conflict.

Colombia's Indigenous peoples have been especially affected by this environmental dimension of the conflict, and they were at risk of being excluded in the peace negotiations and in the redistribution of the territory and resources taken from the FARC. Therefore, the peacebuilding process in Colombia has been characterized by an important territorial dimension, which takes into account the importance of the ecological dimension of the territory and its relation to human beings in the peace process (Cairo 2018;

de Diego 2017; Bautista 2017). This territorial dimension and the participation of local actors in the peacebuilding process are of utmost importance for the Indigenous peoples of Colombia, amounting to 1.5 million people, who take care of 30% of the national protected areas. In Colombia, more than 3 million people have been displaced since the 1990s as a consequence of the conflict, with 34 Indigenous groups at serious risk of cultural or physical extermination (Gómez-Isa 2014).

However, the first law that was enacted regarding the redistribution of land in the peace process in Colombia did not sufficiently consider Indigenous perspectives, or their needs in terms of continued access to their natural resources. Law 1448 (2011) on the internal armed conflict and the reparations entitled to the victims was harshly criticized by Indigenous and Afro-organizations because of their lack of participation in the peacebuilding process (Aponte & López 2013). It was at risk of being declared inapplicable as it was against the principles established in the Constitution, which gave the president of the Republic the power to conduct prior consultation processes and issue decrees regulating the rights of victims of the ethnic minorities, including Indigenous peoples.[7] The government created two ad hoc organizational bodies that could represent the interests of Indigenous peoples and Black communities: the Permanent Round Table for Concertation and the High Level Consultation. Each body drafted legal documents in which they stated their aspirations in the peacebuilding process regarding assistance, care, reparation, and restitution. This process resulted in Decree Laws 4633 and 4635 of 2011: the first for "Indigenous peoples and communities," and the second for "Afro-Colombian, Black, Raizal, and Palenquera communities" (Ruiz Serna 2017).

Aponte and Lopez (2013) argue that the consultation process with Indigenous communities in Colombia regarding the drafting of Law 4633 was not effective for several reasons. For example, it was conducted in an insufficient time frame to allow meaningful participation of all the different communities. In addition, the information, language, and contents were of limited access to the communities; in Colombia, there are more than 64 Indigenous languages, and Spanish is not spoken by all individuals.

Notwithstanding such difficulties in enacting the consultation procedure, Decree Law 4633, the Victims' Act for Indigenous Communities, represents an example of Indigenous peacebuilding, environmental restorative justice,

7 The Constitution of Colombia recognizes Indigenous peoples' rights to self-government in their territories, their legal pluralism, the right to exercise their jurisdictional functions within their territorial jurisdiction in accordance with their own laws and procedures, and the right to administer their resources. In addition, Indigenous peoples are entitled to consultation before any legislative and administrative act that could have an impact on their rights, territories, and resources. The Decree Law 1320 (1998) established the need to specifically regulate prior consultation with Indigenous and traditional Black communities through a specific procedure enabling environmental authorities to exercise their competence in this area.

and an application of the peace ecology approach in law. This law embodies a different ontological representation of the territory, which is not only an object but also a living entity that can therefore suffer harm and is entitled to restoration, restitution, and reparation.[8] By virtue of the principle of *enfoque diferencial* (differential approach), the law develops key concepts according to the Indigenous cosmovision. This differential approach recognizes the importance of protecting Indigenous peoples in armed conflicts, and that such protection entails "a collective exercise of the cultural and spiritual relationship that these have with the territory from its sacred and ancestral character, and which aims to ensure balance and harmony between the elements that comprise it" (Art. 33, Law 4633, 2011).

Law 4633 incorporated the notion of territory as a victim: in Article 45, it recognizes that the territory, understood as a living entity and the basis of identity and harmony according to the Indigenous peoples' own worldview and by virtue of the special and collective bond they maintain with it, *suffers harm when it is violated or desecrated* by the internal armed conflict and its related and underlying factors. Such crucial recognition takes place in accordance with Indigenous cosmovision and territorial sacrality. The law acknowledges the spiritual interconnection of Indigenous peoples with their territory and Mother Earth (Art. 3 and 8).

Regarding agency and who is legally entitled to request reparations for the territory, the law stipulates that, although the territory is a victim, "the holders of rights . . . are indigenous peoples and communities and their individual members" (Art. 3). Therefore, only communities affected by the damage inflicted on the territory can speak for the territory, interpret how the spirits that inhabit the land have been affected by the conflict, and become subjects of restitution and reparation of rights.

If the territory is acknowledged as victim, and therefore reparations are due to the territory, a parallel can be drawn between this conception enshrined in Colombian law and the eco-centric legal paradigm embodied in peace ecology. Such paradigms reflect a holistic conceptualization of harmony between human beings, territory, animals, and all the different manifestations of existence. And reparations for harm should not only include material restoration but also take into consideration the spiritual identity connection of a people with their territory.

Conclusion

This chapter has focused on the synergies and similarities between ERJ, Indigenous peacebuilding, and peace ecology. It has argued that environment

8 The text of the law is available at: https://en.unesco.org/creativity/sites/creativity/files/law1448v18jun20.pdf

and natural resources are – and need to be considered – victims of armed conflict, violence, pollution practices, and degradation. Therefore, if the notion of victimhood is extended to the environment, like in the Colombian case, then environmental reparation and restoration practices should be duly considered in peacebuilding processes. Indigenous cosmovisions represent an example of how ecological harm should not be considered just material damage but spiritual damage that affects the environment as a living entity alongside human beings. Such an approach calls for further investigation regarding the interlinkage between ERJ and peacebuilding, with a view to developing an ecological conceptualization of peace.

Reference List

Al-Alosi, H. & Hamilton, M. (2019) "The ingredients of success for effective restorative justice conferencing in an environmental offending context," *University of New South Wales Law Journal*, 42(4), p. 1460–1488.

Alfred, T. (2009) *Peace, Power, Righteousness: An Indigenous Manifesto*, Ontario: Oxford University Press.

Almassi, B. (2020) *Reparative Environmental Justice in a World of Wounds*, Lanham: Lexington Books.

Amster, R. (2015) *Peace Ecology*, New York: Routledge.

Aponte, J. & López, L. (2013) "El pluralismo jurídico indígena en Ley de Víctimas y Restitución de Tierras: Retorno y consulta previa," *Univ. Estud. Bogotá: Colombia*, 10, pp. 157–176.

Barelli, M. (2016) *Seeking Justice in International Law: The Significance and Implications of the UN Declaration on the Rights of Indigenous Peoples*, New York: Routledge.

Bautista, S. (2017) "Contribuciones a la fundamentación conceptual de paz territorial," *Ciudad Paz-Ando*, 10(1), pp. 100–110.

Bosselmann, K. (2022) "Earth trusteeship and the sovereign state," in: B. Pali, M. Forsyth, & F. Tepper (eds) *The Palgrave Handbook of Environmental Restorative Justice*, Cham: Palgrave Macmillan.

Boyes-Watson, C. (2019) "Looking at the past of restorative justice: Normative reflections on its future," in: T. Gavrielides (ed) *Routledge International Handbook of Restorative Justice*, Abingdon: Routledge.

Cairo, H., Oslender, U., Piazzini Suárez, C.E., Ríos, J., Koopman, S., Montoya Arango, V., Rodríguez Muñoz, F.B., & Zambrano Quintero, L. (2018) " 'Territorial peace': The emergence of a concept in Colombia's peace negotiations," *Geopolitics*, 23(2), pp. 464–488.

de Diego, M. (2017) "¿Qué es la paz territorial?" in: *La paz en el territorio: Poder local y posconflicto en Colombia*, pp. 17–66, Universidad Externado de Colombia.

Forsyth, M., Cleland, D., Tepper, F., Hollingworth, D., Soares, M., Nairn, A., & Wilkinson, C. (2021) "A future agenda for environmental restorative justice?" *The International Journal of Restorative Justice*, 4(1), pp. 17–40.

Gomes, A. (2018) "Lessons from the margin: Indigenous peace ecology," *Organicom*, 15(28), pp. 149–166.

Gómez-Isa, F. (2014) "El desplazamiento forzado de los pueblos indígenas en Colombia," *Anuario Español de Derecho Internacional*, 30, pp. 431–455.

Hamilton, M. (2014) "Restorative justice intervention in an aboriginal cultural heritage protection context: Conspicuous absences?" *Environmental and Planning Law Journal*, 31(5), pp. 352–367.

Hamilton, M. (2021) "Restorative justice in an environmental offending context: Theory and practice," in: M. Hamilton (ed) *Environmental Crime and Restorative Justice: Justice as Meaningful Involvement*, Cham: Springer Nature.
Kyrou, C. (2007) "Peace ecology: An emerging paradigm in peace studies," *International Journal of Peace Studies*, 12(1), pp. 73–92.
Mac Ginty, R. (2008) "Indigenous peace-making versus the liberal peace," *Cooperation and Conflict*, 43(2), pp. 139–163.
Mac Ginty, R. (2011) *International Peacebuilding and Local Resistance: Hybrid Forms of Peace*, Cham: Springer.
McGregor, D. (2021) "Indigenous environmental justice towards an ethical and sustainable future," in: B. Hokowhitu, et al. (eds) *Routledge Handbook of Critical Indigenous Studies*, London: Routledge.
Pereira, R., Sjöstedt, B., & Krause, T. (2021) *The Environment and Indigenous People in the Context of the Armed Conflict and the Peacebuilding Process in Colombia: Implications for the Special Jurisdiction for Peace and International Criminal Justice*: CAPAZ; EDPAL.
Preston, H.J.B.J. (2011) "The use of restorative justice for environmental crime," *Criminal Law Journal*, 35(3), p. 136–153.
Ruiz Serna, D. (2017) "El territorio como víctima: Ontología política y las leyes de víctimas para comunidades indígenas y negras en Colombia," *Revista colombiana de antropología*, 53(2), pp. 85–113.
Schlosberg, D. (2007) *Defining Environmental Justice: Theories, Movements, and Nature*, Oxford: Oxford University Press.
Vayda, A.P. (1976) *War in Ecological Perspective: Persistence, Change and Adaptive Processes in Three Oceanic Societies*, New York: Plenum Press.
White, R. (2014) "Indigenous communities, environmental protection and restorative justice," *Australian Indigenous Law Review*, 18(2), pp. 43–54.
Whyte, K. (2016) "Indigenous experience, environmental justice and settler colonialism," *SSRN*, pp. 157–173.
Williams, K. (2018) *Kayanerenkó:wa: The Great Law of Peace*, Winnipeg: University of Manitoba Press.
Zehr, H. (2008) "Doing justice, healing trauma: The role of restorative justice in peacebuilding," *South Asian Journal of Peacebuilding*, 1(1), pp. 1–16.

Chapter 21

Indigenous Governments Exercising Environmental Law Jurisdiction

Hope for the Future?

Bradford W. Morse

Introduction

This chapter explores examples of Indigenous governments in the United States and Canada exercising their own jurisdiction, as well as jurisdictions conferred upon them by national legislation, to establish, and to enforce, environmental laws that have been consciously designed by them to reflect traditional values and Indigenous traditional knowledge (ITK). Their intention is to respond to the many critical environmental challenges confronting our planet in the 21st century, and particularly tied to the needs of their homelands. Many Indigenous communities are on the front lines of the battle to turn the tide away from environmental destruction, as they readily recognize that drastic action is required before it is too late.

"Empowering Mother Earth through Asserting Indigenous Rights" is a previously authored essay that focused on the success of Indigenous peoples, particularly in Australia and Canada, in asserting aboriginal title, aboriginal rights, or treaty rights to regain more of their traditional territory exclusively through litigation, effective lobbying, or partnerships with national or state/provincial/territorial governments, thereby creating formal comanagement regimes. This strategy was selected where it would best achieve the goal of that Indigenous party in regaining a significant measure of influence or control quickly over how large parts of their traditional homelands are used and protected. The aim was to restore, or at least significantly aid in repairing, the environment of those lands and waters, thereby enabling their rehabilitation and return to healthy conditions as natural ecosystems, and for the benefit of all humankind.

The purpose for this chapter is to examine key examples of how Indigenous peoples, particularly in Canada and the United States, are leading the way to restoring vibrancy to their traditional homelands through the influence of, and actions by, Indigenous governments. The key element is that these situations are ones in which Indigenous governments can protect or rehabilitate ecosystems under their direct control or considerable influence. This chapter also examines how predominantly non-Indigenous or

DOI: 10.4324/9781003440871-25

mainstream governments in both countries can create interesting pathways in which Indigenous governments can enhance their jurisdiction and access sources of funding for environmental remediation and renewal.

American Indian Tribal Governments in the United States

Indigenous peoples, or Native American tribes, as they are more commonly called collectively in the United States, were completely sovereign nations prior to contact with Europeans. The US label "Native Americans" generally excludes Native Hawaiians, whose kingdom was internationally recognized prior to its illegal overthrow by American entrepreneurs with US military support (Morse & Hamid 1990: 407–456), as well as the Inuit and Aleuts of Alaska. All these Indigenous peoples possessed their own forms of government that could differ significantly from each other. For example, some nations selected their leaders (named within their own languages and frequently not translatable as "chiefs") by the clan mothers, some were chosen by elders of both genders, others took office through inheritance, some were selected by spiritual or religious leaders, and some were picked by all adults, leaving those very unhappy with the person(s) chosen having the freedom to depart, which might generate the creation of a new tribe within that language group.

Contrary to popular myth and most Hollywood movies, Indigenous peoples in the Americas were not aimless wanderers, nor did all relocate as seasons changed by following migrating game through their traditional territory. Most nations who lived along the Atlantic or Pacific coasts, and many who relied heavily on farming in the interior, lived primarily in sedentary villages or possessed several villages in different ecosystems. There were, of course, many who did migrate seasonally over traditional routes, and usually in much smaller units during the winter for trapping fur-bearing animals, often only to gather in large numbers in the summer for ceremonies and making key governmental decisions for the whole nation. Concepts of ownership of lands or waters existed, but primarily at the collective or tribal level, rather than by individual owners. Here, too, this did not apply to all, as some nations recognized that particular families or persons might have preferential rights to specific locales for fishing, trapping fur-bearing animals, gathering wild foods and medicines, or for farming.

Europeans came from a completely different mindset, in which fences and other boundary markers denoting private or aristocratic ownership, or lesser leaseholds, was the norm. It is, perhaps, not surprising that seeing no fences, the Europeans assumed that large swaths of territory without fixed permanent villages were free for the taking, even though this was totally false and was quickly disproved by tribal responses to such European settler actions.

Tribes generally had peaceful and mutually beneficial trade relations with the small number of French colonists that entered their territories in the

1600s and 1700s who made minimal requests for some land for trading posts and to be taught to prosper in new lands with different climactic conditions. The goal of France, like Spain originally, was to find a quicker sea route to Asia to displace the Venetians and the Portuguese, who dominated the Asian trade market. The British, however, brought large numbers of immigrants from continental Europe to its colonies in North America from a variety of Protestant sects, as farmers, skilled tradespeople, small business owners, indentured servants released from poorhouses, and prisoners sent to Georgia to serve their sentences. Tribes often vigorously asserted their ownership and sovereignty over their traditional territories to the British governmental representatives, as well as to settlers encroaching on their lands. The British adopted the previous practice of the Dutch of negotiating treaties with significant Indian nations to forge positive economic and military alliances.

England's defeat of France for the second time led to the issuance of the Royal Proclamation of 6 October 1763 by King George III, which declared that the Crown would respect the "Indian tribes or Nations" in the possession of their lands, would block all settlers from trading with Indian nations unless licensed, and that only an imperial, not colonial, government representative could purchase Indian lands within any of the colonies, and then only via a treaty after public negotiations ratified by both parties (Royal Proclamation 1763; Walters 2015). After an invitation by the British special ambassador to Indian Nations across North America to meet him along the Niagara River in July 1764, the Proclamation was presented as a treaty and discussed over several days (Borrows 1997). The Proclamation also declared that all lands west of a boundary to be created by a future survey would remain Indian lands. Many of these promises became major reasons the Indian nations shifted their support from France to Great Britain, and a majority of colonial politicians and residents launched a revolution in 1776. None of that history is relevant today so as to justify national and regional dominant governments in denying either the pre-existence of Indigenous sovereignty or the failure to recognize its continuing existence in either the USA or Canada.

The US Supreme Court came to this realization in *Cherokee Nation v. Georgia* in 1831. Chief Justice John Marshall declared that Indian nations had been independent nations prior to the separate declarations of overarching sovereignty by the British, French, Spanish, and Dutch governments. The arrival of waves of European settlers fundamentally changed the future of the continent. The success of the American Revolution created a new nation – the United States – to which accrued the sovereignty of the British Crown. The US national government asserted that it became the internationally recognized sovereign state with full political control of the 13 former British colonies – as well as over the lands set aside as exclusive Indian territory west of the Royal Proclamation of 1763's boundary. That western boundary was never surveyed and finalized due to the American Revolution. Thus, Chief Justice Marshall stated that Indian nations had become "domestic dependent nations" with direct relations with the US federal government similar to a guardian and its ward (*Cherokee Nation v. Georgia* 1831). Any foreign powers who entered into new treaties with Indian nations would be regarded as committing an act of hostility akin to declaring war against the United States and, as a result, would be responded to accordingly. This view of the law largely remains in place today, except that the president and federal government have been disempowered from entering into new treaties with Indian nations by a rider added to budget legislation passed by Congress in 1871, now recorded in 25 USC 71, although a process exists to obtain federal recognition.

The US Federal Register currently recognizes 574 tribes, pueblos, rancherias, and Alaskan Native villages as well as approximately 66 state-only recognized tribes (2022). The United States holds 56.2 million acres (22.7 million hectares, 227,433 square kilometers, 87,812 square miles) in trust for federal tribes (44 million acres) and individual Indians (11 million acres set aside under the federal Dawes Act, a.k.a. the General Allotment Act of 1887, intended to privatize all reservation land and sell off the balance to non-Indians) or 2.3% of the United States. The US government also holds billions of dollars of tribal money in trust, derived primarily from the sale or lease of reservation lands by the federal government with or without tribal consent.

There are two main strategies in which tribes can assert jurisdiction to regulate the environmental systems that exist within the remaining small residue of their once-massive territories: residual sovereignty and federal legislation. Tribes can exercise their residual sovereignty to enact laws that apply to their reservation lands, whether owned collectively by the tribe or by individual tribal members or tribal businesses. In some situations, tribal law can apply to non-Indian fee lands within the reservation's boundaries. Tribes can also exercise powers made available to them under federal legislation. Congress has passed the following relevant statutes: the Clean Air Act (CAA), the Clean Water Act (CWA), the Safe Drinking Water Act (SDWA), the Toxic Substances Control Act (TSCA), the Emergency Planning and Community Right-to-Know Act (EPCRA), the Comprehensive Environmental Response Compensation and Liability Act (CERCLA), and the Federal Insecticide, Fungicide, and Rodenticide Act (FIFRA).

The US Environmental Protection Agency (EPA) has been assigned the lead role for the federal government to develop, monitor, and implement strategies to fulfill the goals of federal legislation and policies that are intended to improve the environment, whether through preservation of environmentally significant sites or habitats or via remediation of ecosystems that have deteriorated. These federal laws generally allow the EPA to treat eligible federally recognized tribal governments in a similar fashion to state governments, with both being able to carry out the powers granted under the specific federal statute to achieve the legislative goal in relation to state or tribally owned lands. While these seven pieces of legislation do differ from each other, the effect is that federally recognized tribes (thus not applicable to state-only recognized tribes), like state governments, are empowered to carry out federal authority to achieve the goals of each of these statutes that individual tribes seek to achieve.

The EPA also possesses the capacity to provide training and technical expertise to tribal governments as well as project or ongoing program funding, along with delegation of power under the Act to carry out its purposes (see *US v. Mazurie* 1975). It achieves this through provisions enabling the EPA to treat a tribe as equivalent to a state so that it can exercise powers under one or more of these eight federal statutes.

The main requirements for this "treatment as a state" (TAS) status for each tribe is that it be (1) a federally recognized tribe (thereby excluding all state-only recognized tribes), (2) possess a governing body, (3) have appropriate authority to act regarding the specific subject of the federal Act, and (4) have the capacity to carry out the program(s) necessary to achieve the federal statute's mandate (US EPA 2023). As of April 2023, there are 107 federally recognized tribes (out of 574) across the United States that are exercising some element of federally delegated authority concerning major environmental issues that may threaten the residents and visitors to their reservations under one or more of these Acts. This generally encompasses all land within the reservation, including non-Indian-owned fee land within the external reservation boundaries.

Demonstrating the situation's gravity, over 85% of reservations have no standards in effect to assess quality under the CWA; 84 tribes do administer tribal water quality standards that are intended to ensure high-quality water by restricting chemicals added to or leaching into freshwater sources that may render it unsafe to drink, create tumors on or inside fish, or otherwise make clear that only limited quantities of fish can be safely consumed (US EPA 2024a). The EPA has confirmed 84 tribes (15%) as eligible to administer a federal water quality standards (WQS) program under the CWA and approved the WQS plan of 47 of those 84 tribes. The EPA provides a broad array of resources, templates for WQS and water quality certificates, human health ambient water quality criteria and its relationship to daily fish consumption rates, an online reference library of water quality standards (US EPA n.d.), and many more tools to support tribes in their administration and application of these federal statutes. The EPA is consulting currently on a proposed rule to revise the WQS to clarify how state and federal practices "must protect aquatic and aquatic-dependent resources reserved to tribes through treaties, statutes, executive orders, or other sources of federal law, in waters where those rights exist" (US EPA 2024b).

The EPA also has authority to provide funding to tribes under the CWA and the other six federal statutes to carry out developing its own standard setting, lab analyses of test samples, training of staff, and enforcement action regarding any breaches by individuals or corporations. The EPA announced its largest ever investment of over $278M in annual infrastructure funding for tribes and ANVs on 20 June 2023, to enhance clean water, improve drinking water infrastructure, and address emerging contaminants in small and disadvantaged communities (WaterWorld 2023).

There are many tribal initiatives to respond to climate change and to environmental damage, but there is one particularly exciting example. The Confederated Salish & Kootenai Tribes (CSKT) of Flathead Reservation in Montana now manage over 100 federal and state programs on their reservation, as well as 70 tribal programs (CSKT n.d.). They are the first US tribe to own and operate their own hydropower dam and the only local electricity provider. They have repurchased over 245,000 acres of former reservation land since 1944 to possess a 1.2 million acre reservation (1,875 square miles, 4,856 square kilometers), although their original territory was 22 million acres across three states and British Columbia. More recently, the CSKT drafted its first climate change strategic plan in 2013 for its community, which was updated in 2016 (CSKT 2016). A new draft plan was created in August 2020 and released for comment on 13 October 2020, titled CSKT Climate Resiliency Plan (CSKT 2020). It reflected the awareness that climate change was "accelerating so rapidly that many impacts projected to occur many years in the future were already in evidence on the Flathead Reservation in 2019" (CSKT 2020: 15). It was created as a living document to be continually refined and enhanced.

The CSKT selected the planting of whitebark pine trees, which used to be widespread on their lands above the 6,000-foot (1,828-mile) level as a major initiative. The seeds of this tree are vital in sustaining wildlife (over 100 species rely on it), and humans eat the seeds too. An imported fungus from 1909 – white pine blister rust – caused by *Cronartium ribicola*, infested the trees, leading to them becoming more susceptible to mountain pine beetles, which decimated these trees, leaving only 5–8% surviving. Climate warming has compounded the threat by loss of snow earlier than previously, so they decided to restore the forest and rebuild that ecosystem. This reduces snow melting by providing shade cover to allow for the snow to remain longer and melt slower, thereby strengthening the environment. They have been recovering seeds from surviving trees and identified that about 2% of surviving trees have acquired a resistance to the fungus and the mountain beetle. CSKT has been working with the US Forest Service (USFS) to grow seedlings in the USFS's huge greenhouses. The tribe has now begun planting 100,000 seedlings as the next stage of this strategy (CAIHRE 2023). The whitebark pine received US Endangered Species Act protection as a threatened species in 2022 (US FWS), demonstrating the necessity for urgent action.

Indigenous Peoples of Canada

There are three distinct Indigenous peoples now recognized by the Canadian Constitution, as amended: the "Indian, Inuit and Metis peoples" (Canada

Constitution Act 1867–1982). The federal government was given exclusive legislative authority upon the creation of Canada in 1867 through s. 91(24) of what is now called the Constitution Act, which empowered the Parliament of Canada to enact laws for "Indians, and Lands reserved for the Indians."

Parliament exercised that power in 1868 to pass its first statute to exercise control over Indians, titled *An Act providing for the organisation of the Department of the Secretary of State of Canada, and for the management of Indian and Ordnance Lands*. The federal government continued the British policy of negotiating treaties with Indian Nations to form relations and to acquire most of those nations' lands to give to immigrants and existing citizens to create farms and businesses. The Indian Act was passed in 1876 to consolidate a number of colonial and federal statutes to establish stronger federal control over Indian nations by dividing them into small groups, called "bands," with extremely limited lawmaking powers that were controlled by a local federal official called an "Indian agent" (1876). The Indian Act was frequently amended to outlaw traditional ceremonies and compel children to be sent away to Indian residential schools run by four Christian churches (over two-thirds were operated by the Roman Catholic Church), where Indigenous languages and traditional clothing, as well as braids, were suppressed, children were prohibited from speaking to children of the opposite gender, and physical as well as sexual assaults were common (Indian Act 1985). The federal government's goal was to transform Indians into low-skilled Canadian workers, as farmhands and domestic servants, while destroying the cultures, languages, traditional economies, and spiritual beliefs until no distinct Indians remained. This strategy left a legacy of despair and trauma within most Indigenous communities (TRC Canada 2015).

The federal Indian Act, from its early days, was designed to disaggregate Indian Nations into "bands" while restricting them to a tiny land base, impose iron rule via a federal official, suppress Indigenous languages and spiritual beliefs, restrict their movement without permission, and restrict them to passing local bylaws that applied to all people (including non-Indians) while present on that reserve and to the lands themselves. Early lawmaking for the environment meant enacting bylaws (in English or French only) governing "construction and maintenance of watercourses," "water supplies," and "the preservation, protection and management of fur-bearing animals, fish and other game on the reserve" (Indian Act 1985). These bylaws all required the approval of federal officials and the minister responsible for Indian affairs until relatively recently. Regulations passed by the Federal Cabinet under this Act also could deal with depositing garbage on reserves. This structure of the Indian Act, as a vehicle of maximizing federal governmental control via disempowering Indian Nations, continued from 1867 until the 1980s, when it began to undergo significant changes.

A number of relatively new federal statutes have been jointly negotiated with First Nations representatives seeking to expand their jurisdiction to manage their reserve lands without active federal involvement, such as dealing with matrimonial property division (Family Homes on Reserves and Matrimonial Interests or Rights Act 2013), creating comprehensive land codes addressing all aspects of land usage and management, including extensive environmental standards that have been recently updated (*Framework Agreement on First Nation Land Management Act* 2022), and property taxation regimes (First Nations Goods and Services Tax Act 2003; First Nations Fiscal Management Act 2005, which created the First Nations Tax Commission). The latest legislation seeks to align with Articles 26, 29, and 32 of the United Nations Declaration on the Rights of Indigenous Peoples (UNDRIP), in keeping with the Canadian Parliament's passing of the United Nations Declaration on the Rights of Indigenous Peoples Act in 2021.

Over time, the term "Indians" in s. 91(24) was interpreted by the Supreme Court of Canada to include the Inuit via the *Re Eskimo* decision in 1939 (Re Eskimo 1939), and for the Metis in *Daniels v. Canada* in 2016. Thus, the Canadian Parliament can clearly pass laws that affect any of or all the three distinct Indigenous peoples, as well as regional or local aboriginal peoples, so long as such statutes are not violating the "aboriginal or treaty rights" protected by s. 35(1) of the Constitution Act of 1982, which includes all recent treaties, land claims, and self-government agreements. Parliament has not legislated for either the Metis or the Inuit as two distinct peoples, although they are included with First Nations in recent legislation that enables the transfer of jurisdiction over child welfare and family services from provincial or territorial governments to Indigenous governments representing those local or regional peoples (An Act respecting First Nations, Inuit and Métis children, youth and families 2019), creates a national Indigenous Languages Commission to support strengthening or reviving individual Indigenous languages (Indigenous Languages Act 2019), and commits the government of Canada to respect UNDRIP and to ensure that Canadian laws are brought into compliance with its provisions via jointly developed annual action plans (UNDRIP Act 2021).

Canada ended treaty-making with Indigenous peoples in 1924 as a matter of policy, but not by statute, as the United States had done 49 years earlier through the Indian Appropriations Act of 1871 (Rice 1977); however, the *Calder v. British Columbia* decision of the Supreme Court of Canada in 1972 declared that aboriginal title was still part of Canadian common law. As a result, the federal government effectively revived treaty-making but called it "comprehensive land claims," as the then prime minister believed treaties were only between independent nations (Canada Department of Indian Affairs and Northern Development 1973). This policy was later expanded to include negotiating self-government agreements (SGAs) to confirm expanded lawmaking jurisdiction for those Indigenous peoples or nations that negotiated and signed SGAs with Canada and the relevant provincial or territorial public government. As an example, 11 SGAs were negotiated by 11 individual First Nations with Canada and the Yukon Territorial Government from 1993 to 2005 that have relatively common jurisdictional authority, including the power to make laws for their territories over many subjects related to land and their people. While not yet embracing Indigenous governments as sovereign entities, this shift was spurred by the full involvement of Indigenous leaders participating in constitutional conferences with the prime minister and all the premiers from 1983 to 1992.

Governance jurisdictions for the environment reflect legislative control within their significantly expanded "settlement lands" after their individual comprehensive land claim and monetary settlements were reached (i.e., modern treaties) over protection of settlement land, land and water, natural resources, protection of fish and wildlife, planning, zoning and land development, administration of justice, actions that may threaten public health, and the environment. These 11 SGAs are very much the exception rather than the norm among the 640 First Nations in Canada; however, a number of other First Nations in British Columbia and Quebec have settled their aboriginal title claims to acquire a much larger land base, although far smaller than their original territories, along with significantly expanded governance jurisdiction from what had existed during the years of federal suppression of Indigenous sovereignty.

Another key factor at play has been major litigation by Indigenous nations challenging the authority of federal and provincial governments to issue licenses to companies seeking to exploit natural resources within Indigenous territories without their consent. Canadian courts have interpreted the constitutional protection of "aboriginal and treaty rights" in s. 35(2) as confirming that the Crown (later confirmed as both federal and provincial) is in a fiduciary relationship with Indigenous peoples so that it must avoid any sharp dealing (*Guerin v. The Queen* 1984). Similarly, since the Crown is capable of harming aboriginal and treaty rights, it owes a duty to consult with Indigenous peoples whenever considering a course of action that might harm their interests or lands while seeking to avoid such an outcome wherever possible (*Haida Nation v. British Columbia (Minister of Forests)* 2004).

The 2014 decision of the SCC in *Tsilhqot'in Nation v. British Columbia* dramatically changed the landscape even further, as the Court unanimously upheld that aboriginal title existed over the 1,700 square kilometers identified (reflecting a small portion of their territory without third party interests) and that the Crown lost all beneficial interests in those lands but retained legislative capacity to regulate their use so long as it fulfilled its fiduciary duties. The exact parameters of each government's jurisdiction and how their coexistence will evolve remain uncertain today; however, it has meant that the province's freedom to do whatever it wished, ignoring Indigenous interests and goals, is gone. Co-management of lands and waters is, to some degree, inevitable, and the Tsilhqot'in Nation is pushing a strong environmental rehabilitation agenda.

This shifting legal landscape, buttressed by more positive public attitudes and major concerns over climate change and industries seeking a "green" image, has triggered some Canadian governments – and many major companies – to take a far more creative approach to negotiating mutually beneficial agreements recently, which are changing the legal and political landscape. For example, the government of British Columbia recently transferred 20,000 hectares of waterfront and forests (77 square miles), valued at $150M, to the Lake Babine Nation (LBN) as the first step of the 20-year Lake Babine Nation Foundation Agreement, signed by LBN with the governments of British Columbia and Canada in September 2020, along with $43M in funding to pursue a phased approach to implementing aboriginal title, self-government, and other rights (Lake Babine Nation 2020).

Similarly, the Carrier Sekani Tribal Council (composed of seven First Nations) signed the Pathway Forward 2.0 Agreement in 2019 with the governments of British Columbia and Canada that confirms a shared vision for comprehensive reconciliation through sustainable economic development, stewardship, cultural vitality, and respectful governments that will also engage in co-governance. $175M has been provided over five years to develop their involvement in the forest industry, local economy, and culture while "sustainably manag[ing] lands, water, air, and natural resources in a way . . . to meaningfully exercises their Aboriginal title, rights, and interests in their Territories" (Pathway Forward 2.0 Agreement 2019). The Tribal Council represents over 8,400 people, whose traditional territory contains 69,600 square kilometers (26,872 square miles).

The Canadian situation is still somewhat embryonic, as treaties and their wording are being reinterpreted in modern terms to reflect the spirit of the partnership that they promised, rather than constituting mere land purchases by the state. Aboriginal title remains unresolved in the vast majority of British Columbia, as well as in most of all four Atlantic provinces, and pockets elsewhere, while Indigenous rights in beds of waters and subsurface rights remain live issues moving toward appellate courts.

Nevertheless, there is still a very long way from the recognition of "residual sovereignty" that exists in the United States from the "domestic dependent nation" concept of the US Supreme Court's decision of 1831 in *Cherokee Nation v. Georgia*. The Canadian Parliament's passage of the UNDRIP Act in 2021, 18 months after the British Columbia Legislature did the same, does raise hope that we will see significant movement by Canadian governments to recognize the inherent right of Indigenous communities to govern themselves and their exclusively held lands. Such a shift would increase the prospects for more of Canada to receive careful, intergenerational approaches to sustainability so as to restore vitality to portions of Mother Earth.

Reference List

An Act Providing for the Organisation of the Department of the Secretary of State of Canada, and for the Management of Indian and Ordnance Lands (1868) S.C. 1868, c. 42. Available at: www.sac-isc.gc.ca/eng/1100100010196/1618940565216.

An Act Respecting First Nations, Inuit and Métis Children, Youth and Families (2019) S.C. 2019, c. 24.

An Act Respecting Indians [Indian Act] (1985) R.S.C., 1985, c. I-5. Available at: https://laws-lois.justice.gc.ca/eng/acts/i-5/page-1.html.

An Act Respecting Indigenous Languages [Indigenous Languages Act] (2019) S.C. 2019, c. 23.

An Act to Amend and Consolidate the Laws Respecting Indians [Indian Act] (1876). Available at: https://nctr.ca/wp-content/uploads/2021/04/1876_Indian_Act_Reduced_Size.pdf.

Borrows, J. (1997) "Wampum at Niagara: The royal proclamation, Canadian legal history, and self-government," in: *Aboriginal and Treaty Rights in Canada: Essays on Law, Equality, and Respect for Difference*, pp. 155–172, Vancouver: University of British Columbia Press.

Calder v. British Columbia (AG) (1973) SCR 313 [1973] 4 WWR 1.

Canada Constitution Act (1867–1982) *The Constitution Act, formerly called the British North America Act*. Available at: https://laws-lois.justice.gc.ca/pdf/const_e.pdf.

Canada Department of Indian Affairs and Northern Development (1973) *Comprehensive Land Claims Policy*. Available at: https://publications.gc.ca/collections/collection_2017/aanc-inac/R32-81-1987-eng.pdf.

Center for American Indian and Rural Health Equity [CAIHRE] (2023) *Seven Generations* [video], Montana State University. Available at: https://m.youtube.com/watch?v=ZRJDeWW3j_k.

Cherokee Nation v. Georgia (1831) 30 U.S. 1.

Confederated Salish & Kootenai Tribes of the Flathead Reservation [CSKT] (2016) *Climate Change Strategic Plan*. Available at: http://csktclimate.org/downloads/Climate%20Change%20Strategic%20Plan/CSKT%20Climate%20Change%20Adaptation%20Plan%204.14.16.pdf.

Confederated Salish & Kootenai Tribes of the Flathead Reservation [CSKT] (2020) *Draft: CSKT Climate Resiliency Plan*. Available at: http://csktclimate.org/index.php/resources/ongoing-work.

Confederated Salish & Kootenai Tribes of the Flathead Reservation [CSKT] (n.d.) Available at: www.csktribes.org/.

Daniels v. Canada (Indian Affairs and Northern Development) (2016) S.C.C. 12, [2016] 1 S.C.R. 99.

Family Homes on Reserves and Matrimonial Interests or Rights Act (2013) S.C. 2013, c. 20.

First Nations Fiscal Management Act (2005) S.C. 2005, c. 9.

First Nations Goods and Services Tax Act (2003) S.C. 2003, c. 15, s. 67.

Framework Agreement on First Nation Land Management Act (2022) S.C. 2022, c. 19.

Guerin v. The Queen (1984) 2 SCR 335.

Haida Nation v. British Columbia (Minister of Forests) (2004) SCC 73.

Lake Babine Nation (2020) *Lake Babine Nation Foundation Agreement*. Available at: www2.gov.bc.ca/assets/gov/environment/natural-resource-stewardship/consulting-with-first-nations/agreements/lake_babine_nation_foundation_agreement.pdf.

Morse, B. & Hamid, K. (1990) "American annexation of Hawaii: An example of the unequal treaty doctrine," *Connecticut Journal of International Law*, 5, pp. 407–456.

Pathway Forward 2.0 Agreement (2019). Available at: www2.gov.bc.ca/assets/gov/environment/natural-resource-stewardship/consulting-with-first-nations/agreements/cstc_pathways_20_agreement_signed_-_jan_2020.pdf.

Reference as to Whether "Indians" Includes in s. 91 (24) of the B.N.A. Act Includes Eskimo in Habitants of the Province of Quebec [Re Eskimo] (1939) S.C.R. 104.

Rice, G.W. (1977) "Indian rights: 25 U.S.C. § 71: The end of Indian sovereignty or a self-limitation of contractual ability?" *American Indian Law Review*, 5, p. 239. Available at: https://digitalcommons.law.ou.edu/ailr/vol5/iss1/9.

Royal Proclamation (1763) *By the King, A Proclamation*, King George III. Available at: www.sfu.ca/~palys/The%20Royal%20Proclamation.pdf.

Truth and Reconciliation Commission of Canada [TRC Canada] (2015) *Calls to Action*. Available at: https://ehprnh2mwo3.exactdn.com/wp-content/uploads/2021/01/Calls_to_Action_English2.pdf.

Tsilhqot'in Nation v. British Columbia (2014) SCC 44.

United Nations Declaration on the Rights of Indigenous Peoples Act (2021) S.C. 2021, c. 14.

U.S. Environmental Protection Agency [US EPA] (2023) *Tribes Approved for Treatment as a State (TAS)*. Available at: www.epa.gov/tribal/tribes-approved-treatment-state-tas.

U.S. Environmental Protection Agency [US EPA] (2024a) *Actions on Tribal Water Quality Standards and Contacts*. Available at: www.epa.gov/wqs-tech/epa-actions-tribal-water-quality-standards-and-contacts.

U.S. Environmental Protection Agency [US EPA] (2024b) *Revising the Federal Water Quality Standards Regulations to Protect Tribal Reserved Rights*. Available at: www.epa.gov/wqs-tech/revising-federal-water-quality-standards-regulations-protect-tribal-reserved-rights.

U.S. Environmental Protection Agency [US EPA] (n.d.) *Tribal Air and Climate Resources*. Available at: www.epa.gov/tribal-air.

U.S. Federal Register (2022) vol. 86, No 18, 7554–8.

U.S. Fish & Wildlife Service [US FWS] (2022) *Whitebark Pine Receives Endangered Species Act Protection as a Threatened Species*. Available at: www.fws.gov/press-release/2022-12/whitebark-pine-receives-esa-protection-threatened-species.

U.S. v. Mazurie (1975) 419 US 544.

Walters, M. (2015) "The Aboriginal Charter of Rights: The Royal Proclamation of 1763 and the Constitution of Canada," in: T. Fenge & J. Aldridge (eds) *Creating Canada: From the Royal Proclamation of 1763 to Modern Treaties*, Montreal and Kingston: McGill-Queen's University Press.

WaterWorld (2023) *EPA announces $278M for Tribes', Alaska Native Villages' Water Infrastructure*. Available at: www.waterworld.com/drinking-water/infrastructure-funding/press-release/14295804/epa-announces-278m-for-tribes-alaska-native-villages-water-infrastructure.

Chapter 22

Find the Law in Your Heart

The Role of Non-Local Intuition in Legal Reasoning

Orsolya Bányai

Background and Basic Assumptions

We are collectively experiencing difficult times on this planet. Our civilization is very close to reaching its limits, and radical transformation is needed to be able to continue. According to system science, when such a big problem needs to be solved, it is not enough to use well-known sources to find a solution, but rather previously marginalized information (Fodor 2019: 38). This chapter aims to represent a still-marginalized understanding of reality, but one with a high potential for inducing meaningful change with a new, fresh legal approach to current eco-social problems. This will be done in light of the values of the Earth Charter (2000), which is a non-binding international declaration on the vision of basic values to provide an ethical foundation for the emerging world community. It highlights the interrelated nature of ecological integrity, social and economic justice, democracy, nonviolence, and peace. Its preamble says: "Our environmental, economic, political, social, and spiritual challenges are interconnected, and together we can forge inclusive solutions" (Earth Charter 2000). This chapter advances such an integrative legal approach which supports a deep transformation of our society in order to realize respect and care for the community of life.

Previously, the focus of my research was on the very significance of individual responsibility, based on the theory of quantum social change and critical mass (Bányai 2022). It argued that personal transformation toward an integrative, holistic worldview, alongside grassroots movements which hold that integrative worldview, has the ability to attain transformation of our whole society. In this process, any kind of legislation has a secondary, but still important, role. Nevertheless, considering that our culture today is on the verge of ecological collapse, it is time to start talking about new legal frameworks for a renewed society. This kind of discussion is part of the transformation itself.

The following discussion will consider the importance of integrative worldview (Bányai 2020; De Witt 2014; Van Egmond & De Vries 2011), especially as it is represented by the integral philosophy of Ervin László (2014).

DOI: 10.4324/9781003440871-26

According to the integrative worldview of this chapter, (1) one part of human consciousness extends beyond space and time (therefore, it has a transcendental[1] aspect); (2) all individual and transcendental consciousness is part of a universal, collective consciousness on which our reality, that is, what we are experiencing, depends, as consciousness determines reality; (3) between universal and individual consciousness, information flows in real time through non-local intuition; and therefore, (4) all living entities are part of the web of life, interconnected with each other at different energetic, physical, chemical, biological, and even transcendental levels. This transcendental level cannot be ignored, because the other levels within space and time build upon this. This worldview, previously called "Eternal Order" (Bányai 2019, 2020, 2021), is based on the results and interpretations of different scientific disciplines as quantum mechanics, consciousness research, biology, epigenetics, cosmology, psychology, etc. (Dossey 2017; Schwartz et al. 2017; Walach 2020).

From a legal point of view, the four aforementioned elements of this alternative reality can be considered as eternal universal laws that human law should take into account. This Eternal Order has two dynamic aspects. The first is that the continuous movement of the reality around us and within us can be considered as a universal movement of the cosmos (*Rta* in traditional Hindu law). The second dynamic aspect is the constant information transmission between individual and collective consciousness through the so-called non-local intuition and acting according to that impulse at the individual level. Through this non-local intuition, eternal laws are being transformed to the language of everyday life. Thus, they become an excellent compass in life to help show us the direction, especially in those circumstances where rational decision-making cannot help.

This chapter is going to focus on the dynamic aspect of the Eternal Order at the individual level, what symbolically – and perhaps literally as well – manifests itself through the heart. The heart is crucial in both Christian and Hindu spiritual traditions. As the place of conscience, it has significance in traditional natural legal theories as well. Furthermore, modern scientific research also suggests that our heart is the organ responsible for non-local intuition. Finally, there are personal experiences, including my own. With its help, I navigated safely through the most complex and difficult situations of my life. Of course, I had to learn how to use it together with my rationality, but it helped make it possible not just to live in survival mode but to integrate life experiences and realize an inner personal development. It helped me reconnect both nature and society in a respectful, meaningful, and intimate way.

1 In this chapter, the term "transcendental" means "beyond space and time." It is distinguished from the term "immanent," which means "within space and time."

Is the Law in Our Heart?

A group of researchers claims that the heart is not only a pump or neurological center (see heart–brain) but also the place where non-local intuition occurs (Bradley 2007; McCraty 2015; La Pirra & Gillin 2006; Radin 2008; Rezaei et al. 2014.) According to McCraty, *non-local intuition* (2015: 47)

> refers to the knowledge or sense of something that cannot be explained by past or forgotten knowledge or by sensing environmental signals. It has been suggested that the capacity to receive and process information about non-local events appears to be a property of all physical and biological organization and this likely is because of an inherent interconnectedness of everything in the universe. Examples of non-local intuition include when a parent senses something is happening to his or her child who is many miles away, or the repeated, successful sensing experienced by entrepreneurs about factors related to making effective business decisions.

This understanding is based on a large body of documented rigorous scientific research on non-local intuitive perception, which demonstrated that the heart receives and processes information about a future event before the event actually happens. It seems that our heart is coupled to a field of information not bound by the classical limits of time and space (Radin 2008; Rezaei et al. 2014). There is a growing number of research on how this non-local intuition functions from both a neuroscientific (Frecska 2012) and a decision-making point of view (Bradley 2007; Hiley & Pylkkänen 2005; Marcer & Schempp 1997, 1998; Mitchell 2000; Rauscher & Targ 2001; Sinclair 2011; Tiller et al. 2001). Research shows that entrepreneurs (La Pirra & Gillin 2006), teachers (Sipman et al. 2019), and psychologists (Frecska 2016; Shapiro & Marks-Tarlow 2021) use such intuition. Nita directly suggests to apply such intuition to employees of public administration (2014). This kind of intuition should be distinguished from implicit knowledge, energetic sensitivity, or any kind of local intuition (McCraty & Zayas 2014).

It seems that people get continuous intuitive information from the quantum field that is beyond space and time, but many are not aware of it. And actually, the ability can be improved: It is like any other skill in the sense that it must be practiced in order for one to become proficient at it (Sowerby 2001). Non-local intuition seems to require "becoming aware" of intuitive signals, because these intuitions emerge subtly (Sipman et al. 2019). Meditative exercises and mindfulness, when done frequently, can help in this regard (Dane 2011). Studies on intuition suggest those people who were in a coherent state (emotionally, mentally) prior to non-local intuition experiments were significantly more attuned to the information from the heart (McCraty et al. 2004). However, Frecska also warns that to learn how to use such intuition consciously needs time, because our "perceptual-cognitive" mind is so unprepared, being incapable of distinguishing between what is personal and what is transpersonal (Frecska et al. 2016: 164).

The results of using this kind of intuition are more effective communication skills, the ability of better social cooperation, and a positive attitude toward life (McCraty & Zayas 2014; Bradley et al. 2011). People who use non-local intuition are able to better respect and love people, animals, plants, and even non-animate surroundings (McCraty et al. 2004).

> It is no coincidence that Francis of Assisi does not pray for nature in abstract, but what he can emotionally touch on and relate to is particular objects of nature that are personified (like "brother sun" and "sister moon") in his Cantico del sole.
>
> (Deák 2022: 6)

Another important result of non-local connection can be individual and social resilience (McCraty et al. 2004). *Resilience* is the capacity to prepare for, recover from, and adapt in the face of stress, adversity, trauma, or challenge (McCraty 2015: 8). Non-local intuition can give us strength and creativity to overcome low points in life (Raftopoulos 2011; Ebrahimi et al. 2012). Finally, non-local intuition is used mostly in situations where information is limited, there were time constraints, or there was a great deal of complexity (Allinson & Hayes 1996; Sharda 2016). It is observed that as problem complexity increases, human capacity for rational analysis decreases and the mind searches for heuristic and intuitive ways to reduce cognitive demands (Davis 2004b). Indeed, when rationality cannot be applied, it is best to try to reach non-local intuition. This is because heart intuition is based on presence and real-time information and gives the possibility to solve the problem in the highest level of creativity. It also enables us to discover new situations, as non-local intuition is not based on already-learnt past patterns. Doing this, our personal worldview will be challenged and formed by new experiences and give us the possibility to reach a more integrated and coherent personality and worldview. Thus, the use of non-local intuition supports the evolution of an integrative worldview by inner inspiration.

To support the preceding argument, it is also important to review the Bible and the Bhagavad-Gíta, as scriptures of the two most popular Western and Eastern religions of the world. According to the Bhagavad-Gíta (BG 18: 61–62):

> The Supreme Lord is situated in everyone's heart, O Arjuna, and is directing the wanderings of all living entities, who are seated as on a machine, made of the material energy. O scion of Bharata, surrender unto Him utterly. By His grace you will attain transcendental peace and the supreme and eternal abode.

In the Bible, in the gospel of Luke it is written: "[T]he kingdom of God is within you" (Luke 17:21). In the prophet of Jeremiah, the Lord says, "I will put My law in their minds, and write it on their hearts; and I will be their God, and they shall be My people (Jeremiah 31:33)." Pope Benedict XVI refers to Saint Paul in the Letter to the Romans as an example that the listening heart is open to the language of being (Pope Benedict XVI 2011):

> When Gentiles who have not the Law [the Torah of Israel] do by nature what the law requires, they are a law to themselves . . . they show that what the law requires is written on their hearts, while their conscience also bears witness..
>
> (Romans 2:14–15)

Thus, we can see directly that sacred scriptures support scientific understanding. Of course, in these scriptures, not the heart itself is important, but the fact that anyone can be attuned to God directly within themselves. However, as will be discussed later in more detail, this connection comes hand in hand with its well-known risks. Therefore, regarding heart-based intuition, both Christian and Vedic theologies emphasize the importance of other authoritative sources.

From a legal point of view, non-local flow of information (and energy) can be considered as a dynamic universal law in our life. This dynamic universal law, which reveals itself through our heart, similarly to the movement of the ocean, is constantly changing, but the drop of the ocean is always able to follow the different streams and waves of the ocean. Other drops of the ocean potentially still receive the signals from the non-local consciousness and can act in harmony. Our natural task is to find and follow this flow in our life, or alternatively, we can decide not to follow, but in this case, we would experience separateness, trying to rival something with which we are a part. This flow can be considered as the law of everyday life. Of course, understood not in a sense of modern legal positivism but as a constantly changing equilibrium in life.

Non-Local Intuition in Hindu Legal Traditions and Western Classical Natural Law

This dynamic universal law in our life can be considered as one aspect of the *dharma* in traditional Hindu law, or the expression of cosmic *Rta* in human life. *Rta* most often means "truth" or "natural order" and refers both to the true and everlasting movements of the cosmos (Davis 2004a) and literally means "the cause of things" (Koller 1972: 133). By using a metaphor, *dharma* would be our individual voice in the ongoing cosmic musical performance. In traditional Hindu law, complex meanings of *dharma* exist (Koller 1972), and here, *dharma* is used in two ways. The first is called *moksadharma* (Koller 1972; Davis 2004a), or the dynamic universal law which can be understood through our heart. The second, lower *dharma* is being in accord with the manifested reality of ordinary experience and its aim to support *moksadharma*. This "social" *dharma* means different things in traditional Hindu law, especially different legal and religious duties, or norms on the basis of social position or life stages, which can sound strange for those who are not familiar with Indian culture. But because man-made law closely relates to this (social) *dharma*, I simply refer to *dharma* as the synonym of *man-made law*.

Thus, man-made law should be designed in a way to support the realization of *moksadharma*, or in other words, to be able to live in full harmony with our heart. But man-made law in our era is not designed to help the realization of *moksadharma*; thus, there is a heavy burden on individuals to find the way to it. In this journey, one important tool is non-local intuition, which can be extremely helpful in difficult circumstances, where usual answers and rationality are not enough. But it also has its risks, especially because people generally do not have a full integrative worldview, with no coherent personality of appropriate and good qualities (e.g., patience, wisdom, humility, etc.), or trained in how to live a transcendental way of life. Therefore, people should rarely make mistakes when following their intuition, or in case they trust in it too enthusiastically and somehow make errors, they should learn to accept the consequences of the mistakes in their life (moral, legal, etc.) and try to learn from them. It is important to consult with those who have more knowledge and experience in this regard, and written traditional scriptures are also helpful to find the equilibrium (i.e., the *moksadharma*) in a particular situation.

In classical Hindu law, the *Vedas*, the oral tradition called *smrti*, and customs were the privileged sources of *dharma*, and non-local intuition, called *atmatusti*, meaning "what is pleasant to one's own self," was the last consideration (Davis 2010). Here, the "self" refers to the *atma*, the pure consciousness within us. The pleasure of the *atma* is intrinsic, within us, but we are not usually aware of it. The strong feeling of pleasure thus can be the sign of what the *moksadharma*, or the law of the heart, is in a particular situation. But this kind of pleasure can only be reached by non-local intuition and should not be mixed with the immanent weak sense of emotional satisfaction. Thus, *atmatusti* was only accepted as the source of law if it was used by qualified wise persons, like sages, because the mental satisfaction of such great and wise persons is of such tremendous power that even *adharma* (not-*dharmic* "wrong") becomes *dharma* ("right"), or *dharma* becomes *adharma* (Davis 2007). So the pleasure of the self can be authoritative; it can bend reality, but only according to people with exceptional qualities. Therefore, in such extraordinary cases, even the *Vedas* and other outer sources of *dharma* will fully support such *atmatusti*. At this level, there is no longer a difference between inner and outer sources of *dharma*.

The next step is then to focus on the question whether this law of the heart, as it is interpreted by the classical Hindu law, is equal with the Western term "conscience." Ultimately, it depends on what *conscience* means, because it has been defined in different ways over the centuries. But it is true that the classical Greek and Scholastic legal philosophical traditions accepted one aspect of conscience (*synderesis*) as God's Word implanted in us, or the intuitive perception of eternal law (*partitipatio legis aeterna in rationali creatura*). According to St. Paul, for example, it plays a directive role before action takes place (Eardley 2021). In light of modern science and traditional Hindu law, only that aspect of conscience can be considered as the law of individual action, which is based on non-local intuition. It is not equal with any local intuition, moral sense, or some sense of rationality. Thus, only with these restrictions can *synderesis* be considered as *atmastuti*, or the law of the heart.

Nevertheless, Davis warns us that from the classical Hindu legal perspective, *atmastuti* traditionally is not the compass of everyone but, rather, an inner sense of the *moksadharma* based on sophisticated educational training (Davis 2007: 294). Again, unfortunately, it is true. However, today, when human society as a collective does not take into account universal order and is so far from that natural order (*Rta*), only individuals can find the way of harmony. For individuals in such a chaotic social environment, only our intuitive heart is left to find the way out. This conscience must be, or at least is better to be, followed by us. It is a conscience that is above immanent moral or ethical rules, above good and bad, and if we ignore it, it brings natural negative consequences. It is like a building being constructed where gravity is not taken into account – nothing good can result. Our collective social and environmental problems, similar to such a tragic construction, can be seen as the consequences of the fact that most people ignore eternal universal order and their own individual reflections of the heart.

Non-Local Intuition in Legal Reasoning and Rulemaking

For our shared future, people need to be trained in how to be able to live in harmony with our heart, and states should take responsibility for this by ensuring supportive conditions for such improvement. For this purpose, man-made law should integrate in its policies the aim to raise the level of consciousness of the people in order for everyone to be able to live in full harmony with their heart, which, as mentioned, results in many gifts in our life: resilience, intrinsic pleasure, positive attitude to others, etc. But do not forget, this is still a vision.

As for the present, it must be understood that people are on different levels of consciousness, and they also have different worldviews. And generally speaking, the state, in its current form, is not able to give such support. However, a possible step forward would be for those who already have an integrative worldview, with some experience on heart-based intuition, to apply non-local intuition with their particular fields of work, whether as lawyers, politicians, or bureaucrats, responsible for rulemaking, jurisprudence, or enforcement. This can give them the ability to interpret existing legal texts in a way which helps respect both the Eternal Order and ecological integrity.

Non-local intuition and intellect together do not only support open-mindedness and wisdom but also give strength and power to make hard, but heart-based, decisions which are necessary for a more resilient society and environment. These decisions may come with difficulties and conflicts, but with the help of non-local intuition and rationality together, they can be successfully overcome. Even the risks that come from blindly following our heart can be minimized by this solution. This does not mean that a decision made on an intuitive base cannot be accepted by our rationality but that non-local intuition and rationality in an ideal case support each other. With this approach, current structures at both the micro and macro levels may become renewable and flexible enough to support individual and social transformation that realizes ecological integrity.

Reference List

Allinson, W.C. & Hayes, J. (1996) "The cognitive style index: A measure of intuition-analysis for organizational research," *Journal of Management Studies*, 33(1), pp. 119–135.

Bányai, O. (2019) "The foundation of an upcoming civilization able to reach its fulfillment within the ecological limits of the Earth: The eternal order," *World Futures*, 75(5-6), pp. 298–323. Available at: https://doi.org/10.1080/02604027.2019.1591812.

Bányai, O. (2020) "Quantum mechanics and law: What does quantum mechanics teach us?" in: L. Westra, K. Bosselmann, & M. Fermeglia (eds) *Perspective on Ecological Integrity in Science and Law*, pp. 147–157, Cham: Springer.

Bányai, O. (2021) "The theory of worldviews from an environmental law perspective," *World Futures*, 77(8), pp. 613–635.

Bányai, O. (2022) "A theory of quantum social change: Individual responsibility for ecological sustainability," in: L. Westra, K. Gwiazdon, & G. Garver (eds) *The Law and Politics of Ecological Integrity*, Cham: Springer, forthcoming.

Bradley, R.T. (2007) "The psychophysiology of intuition: A quantum-holographic theory of non-local communication," *World Futures*, 63(2), pp. 61-97.

Bradley, R.T., Gillin, M., McCraty, R., & Atkinson, M. (2011) "Non-local intuition in entrepreneurs and non-entrepreneurs: Results of two experiments using electrophysiological measures," *International Journal of Entrepreneurship Small Business*, 12(3), pp. 343–372.

Dane, E. (2011) "Capturing intuitions 'in flight': Observations from research on attention and mind-fulness," in: M. Sinclair (ed) *Handbook of Intuition Research*, pp. 217–226, Cheltenham: Edward Elgar.

Davis, D.R. Jr. (2004a) "Traditional Hindu law in the guise of postmodernism: A review article," *Michigan Journal of International Law*, 25(3), pp. 735–749.

Davis, D.R. Jr. (2007) "Atmatusti as a source of dharma," *Journal of the American Oriental Society*, 127(3), pp. 279–296.

Davis, D.R. Jr. (2010) *The Spirit of Hindu Law*, Cambridge: Cambridge University Press.

Davis, S. (2004b) "The myth of the rational decision maker: A framework for applying and enhancing heuristic and intuitive decision making by school leaders," *Journal of School Leadership*, 14(6), pp. 621–652.

Deák, D. (2022) "Physical, Social, Ecological, and Spiritual Homes, and Their Legal Significance," *World Futures*. Available at: https://doi.org/10.1080/02604027.2022.2101856.

De Witt, A. (2014) "The integrative worldview and its potential for sustainable societies," *Worldviews*, 18(3), pp. 191–229.

Dossey, L. (2017) "Consciousness and TDVP: Welcome to a new world," *Explore*, 13(4), pp. 233–237.

Eardley, P. (2021) "Medieval Theories of Conscience," in: E. Zalta (ed) *The Stanford Encyclopedia of Philosophy*. Available at: https://plato.stanford.edu/archives/spr2021/entries/conscience-medieval.

Earth Charter (2000) Available at: https://earthcharter.org/library/the-earth-charter-text/.

Ebrahimi, A., Keykhosrovani, M., Dehghani, M., & Javdan, M. (2012) "Investigating the relationship between resiliency, spiritual intelligence and mental health of a group of undergraduate students," *Life Science Journal*, 9 pp. 67–70.

Fodor, L. (2019) *A falu füstje. A települési önkormányzatok és a környezet védelme a 21. Század eleji Magyarországon* [The smoke of the village. Municipalities and the protection of the environment in Hungary of the early 21st century], Budapest: Gondolat Kiadó.

Frecska, E. (2012) "Non-locality and intuition as the second foundation of knowledge," *NeuroQuantology*, 10(3), pp. 537–546.

Frecska, E., et al. (2016) "Non-locality and the shamanic state of consciousness," *NeuroQuantology*, 2, pp. 155–165.

Hiley, B.J. & Pylkkänen, P. (2005) "Can mind affect matter via active information?" *Mind and Matter*, 3(2), pp. 7–27.

Koller, J.M. (1972) "Dharma: An expression of universal order," *Philosophy East and West*, 22(2), pp. 131–144.

La Pirra, F. & Gillin, M. (2006) "Non-local intuition and the performance of serial entrepreneurs," *International Journal of Entrepreneurship and Small Business*, 2006/3(1), pp. 17–35.

László, E. (2014) *The Science and Akashic Field. An Integral Theory of Everything*, Rochester: Inner Tradition.

Marcer, P. & Schempp, W. (1997) "Model of the neuron working by quantum holography," *Informatica*, 21, pp. 519–534.

Marcer, P. & Schempp, W. (1998) "The brain as a conscious system," *International Journal of General Systems*, 27, pp. 231–248.

McCraty, R. (2015) *The Science of the Heart. An Overview of Research Conducted by Heart Math Institute*, 2nd ed., Boulder Creek: HeartMath Institute.

McCraty, R., Atkinson, M., & Bradley, R.T. (2004) "Electrophysiological evidence of intuition: Part 2. A system-wide process?" *Journal of Alternative and Complementary Medicine*, 0(2), pp. 325–336.

McCraty, R. & Zayas, M. (2014) "Intuitive intelligence, self-regulation, and lifting consciousness," *Global Advances in Health and Medicine*, 3(2), pp. 56–65.

Mitchell, E. (2000) "Nature's mind: the quantum hologram," *International Journal of Computing Anticipatory Systems*, 7, pp. 295–312.

Nita, M.A. (2014) "From emotional intelligence to spiritual intelligence in public administration," *Juridical Current*, 56(1), pp. 165–181.

Pope Benedict XVI (2011, September 23) *The Listening Heart: Reflections on the Foundations of Law*. Available at: www.abc.net.au/religion/the-listening-heart-reflections-on-the-foundations-of-law/10101114.

Radin, D. (2008) "Testing non-local observation as a source of intuitive knowledge," *Explore*, 4(1), pp. 25–35.

Raftopoulos, M. (2011) "It's that knowing that you are not alone': The role of spirituality in adolescent resilience," *International Journal of Children's Spirituality*, 16(2), pp. 151–167.
Rauscher, E. & Targ, R. (2001) "The speed of thought: Investigation of a complex spacetime metric to describe psychic phenomenon," *Journal of Scientific Exploration*, 15, pp. 331-354.
Rezaei, S., Mirzaei, M., & Zali, M.R. (2014) "Non-local intuition: Replication and Paired-subjects enhancement effects," *Global Advances in Health Medicine*, 3(2), pp. 5–15.
Schwartz, G., et al. (2017) "The academy for the advancement of postmaterialist sciences: Integrating consciousness into mainstream science," *Explore, The Journal of Science and Healing*, 14(2).
Shapiro, Y. & Marks-Tarlow, T. (2021) "Varieties of clinical intuition: Explicit, implicit, and non-local neurodynamics," *Psychoanalitic Dialogues*, 31(3), pp. 262–281.
Sharda, N. (2016) "How do spirituality, intuition and entrepreneurship go together?" *Philosophy of Management*, 15, pp. 65–82.
Sinclair, M. (2011) "An integrated framework of intuition," in: M. Sinclair (ed) *Handbook of Intuition Research*, pp. 3–16, Cheltenham: Edward Elgar.
Sipman, G., Thölke, J., Martens, R., & McKenney, S. (2019) "The role of intuition in pedagogical tact: Educator views," *British Educational Research Journal*, 45(6), pp. 1186–1202.
Sowerby, D.F. (2001) *The Light of Inner Guidance: A Heuristic Study of the Recognition and Interpretation of Intuition*, Dissertation, Institute of Transpersonal Psychology, Palo Alto, CA.
Tiller, W.A., Dibble, W.E. Jr., & Kohane, M.J. (2001) *Conscious Acts of Creation: The Emergence of a New Physics*, Walnut Creek: Pavior Publishing.
Van Egmond, K. & De Vries, B.J.M. (2011) "Sustainability: The search for the integral worldview," *Futures*, 43(8), pp. 853–867. Available at: https://doi.org/10.1016/j.futures.2011.05.027.
Walach, H. (2020) "Naturalizing religion, spiritualizing science: The role of consciousness research," *Journal of Consciousness Studies*, 27(7–8), pp. 165–194.

Chapter 23

The Earth Charter Movement for Social Democracy and an Ecological Economy

Peter Venton

Introduction

By 2020, the 2000 Earth Charter Vision for democracy and stewardship of the Earth's environmental and ecological systems had completely failed in the United States of America (US; America). The US capitalist economy was simply broken. The increase in US GDP that had been powered by energy from burning fossil fuels since 1970 had increased greenhouse gas emissions to the point of a climate crisis which could only be averted with a radical downsizing and restructuring of the US economy – according to the United Nations. Over the same period, a huge increase in the inequality of personal wealth among Americans occurred. The share of personal wealth of the richest 10% of Americans increased from 28% in 1970 to 50% in 2010, while the share of wealth for the next 40% of wealthiest Americans had declined to 30%, and the share of the bottom half of wealthy Americans to 20% (Piketty 2014: 249, 324).

In 2021, I was introduced to the book titled *Doughnut Economics: Seven Ways to Think Like a 21st Century Economist* written by economist Kate Rayworth, who teaches at Oxford University's Environmental Change Institute. The book features a diagram of two rings in the shape of an American-style doughnut: an outer ecological ceiling ring depicting limits on the Earth's capacity to support current human population and species, and an inner social foundation ring for a democracy. The book comprises prescriptions by imagined 21st-century economists for first realizing a democracy, which then manages and controls the capitalist economy to render it an "ecological economy." This sounded to me like the vision of the Earth Charter.

The ecological ceiling ring has several elements, one of which is climate change, which has "transgressed" the top of it. To correct this transgression, the economists propose a carbon tax and regulation of greenhouse gas emissions. They also propose regeneration policies to reduce the material waste from capitalist production that leaves material footprints and regeneration policies for biological waste by recycling it through the living Earth. The economists also propose a series of policies for management of the economy

DOI: 10.4324/9781003440871-27

without any growth in GDP. One of these is to reduce the workweek to maintain full employment in the wake of technological change.

The social foundation ring includes several elements of the material basics of human life plus principles of equality and social justice in a democracy. Here the economists prescribe designs to redistribute wealth and income for more economic equality. These include special tax rates on personal wealth of the 1% rich, more progressive tax rates on employment income and corporate income as well as private wealth, the end of "faux" intellectual property patents, and new kinds of businesses, such as cooperatives and "stakeholder capitalism," where labor shares some of the corporate profits. And there are measures for reducing global inequity between high-income countries and less-developed countries in the Global South. All the foregoing designs are expected to correct the transgressions from the two rings so as to bring human society into the "safe and just space" between the rings of the Doughnut (Rayworth 2017: 38).

From the Earth Charter to Doughnut Economics

In his prologue to the book *The Earth Charter, Ecological Integrity and Social Movements*, J. Ronald Engel wrote that North American theological ethicist Douglas Sturm proposes that the Earth Charter is a summons to extend a first axial age that was grounded in a moral system for radical democratic human freedom, equality, and solidarity. The first axial age was a vision of the sacredness of human life and a universal human community announced by the prophets (Confucius, Lao-Tzu, the Upanishads, the Buddha, Zarathustra, Plato, Isaiah, Jeremiah, and Jesus of Nazareth), and it was extended in a second axial age, where humans joined other species in solidarity in a social democracy, but with the purpose of preserving the planet's ecological and environmental systems. In other words, it contemplated an ideal social democracy that included other species and added the paramount purpose of stewardship over the planet's ecological system and environmental commons (Engel 2014: xvi).

Fourteen years after the launch of the Earth Charter in 2000, Engel briefly describes a downward spiral of accelerated rates of increases in greenhouse gas emissions and climate change, increasing pollution, accelerating loss of biological diversity, increasing difference in per capita incomes among countries, expanding world population, and rising military conflict casualties. He finds it irrefutable that the primary agencies responsible for the downward spiral are corporate economic interests in alliance with government and military interests (the so-called "military–surveillance–police state industrial complex") that have successfully seduced the masses of world citizens. They have built a hierarchy of wealth, power, and privilege that now effectively governs our world civilization. The ideologies of these groups include the fantasy of unlimited expansion of economic activity and population in a world without limits and the dogma that unregulated private free market organization of economic life is the only kind that is practical regardless of the human or ecological consequences (Engel 2014: xxi).

On Capitalism

J. Ronald Engel's stark assessment is about capitalism, which reminded me of the Munk Debate on capitalism held in Toronto on 4 December 2019. The motion was: "Be it resolved: The capitalist system is broken. It's time to try something different" (Griffiths 2019: 40). Greek economist and former politician Yanis Varoufakis led off the debate arguing for the motion with the following statements:

> The problem with capitalism is that it is particularly inefficient at using the fantastic technologies and wealth that it produces. . . . and really, existing capitalism is profoundly anti-democratic. We have captains of industry and masters of finance who accumulate war chests with which they effectively buy politics, buy campaigns and capture regulators. Ecological destruction is an essential aspect of this technostructure as markets were never designed to protect public goods like the environment. They were only designed to create private goods. There is no way that this corporatized, financialized capitalism that we live in can ever value the scarce resources of our nature, of our environment. This capitalism will assign a near-zero price to it and therefore deplete it before it is gone.
>
> (Griffiths 2019: 48–49)

Doughnut Economics

On 22 November 2021, I read an article on the Common Dreams website in Maine USA titled "Can a Doughnut Heal Our World?" by American philosopher and Theravada monk Bhikkhu Bodi. Bodi wrote that the human community faces two momentous challenges today that we must address: One is to establish the social and economic conditions necessary for everyone on the planet to flourish, to live with dignity and purpose and fulfill their life's potential. The other is to safeguard the natural environment from the callous harm caused by unconstrained extraction and consumption. He suggested that British economist Kate Rayworth has proposed an alternative economic model in a clear visual image which illustrates the goalposts we must pursue in order to jump over the hurdles. She has designed a model she calls "Doughnut economics" (Bodhi 2021: 1–3).

The Doughnut

The diagram of the Doughnut features two concentric segmented rings in the shape of an American-style doughnut (Raworth 2017: 44). The outer dark-green *ecological ring* has nine segments described as area boundaries that represent the limits for sustaining the Earth's ecological environment for humans in future millennia. The names of these areas are listed in Table 23.1.

Four of the area boundaries – climate change, nitrogen and phosphorous loading, land conversion, and biodiversity loss – have been transgressed. Three others – ozone layer depletion, ocean acidification, and freshwater withdrawals – have not yet reached their limit. Two others – chemical pollution and air pollution – have not been quantified yet (Rayworth 2017: 258).

The inner *social foundation* ring has 12 segments that represent the basics of life on which no one should be left falling short. These include sufficient food, clean water, and decent sanitation; access to energy and clean cooking facilities; access to education and to health care; decent housing, a minimum income, and decent work; and access to networks of information and to networks of social support. Furthermore, it calls for achieving these with gender equality, social equity, political voice, and peace and justice. They are all included in the United Nation's Sustainable Development Goals, agreed to by 193 member countries in 2015, and the vast majority of these goals are to be achieved by 2030 (Rayworth 2017: 39). Statistical indicators of shortfalls in each of the 12 elements of the social foundation are also presented (Rayworth 2017: 255). They are percentages of countries in the world falling short of indicators as distinct from high-income countries that are the focus of the book.

Political voice includes the right and capacity of citizens to engage in public debate with respect to the development of legislative policies and to hold the executive of the government accountable to its administration and actions (Rayworth 2017: 67). It comprises institutions that give many people a say in decision-making that offset the voice of the privileged few and reduce the threat of authoritarian-led states. To avoid tyranny of the state and the tyranny of free markets, democratic politics are key, as they reinforce the foundational role played by society in generating the civic engagement needed for participation and for holding political representatives in the executive of the government accountable (Rayworth 2017: 74).

Table 23.1 Ecological ring area boundary names

1	Pollution
2	Biodiversity loss
3	Land conversion
4	Freshwater withdrawals
5	Nitrogen and phosphorous loading
6	Chemical pollution
7	Ocean acidification
8	Climate change
9	Ozone layer depletion

Seven Prescriptions for 21st-Century Economists

Change the Goal From GDP to the Doughnut

The first prescription is to change the goal of GDP to the Doughnut. *GDP* is a measure of economic output of goods and services of the traditional economic system within the state. It has been the *de facto* paramount goal of nations at least since the 1960s. The Doughnut includes the 9 elements (goals) of the "ecological ceiling" plus the 12 elements (goals) of the "social foundation," which represent, respectively, the goals of an ecological economy and a social democracy.

See the Big Picture

The big picture of three concentric circles is not shown for lack of a copyright. It shows how the smallest ring circle, labeled "economy," is comprised of four quadrants: the state, the commons, the markets, and the households. The economy is embedded in a larger ring, labeled "society," which, in turn, is embedded in an even larger ring, labeled "Earth." *Solar energy* is a line of arrows that penetrate the center of the entire diagram of concentric circles from the left to the right and pass through the center, generating waste and heat in the economy that exits right outside the "Earth," presumably as greenhouse gas emissions.

My interpretation of the diagram is as follows. If the economy breaches society, it means that society does not have democratic control over the economy. If society breaches the ecological ceiling, it will not be "safe." If society breaches the social foundation, it will not be a just society. Further, the big picture diagram is not only about the state; it is also about the world, and hence global. In other words, there are two economies: "nation-state" and "global." The big picture is thus about the World Federalist Movement for global governance over the commons and the global economy, as presented in the book *Global Democracy: The Key to Global Justice*, published by Democracy Without Borders in Berlin (Gilad & Freeman 2022).

Nurture Human Nature

There is growing use of monetary incentives in policies at ending human deprivation and ecological degradation. These extrinsic values tend to crowd out the intrinsic moral values. Twenty-first-century economists believe that humans have to learn more about our intrinsic values and morals – as distinct from extrinsic market prices in the neoliberal capitalist system. As French economist Thomas Piketty puts it, the price system knows neither limits nor morality (Piketty 2014: 6). The idea of intrinsic values is a central theme of Canadian economist and former governor of the Banks of Canda England Mark Carney's book *Value(s) Building a Better World for All* (Carney 2021).

Carney argues that putting a price on every human activity erodes certain moral and civic goods and that we have moved from a market economy to a market society which is now undermining our basic social contract of relative equality of outcomes, equality of opportunity, and fairness across generations (Carney 2021: 5). These values are obviously principles of democracy that can be found in the social foundation of the Doughnut.

Twenty-first-century economists are integrating the academic disciplines of ecology, psychology, and behavioral sciences into the traditional academic disciplines of economics, political science, and sociology. These economists are expected to introduce "nudges" that tap into underlying norms of the citizenry to change individualistic, self-interested citizens into more social and reciprocating citizens (Raworth 2017: 105). So here is the essence of the tacit social contract in which citizens give up some of their interests in return for reciprocity in the form of government benefits and policies that protect not only them but also the majority of their fellow citizens.

Get Savvy With Systems

There are two complex systems to understand: the *laissez-faire* economic system and the much different ecological system. On top of this, the ecological system intersects with the economic system – an intersection that adds to the overall complexity of systems analysis of the two (Rayworth 2017: 118). As Derek Paul writes in his book *A Leap to an Ecological Economy*, the market pricing system in the neoliberal capitalist economy is hugely different from and antiethical to the intrinsic valuation of natural resources and ecological systems.

The *laissez-faire* capitalist economic system is prone to overproduction, less-than-full employment, massive over-financialization of the economy, and periodic instability that requires extra GDP to "even things up," as Rayworth explains in the title of Chapter 7, "Be Agnostic about Growth." And economic growth impinges severely on the environmental system, as illustrated in the diagram of the Doughnut.

A systems perspective makes clear that the prevailing direction of global economic development is caught in the twin dynamics of growing social inequality and deepening ecological degradation. These trends echo the conditions under which earlier civilizations, such as the Easter Islanders and the Greenland Norse, collapsed, as Jared Diamond explained in his 2002 book *Collapse: how societies choose to fail or survive* (Rayworth 2017: 132, citing Diamond 2002 in note 42).

More recently, the 1972 MIT study "The Limits of Growth" created one of the first dynamic computer models of the global economy, known as World 3. It explored a range of economic scenarios up to 2100, taking five factors they saw as determining and ultimately limiting output growth: population, agricultural production, natural resources, industrial production, and pollution. According to their projections for the business-as-usual model, as global population and output expanded, nonrenewable resources, such as oil, minerals, and metal, would be depleted, leading to a drop in industry output and food production, ultimately resulting in famine, a large fall in human population, and greatly reduced living standards for all. Recent data comparison with the 1972 model fund shows that the global economy appears to be tracking the business-as-usual scenario, and that doesn't end well. If we are to avoid such a fate for our global civilization, we clearly need a transformation for a "regenerative and distributive economy." We now turn to the review of Thomas Piketty's innovative analysis of capitalism in his book *Capital in the Twenty First* Century (Venton 2018), followed by Kate Rayworth's comments on designs for redistribution.

Thomas Piketty on Design to Distribute

Piketty's innovation is to separate the analysis of economic distribution into two components of income and wealth and their interrelationships. It also has a more sophisticated historical analysis done in the 20th and, indeed, the 19th century, assisted by a massive historical database. The review is organized under several issues or questions.

Why Capitalist Economic Systems Do Not Self-Correct

A common 20th-century justification for the capitalist market economies in circumstances of growing inequality of wealth and income is that the market will correct by itself in the sense that the rate of return on capital will fall with the result that inequality will be reversed or that technological change will somehow make things better. That is an illusion, because it is political forces that are central to ensuring that the wages of human capital will triumph over the income from capital. Furthermore, technology and markets will not necessarily prevent growing inequality because of the caprice of technology. Some have argued that proper investment in training and skills and the diffusion of knowledge will work in labor markets to enhance social mobility and reduce unequal income to a degree. Piketty concludes, with respect to the United States, that there is little correlation between spending on public education and social mobility.

An Empirical Law Explaining Why the Inequality of Wealth Keeps Increasing

Inequality of wealth in developed capitalist countries increases indefinitely over time in circumstances where the rate of return on capital (i.e., wealth) is greater than the rate of growth in national income.

Why Is It Undemocratic When Inherited Wealth Grows to Dominate the Economy?

When the rate of return on capital significantly exceeds the growth rate of income, it logically follows that inherited wealth grows faster than income. Democratic societies view that inequality should be based on effort and merit, rather than kinship, which requires no effort. Inherited wealth based on kinship is not meritorious. Furthermore, rentiers are the enemy of democracy (Piketty 2014: 422).

Inequality of Income From Labor Can Be Reduced With Little Loss in Economic Output

Inequality of income from labor was limited in the mid-20th century in the United States, for example, when top marginal income tax rates were 94%. In that situation, senior executives had little incentive to bargain with directors for higher income because most of the income would be taxed away. The result of the massive reduction of top marginal income tax rates largely from 1965 to 1990 resulted in an explosion of income of senior executives over this period. Moreover, Piketty's analysis of 3,000 firms in 14 countries concluded that higher compensation of senior executives was largely unrelated to their personal productivity and, hence, GDP. Therefore, he suggested top marginal rates of 80% would not impede economic growth.

Inequality in the Global Capitalist System Can Be Fixed With a Modest Progressive Tax on Capital

In Canadian dollars, for example, suggested rates of tax on wealth are 1% on wealth between $1.4 million and $7.0 million, and 2.0% on wealth in excess of $7 million. The revenue from these rates is equivalent to a tax of 10% and 20% on income from capital for the next ten years. Assuming a 5% rate of return on wealth, these rates would generate $700 to $35,000 annually for the first bracket, and $70,00 and up for the second bracket. A billionaire would pay $10 million per annum.

Democratic Control Over Capitalism in the Global Economy Has Been Lost

To gain control, it must ideally organize a progressive global tax on capital, coupled with a very high level of international financial transparency, to avoid tax evasion. Such a tax would avoid the "endless inegalitarian spiral and control the worrisome dynamics of global capital concentration" (Piketty 2014: 515, 516). The international tax is technically feasible.

Addressing Climate Change Is a Risky Proposition With Potentially Huge Returns

According to British economist Nicholas Stern, heading off climate change justifies spending five points of global GDP a year right now to attempt to mitigate climate change in the future. This represents spending by the rich countries on a far vaster scale than the 1 to 1.5% of GDP that they have been spending over the last several decades (Piketty 2014: 569, 654).

Kate Rayworth on Design to Distribute

As noted earlier in this chapter, the central purpose of economic redistribution is to reduce inequality to a moderate level. This is a key policy for restoring the democracy that the capitalist economic system has destroyed. Moderate inequality is a key element of the common good of the democracy. Social welfare is also positioned as a key element by Richard Wilkinson and Kate Picket in their 2009 book *The Spirit Level* (Wilkinson & Pickett 2009). Beyond certain levels of economic income, there is little evidence that further increases bring materially more happiness (Rayworth 2017: 146). So there is no collateral damage to the economy from the redistribution of income through taxation. The main policy for reducing inequality of income includes higher top marginal tax rates on employment and corporate income. But instead of focusing on redistributing income, 21st-century economists will also seek to redistribute wealth (Rayworth 2017: 173).

While redistributive policies for income can be life-changing for those who benefit from them, they still may not get to the root of economic inequalities because they do not focus on redistributing the wealth that generates the income. Distributive design has an unprecedented opportunity to transform the dynamics of wealth ownership of land, to create money, to run enterprises, to employ technology, and to share knowledge (Rayworth 2017: 151). These include taxation of the value of land; land owned by the community; controlled low interest rates for investors, financiers, and commercial bankers, who are the largest rentiers. Investors can be controlled by central banks that require reserves of 100%, or state-owned banks, where profits are shared with the public or lower interest rates for rentiers – which Keynes called "functionless investor." Enterprises that allow labor to own their shares – an example of "shareholder capitalism" (Rayworth 2017: 196). Not-for-profit benefit companies or community interest companies (Rayworth 2017: 160). Monopolistic owners of robots and internet companies like Google, YouTube, Apple, Facebook, eBay, PayPal, and Amazon, digital monopolies at the heart of the network world, should be broken up or regulated or taxed on their excessive unfair profits (Rayworth 2017: 169), and owners of ideas through patents may be replaced by open-source design that promises large social benefits (Rayworth 2017: 166).

In conclusion, Rayworth believes that 21st-century economists will design distributive flow into economic institutions and harness the market, the commons, and the state alike to make it happen. In this regard, they will work with bottom-up networks that are already driving a revolution in redistribution (Rayworth 2017: 173).

Create to Regenerate

Creating to regenerate is a process of designing products which contain materials in a way that enables the materials to be reused after the products have become obsolete or have finished their useful lives. Planned obsolescence is one of the features of the neoliberal capitalist system, about which consumers complain, but it is also a problem for an ecological economy, as explained by retired physicist Derek Paul in his book *A Leap to an Ecological Economy* (Paul 2020). Paul states that "[t]he case has been made in the literature . . . that humanity needs to reduce its material expectations because its ecological footprint now exceeds its bio-capacity. Re-use and recycling are essential processes for a truly ecological economy" (Paul 2020: 72–73). As an example, Paul cites the "re-use of materials in computers" as an obvious step toward sustainability (Paul 2020: 69–70).

Creating to regenerate is a process for redesigning materials for reuse in a "circular butterfly economy," which is depicted in a diagram on page 188 and titled "The butterfly economy: regenerative by design." The butterfly circular economy has two wings, technical and biological. On the biological wing, all nutrients are eventually consumed and regenerated through the living Earth. An example is in the production of coffee, where leftover coffee grounds have nutrients that are ideal for growing mushrooms that can be used as feed for cattle, chickens, and pigs and so are returned to the soil as manure (Rayworth 2017: 188).

On the technical wing, products are made using technical nutrients, such as metal and synthetic fiber, that do not normally decompose (Rayworth 2017: 189). In today's technical wing of the circular butterfly economy, mobile phones, which are chock-full of gold, silver, cobalt, and rare Earth materials, are typically used for just two years. New computers could be designed for easy collection and disassembly that would lead to the refurbishment and resale of all their material parts at the end of their useful lives and could be reused in other products (Rayworth 2017: 189).

The problem with pursuing regenerative industrial design is that it does not work very well when it is based on free market prices for consumer and production goods and finance in the neoliberal capitalist economy. Economist Mark Carney has explained at length that market prices do not represent intrinsic values (Carney 2021). The shift to regenerative economic design can be monitored only if it is backed up by metrics that measure the sustainability values of its mission. One example cited was the living metrics for the city of Oberlin, Ohio, United States. For businesses, three initiatives for metrics were listed: the Economy for the Common Good, B Corp's Impact Reports, and the MultiCapital Scorecard (Rayworth 2017: 204, note 64).

It is time for economists to step up to the design table and take a seat alongside those innovative architects, industrial ecologists, and product designers who are spearheading the revolution in regenerative design. The role of the economists is advising and managing on pricing for regeneration rather than pricing for profits. Together they will unleash the extraordinary potential for the circular economy and regenerative design (Rayworth 2017: 205).

Be Agnostic About Growth

What Rayworth means by being agnostic about growth is to design an economy that promotes human prosperity regardless of whether or not the design will increase GDP (Rayworth 2017: 209). Environmental economist Peter Victor's book *Managing Without Growth: Slower by Design, Not Disaster* (Victor 2008) provides insights into managing growth for mitigating climate change and improving the social foundation of the Doughnut at the same time.

Victor had discovered that during the period of just over 20 years from 1971 to 2004, GDP per capita in the United States grew by 105%, while the genuine progress indicator (GPI), which is taken as an indicator of the common good of the United States, grew by only 5% (Victor 2008: 129). That is because the GPI incorporates negative changes related to the increase in economic inequality and increased pollution of the Earth's atmosphere.

Victor presented a set of policies for reducing greenhouse gas emission through slower economic growth in Canda over a hypothetical projected 30-year period, from 2005 to 2035 (Venton 2020). These included:

1. Eliminate poverty through tax changes to redistribute after-tax income.
2. Reduce the workweek by 15%, for more leisure and less consumption.
3. Invest more in public infrastructure to reduce crowding, and less in the production of private status goods.
4. Introduce a tax on carbon and regulate carbon emissions, as both are complementary.

To address the common fear of economists that lower growth will reduce full employment, which is an important element of the social foundation, John Maynard Keynes anticipated that, as technology increased labor productivity, the typical working week would shorten; he famously predicted that a 15-hour workweek would suffice in the 21st century, and society would then endeavor to make what work there is still to be done as widely shared as possible (Rayworth 2017: 236, note 61).

Rayworth notes:

> [T]he UK's New Economics Foundation plan to shorten the standard paid work week in high income countries to 21 hours as a means of tackling both unemployment and over work. To make this work social policy expert Anna Coote behind the proposal said we would have to get rid of perverse incentives in tax and insurance systems so that employers are encouraged, rather than penalized, for taking on more workers.
>
> (Rayworth 2017: 236, note 63)

This is another possible design for getting people into the safe and just space of the Doughnut.

Conclusion

Doughnut economics sets out an optimistic vision of humanity's common future: a global ecological economy that is created by a combination of distributive and regenerative design and a substitution of the goals of human flourishing in a social democracy for the current capitalist goal of endless increase in GDP and an unsustainable extreme concentration of wealth. This is the way forward toward J. Ronald Engel's Earth Charter Vision. I would venture to suggest that the success in achieving this outcome will depend on how successful 21st-century economists are at nudging human behavior to adopt morals that will blunt the power of markets and collaborate to gain democratic control over the small hierarchy of wealth, power, and privilege that now governs them.

Reference List

Bodhi, B. (2021) "Can a doughnut heal our world?" *Common Dreams*. Available at: www.commondreams.org/views/2021/11/22/can-doughnut-heal-our-world.

Carney, M. (2021) *Value(s) Building a Better World for All*, Toronto: Penguin Random House.

Engel, J.R. (2014) "Prologue: Summons to a new axial age-the promise, limits and future of the Earth Charter," in: L. Westra & M. Vilela (eds) *The Earth Charter, Ecological Integrity and Social Movements*, Abingdon Oxon: Routledge.

Gilad, O. & Freeman, D. (2022) *Global Democracy: The Key to Global Justice*, Berlin: Democracy Without Borders.

Griffiths, R. (2019) *The Future of Capitalism: Katrina Vanden Heuvel, Yanis Varoufakis Vs. Arthur Brooks, David Books*, Toronto: Anansi Press.

Paul, D. (2020) *A Leap to an Ecological Economy*, 2nd. ed., Montreal: Maison Hilton.

Piketty, T. (2014) *Capital in the Twenty-First Century*, trans. by Arthur Goldhamer. Cambridge, MA: The Belknap Press of Harvard University Press.

Raworth, K. (2017) *Doughnut Economics: Seven Ways to Think Like a 21st Century Economist*, White River Junction Vermont: Chelsea Green Publishing.

Venton, P. (2018) *Capital in the Twenty-First Century* by Thomas Piketty: A Review. Available at: https://puwashgroup.ca.
Venton, P. (2020) "The Political Economy of Managing Without Growth," in: L. Westra, K. Bosselmann, & M. Fermeglia (eds) *Ecological Integrity in Science and Law*, New York: Springer Nature.
Victor, P. (2008) *Managing Without Growth: Slower by Design, Not Disaster*, Northampton: Edward Elgar.
Wilkinson, R. & Pickett, K. (2009) *The Spirit Level*, London: Penguin.

Chapter 24

Earth Trusteeship

Unraveling the Mysteries of an Ecological Trust

Justin Sobion

Introduction

Earth trusteeship[1] is defined as the institutionalization of the duty to protect the integrity of the Earth's ecological systems. The concept of maintaining the integrity of ecological systems has existed in international environmental law agreements for over 40 years. Modern scholarship has advanced the prospect of ecological integrity being an eco-centric ethical paradigm that is rooted in reasonableness and natural law (Taylor 2020: 231; Bosselmann 2017: 99). Following this, academics who have studied the notion of ecological integrity argue that the integrity of Earth's life support systems should be promoted as a basic norm (or *grundnorm*) of international law (Kim & Bosselmann 2015: 205).

This chapter introduces the reader to the history leading up to the launch of the Hague Principles which formally established the concept of Earth trusteeship. By critically examining the Background Note and Principle 1 of The Hague Principles, this chapter seeks to identify who are the trustees, the beneficiaries, and the trust *res* under the Earth trusteeship model. This chapter explores the notion of holding the Earth in trust for future persons, and the fact that such an endeavor is already supported in international law, constitutional law, judicial decisions, and scholarly contributions. As a relatively new concept, Earth trusteeship captures an ethical vision or a "new consciousness of law" – one that was envisioned by Nobel laureate Albert Schweitzer (Cicovacki 2009: 111). In the end, I attempt to unravel the mystery of this "ecological trust" and its possible role in future UN reform and the ICJ Advisory Opinion on climate change.

An Introduction to The Hague Principles

The text containing the Hague Principles for a Universal Declaration on Responsibilities for Human Rights and Earth Trusteeship (Hague Principles)

1 For more on Earth trusteeship and the text of The Hague Principles, please visit: www.earth-trusteeship.world.

DOI: 10.4324/9781003440871-28

is the result of a collaboration among representatives from a wide spectrum of environmental, Indigenous, and human rights organizations which came together under the auspices of the Earth Trusteeship Initiative (ETI). The collaboration commenced with an Earth Trusteeship Symposium held at Utrecht University in June 2017. This meeting was followed by further consultations and drafts around human rights and responsibilities which were produced at an Earth trusteeship gathering of about 80 delegates in The Hague exactly one year later. The text containing the Hague Principles was eventually finalized in September 2018, before the convening of its inaugural launch in December that year.

The Hague Principles and the ETI were formally launched on 10 December 2018 at the opening of the Earth Trusteeship Forum, held at the Peace Palace, The Hague, Netherlands. The launch of the Hague Principles was strategically timed to coincide with the commemorations of the 70th anniversary of the Universal Declaration of Human Rights (UDHR) (van Willenswaard 2020b: 308). The text containing the Hague Principles was inspired by the Earth Charter (2000), which was launched 18 years earlier at the same venue (Bosselmann 2018).

During our everyday speech, we often speak about human rights but seldom about human responsibilities. The Hague Principles teach us that human rights are fundamental, but not as fundamental as the responsibilities which we and our political institutions have to the Earth (Bosselmann 2019). If we do not have clean air to breathe, fresh water from the lakes, and food from plants to nourish us, then our lives will swiftly dissipate. Our right to life – the cornerstone of all human rights – will be immediately extinguished. The core message of the Hague Principles is that humans are members of the community of life (or Earth community), and it is this membership which defines what relationship we have with others (Bosselmann 2019). The intention is that the Hague Principles act as a "guide" for drawing up a Universal Declaration on Responsibilities for Human Rights and Earth Trusteeship.

The Content of the Hague Principles

In terms of structure, the text of the Hague Principles can be divided into four distinct parts: (1) the Background Note, (2) the Preamble, (3) the Three Principles, and (4) the Annex. These four parts will be discussed next.

The Background Note

The Hague Principles commence with a Background (or Background Note), comprising four paragraphs, which sets the tone of the intention of the drafters. The Background Note commences, and culminates, with some laudable language concerning the UDHR. In the opening paragraph, the

Background Note celebrates the 70th anniversary of the UDHR and refers to it, and subsequent human rights covenants, as "precious manifestations of the human spirit." The Background Note continues: "The recognition of equal and unalienable rights of all human beings is an indispensable prerequisite for achieving just and sustainable societies." The Background Note recognizes that "we" (human beings) belong to the community of life (or Earth community). As part of this community, we also have rights and responsibilities "for each other, for future generations, for all living beings and the Earth." Notwithstanding this, the Background Note acknowledges that we live in a different world today compared to when the UDHR was adopted in 1945. In the second paragraph, the Background Note heralds this bleak warning:

> The Earth community is in grave danger. Over the last 70 years the human population has tripled and disparities in economic wealth have greatly increased. Moreover, overall consumption of Earth's natural resources by humans has grown at such a rate that the entire Earth system with its interconnected components – the geosphere, the hydrosphere, the atmosphere, and the biosphere – are all now at risk.

Consequently, the Background Note asserts that under the present living conditions, "humanity is threatening the very living conditions that make the enjoyment of civil, political, cultural and economic rights possible." The second paragraph continues:

> Human rights must therefore include responsibilities for the Earth community and the entire Earth system. *These imply obligations and new opportunities for all human beings to act as Earth trustees* [emphasis added].

In the penultimate paragraph, the Background Note explains that over the last 70 years, people and organizations in many countries and cultures have called for the recognition of human responsibilities. The Background Note makes mention of the Annex, which provides a list of 17 instruments that acknowledge human responsibilities. Those instruments listed in the Annex include the American Declaration of the Rights and Duties of Man 1948, the Rio Declaration on Environment and Development 1992, and the Universal Declaration of the Rights of Mother Earth 2010. The final paragraph of the Background Note declares that while the drafters celebrate and embrace the UDHR, they call upon all peoples, the UN, and the international community of states to not only reaffirm their responsibilities for human rights but to also "recognize their responsibilities to the Earth community and the Earth system and to adopt *Earth Trusteeship as an active and all-encompassing way forward*" (emphasis added).

The Preamble

The Preamble of the Hague Principles confirms that it was a group comprising "citizens of nation-states, actors in the world economy and members of global civil society," which was responsible for the drafting and agreeing of the text. This group acknowledges that the well-being of human beings depends on the health and well-being of other beings and ecosystems. The Preamble continues:

> Considering that the totality of beings and ecosystems on Earth forms a community of life (the "Earth community"), [and] Realizing that just as human beings have rights that suit their needs, other beings have the right to exist and flourish according to their specific needs.

The Preamble closes on the understanding that a "new, more mindful, and appropriate relationship with Earth and Nature is necessary for the flourishing of all beings." As a result, the drafters of the text agreed that the Hague Principles be a "guide" for drawing up a Universal Declaration on Responsibilities for Human Rights and Earth Trusteeship. The use of the word "guide" suggests that the Hague Principles was not intended by the drafters to be a legally binding document but, rather, a roadmap for using Earth trusteeship as a tool for future ecological governance.

The Three Principles

The Hague Principles contain three core principles. The first two principles, Principle 1 (Responsibilities for Earth) and Principle 2 (Responsibilities within the Community of Life), can be classified as eco-centric principles. This eco-centric language found under Principle 1 and Principle 2 is lacking in the UDHR text. Principle 3 (Responsibilities for Human Rights) asserts, *inter alia*, that a state is a trustee of the human rights of its citizens and all human beings (Principle 3.2). Principle 3 is more mundane in that it prescribes an anthropocentric "rights-focused" approach where not only the state but also all human beings are responsible for the protection of human rights (Principle 3.1). This objective is similar to that of the UDHR, where, in the Preamble, member states have pledged to achieve the promotion of universal respect for human rights. The Hague Principles, when read as a whole, illustrates that a state's trusteeship responsibilities do not end with human rights. States also have parallel ecological responsibilities which cannot be isolated from their human rights responsibilities (Bosselmann & Taylor 2019: 281). The ability to connect biocentrism and anthropocentrism in one single text remains the most outstanding feature of the Hague Principles (Sobion 2023: 133). This is in stark contrast to our traditional legal system, where it has been submitted that human rights and environmental rights still focus on human well-being

(UN Secretary-General Report – Harmony with Nature 2017: 12). In order to comprehend the future role of the trustees and the intended beneficiaries under Earth trusteeship, this chapter would briefly examine the concept of a trust and whether it reconciles with Principle 1 of the Hague Principles.

Reconciling the "Trust" and Earth Trusteeship

Trusts are rightly regarded as one of the hallmarks of legal systems of the common-law family (Johnston 1988: 1). The conceptual starting point of a "trust" is "a confidence reposed in some other" (Moffat 2009: 1). From this point of view, a "trust" in English law is, in some measure, a translation of the word "trust" as used in ordinary speech (Moffat 2009: 1). When one speaks of "trusteeship," it refers to the office, status, or function of a trustee – the person in whom the confidence is reposed (Garner 2019: 1826). Generally speaking, a *trustee* is someone who acts on behalf of another. When one thinks of "a trustee," other synonyms can come to mind, such as *guardian*, *caretaker*, or *custodian*. The duties of a trustee can range from holding and preserving property on behalf of someone else to managing the affairs of a minor until she/he attains the age of majority. Trusteeship, therefore, is a form of governance that requires a person or entity to act (as a "trustee") on behalf, and for the benefit, of another person or entity ("beneficiary") (Bosselmann 2020: 242).

The concept of Earth trusteeship is formally established under Principle 1 of the Hague Principles, which reads:

> 1.1 All human beings, individually and collectively, share responsibility to protect Nature, of which we are an integral part, the integrity of Earth's ecological systems and Earth as a whole, home of all living beings.
>
> 1.2 Each state individually, and the international community of states collectively, acknowledge that they have, and share, responsibilities for Nature, in cooperation and in alliance with their citizens as *equal trustees of Earth and the integrity of Earth's ecological systems* [emphasis added].

Principle 1.1 confirms not only our deep human connection with nature but also our natural duty to protect it (including the Earth and Earth's ecological systems). Having established our responsibility to nature, Principle 1.2 highlights that both states and human beings share responsibilities as "equal trustees" of the Earth. The use of the phrase "equal trustees" symbolizes a joint, democratic, and participatory process in the ecological management of the planet. The corpus or trust *res* (the body or the subject) of the trust is identified under Principle 1.2 and consists of three elements: (1) nature, (2) Earth, and (3) Earth's ecological systems. The trust *res* consists of ecological matter.

Principle 1.2, therefore, is the cornerstone of Earth trusteeship. It creates a specialized form of a trust, something which is akin to an ecological trust.

It can be argued that even with the presence of Principle 1.2, the concept of Earth trusteeship and what it sets out to achieve is not adequately defined. According to the Background Note, "[h]uman beings must therefore act as Earth trustees individually and collectively through new arrangements of Earth Trusteeship at all levels." But what exactly is *Earth trusteeship*, and how are these "new arrangements" to be realized?

Earth trusteeship is the institutionalization of the duty to protect the integrity of the Earth's ecological systems (Bosselmann 2021: 36). The idea of protecting the integrity of Earth's ecological systems is not novel in academic scholarship and law. The duty to protect the integrity of ecological systems is expressed in no fewer than 27 international environmental agreements (Bosselmann 2023) over the period of more than 40 years (from the Preamble of the 1980 Convention on the Conservation of Antarctic Marine Living Resources to the recent High Seas Treaty). Articles 7(f) and (h) of the High Seas Treaty (2023) provide that states parties shall be guided by the "ecosystem approach," which "builds ecosystems resilience" through the maintenance and restoration of "ecosystem integrity" within the oceans. By maintaining the ecological integrity of the Earth, we are essentially maintaining the integrity of Earth's life support systems (Kim & Bosselmann 2015). For Earth trusteeship (and the duties that are attached to it) to be institutionalized would require some norm-setting activity which integrates the concept into our way of thinking, our culture, and our philosophy of life. From this perspective, using the Hague Principles as a guide is a step in the right direction. It is important to note that in the given definition, Bosselmann combines all three subjects of the trust (nature, the Earth, and Earth's ecological systems) under one broad heading: Earth's ecological systems. By so doing, greater emphasis is placed on the ecological nature of the trust, thereby justifying Earth trusteeship as an ecological trust.

Having defined who the trustees are and what is the trust *res* is, a natural, yet bold, question that arises is, "Who owns the Earth?" In response to this, a staunch advocate for Earth trusteeship would claim, "We all do. Or more correctly, none of us does." It would be outrageous for the proponents of Earth trusteeship (or anyone else, for that matter) to accept that human beings and/or states own the Earth. The Earth, like the atmosphere or the oceans, has been long cited as being *res nullius* (nobody's thing) and, therefore, incapable of being owned by anyone. Van Willenswaard, one of the stalwarts for Earth trusteeship, presents a new paradigm of ownership which is diametrically opposed to the conventional Western property model. He labels this new paradigm as "ownership" for the benefit of others, which includes the well-being of future generations (van Willenswaard 2016: 206). In his reflections, van Willenswaard considers Earth trusteeship as a governance and legal concept that emphasizes a "joy of duty" and a "powerful realization of altruism" (van Willenswaard 2020a). The idea that a trustee's obligation is grounded in altruism has also been supported by a leading scholar, Peter Birks. According to Birks, a trustee's obligation requires the highest degree of altruism (Birks 2000: 14, 29). Birks defined *altruism* as action in the interests of another or the disposition to act in the interests of another (Birks 2000: 14). So even if Principle 1.2 asserts that humans and states are joint trustees for nature, the Earth, and Earth's ecological systems, this does not mean that they own these matters that are the subject to the trust. As trustees, they merely hold and preserve the *corpus* of the trust for the benefit of others.

Earth trusteeship therefore requires us to rethink the way in which we view traditional property rights (van Willenswaard 2020b: 265–267). By so doing, Earth trusteeship does not intend to dispense with traditional private ownership but to add an essential dimension of service based on intrinsic ownership obligations for natural spaces which are meant to benefit all – such as the Earth, nature, and the global commons (van Willenswaard 2020b). This selfless nature of the trustee's office should be an integral feature of Earth trusteeship. For it to work, Earth trusteeship requires the trustees to act selflessly and in the interests of others when caring for the Earth – an object which does not belong to them.

Who Are the Beneficiaries of the Trust?

Although Principle 1.2 of the Hague Principles designates the trustees, the text remains unclear as to who the beneficiaries are. In interpreting who the beneficiaries are, the Background Note can assist:

> We all belong to the community of life, the Earth community, which determines what rights and responsibilities we must recognize and honour *for each other, for future generations, for all living beings and the Earth, our home* [emphasis added].

The phrase "future generations" – which is mentioned only once within the entire Hague Principles text – could be interpreted as minors or generations unborn (or both). However, since the words "each other" precede "future generations," a strict statutory interpretation of the sentence conveys that the former alludes to the present generation (which includes minors), and the latter only refers to those who are unborn. Using the Background Note as the reference point, and reading the Hague Principles in its entirety, human beings have trusteeship responsibilities for each other (present generation), for future generations (the unborn generations), for all living beings (living non-humans), and for the Earth (which is considered a living non-human). These are the assumed beneficiaries of Earth trusteeship. By holding the Earth in trust for those mentioned, all would benefit – including the Earth itself. This makes the Hague Principles more interesting and perhaps more eco-centric than originally imagined.

Why Should We Hold the Earth in Trust for Future Persons?

The concern for future generations is a familiar feature in international and environmental law (Bosselmann 2013: 107). Our responsibility to conserve and protect the environment or the climate for the benefit of future generations is highlighted in a number of treaties and international law instruments, including the Stockholm Declaration 1972, the Rio Declaration 1992, the World Charter for Nature 1982, the Convention on Biological Diversity 1993, and the United Nations Framework Convention on Climate Change (UNFCCC) 1994. The Earth Charter, which inspired the Hague Principles, declares (in its Preamble) that "[t]he protection of Earth's vitality, diversity, and beauty is a sacred trust" and (at Principle I.4) that the Earth's "bounty and beauty" must be secured "for present and future generations." This language found in the Earth Charter represents the epitome of what Earth trusteeship seeks to fulfil in terms of planetary and ecological governance.

There are judicial decisions that subscribe to the philosophy of holding the Earth on trust for others. The South African Constitutional Court, in *Fuel Retailers Association of Southern Africa v. Director-General Environmental Management*, held that "[t]he present generation holds the earth in trust for the next generation. This trusteeship position carries with it the responsibility to look after the environment." Similarly, the High Court in India, in *Miglani v. State of Uttarakhand*, acknowledged that because the past generation handed over the Earth to the present generation in its pristine glory, the present generation is morally bound to reciprocate this to the future generation. The High Court, in *Miglani v. State of Uttarakhand*, touched upon the theory of intergenerational equity, which is widely discussed in Edith Brown Weiss's influential text *In Fairness to Future Generations*. This theory postulates that each generation is entitled to inherit a planet at least as good as that of previous generations. In other words, all generations are entitled to at least the minimum level that the first generation in time had (Weiss 1989: 24–25).

On this subject of future generations, the scholarly work of Behrens, a South African professor in bioethics, is also relevant. Behrens explains that the present generation should express gratitude to its predecessors for preserving the environment on its behalf, by emulating its predecessors and preserving the environment for future generations. This sense of duty stems from the notion of moral obligations toward an unidentifiable, contingent group of future persons (Behrens 2012: 179). Since future generations have no voice to speak for themselves, it is the conscience of the present generation that needs to keep their welfare constantly in view (Weeramantry 2009: 254). As Behrens asserts, "no sound environmental ethic can now ignore the question of what (if anything) we owe to posterity" (Behrens 2012: 180). Behrens admits that although the notion that we owe something morally to unborn generations is almost obvious by Africans, it could be contested in Western philosophy (Behrens 2012: 181). Behrens strongly relies on the scholarly contribution of Kwasi Wiredu – a prominent Kenyan philosopher – to support his views (Behrens 2012: 181). Wiredu wrote:

> Of all the duties owed to the ancestors none is more imperious than that of husbanding the resources of the land so as to leave it in good shape for posterity. In this moral scheme the rights of the unborn play such a cardinal role that any traditional African would be nonplussed by the debate in Western philosophy as to the existence of such rights.

Weiss accepts that there are cultural traditions that recognize human obligations to conserve the planet for future generations. Such cultural traditions are deeply rooted in, for example, Islamic law, in African customary law, and in Asian nontheistic traditions (Weiss 1989: 18). Weiss also invokes the socialist legal tradition of Marx, who shared the view that communities are only possessors (not owners) of the Earth, with obligations to protect the Earth for future generations (Weiss 1989: 19–20). In Weiss's *In Fairness to Future Generations*, she recognized that we all share one Earth and have a fiduciary obligation to conserve it for future generations (Weiss 1989: 2–3). Recognizing that global security is at stake, Weiss revisited the scope of her original planetary trust to explicitly encompass other diverse and increasing threats to the planet, including nuclear weapons, biotechnology, cybertechnology, artificial intelligence, and possible pandemics (Weiss 2020: 449–450).

Holding the Earth in trust for future generations is also in line with the Buddhist philosophy. Weeramantry, in his extrajudicial writings, shares the Buddhists' ethical perspective, which calls upon us to re-examine what kind of world we have inherited, what we are responsible for, and what we will pass on to the coming generations (Weeramantry 2009: 138). This ethical perspective has been enshrined in the Constitution of the Kingdom of Bhutan (2008), where Buddhism is the official religion of the state. Article 5(1) of the Constitution prescribes:

> *Every Bhutanese is a trustee of the Kingdom's natural resources and environment for the benefit of the present and future generations* and it is the fundamental duty of every citizen to contribute to the protection of the natural environment [emphasis added].

This trusteeship duty must be read in conjunction with Article 8(2) of the Constitution, which expresses that every Bhutanese has the duty to "preserve, protect and respect the environment." The country of Bhutan, therefore, provides a constitutional example where the citizens are the trustees of the natural resources and environment, while individuals who are living (present generation) and those who have abstract claims in the future (future generations) are the beneficiaries. To prevent the degradation of the country's ecosystems, the state is obliged to ensure that a minimum of 60% of Bhutan's total land is under forest cover (Art. 5(3)). Article 1(12) of the Constitution goes on to provide a generic list of natural resources, which include minerals, rivers, lakes, and forests. These natural resources are vested in the state and regulated by the law. An interpretation of Bhutan's Constitution indicates that while the state is the ultimate "owner" of the natural resources, citizens play a critical intermediary role as trustees for the present and future generation.

A second constitutional example is found in Papua New Guinea, where the Constitution uses a similar trusteeship model, albeit with one noticeable difference (Constitution of the Independent State of Papua New Guinea 1975). The Preamble of the Constitution of Papua New Guinea declares – as its fourth goal – that "[w]e" (the people) conserve the natural resources and environment "for the collective benefit" of all. The Preamble further calls for the people to make "wise use" of the natural resources and the environment (on the land or seabed) by way of a trust for future generations. Unlike Bhutan's, the Constitution of Papua New Guinea does not make a distinction between the owner and the trustees of natural resources and the environment. Rather, Papua New Guinea perceives all persons (the state and its citizens) as trustees. This single characteristic makes it closer to Principle 1.2 of the Hague Principles. Notwithstanding this subtle difference, the examples of Bhutan and Papua New Guinea demonstrate a form of Earth trusteeship at work.

The argument pertaining to trusteeship and intergenerational equity has also reached the pinnacle of the UN. According to a General Assembly Report issued by the Secretary-General (2013: 12),

> it has been argued that the notion that each generation holds the earth as a trustee or steward for its descendants strikes a deep chord with all cultures, religions and nationalities.

It is possible to mount an argument that since we are morally responsible for future persons, why cannot we also be responsible for future generations of living beings who are not persons? Behrens goes so far to suggest that a strong case could be made for taking future generations of all morally considerable entities into account (Behrens 2012: 179). Such a statement may be consistent with Principle 2 of the Hague Principles, which recognizes our responsibilities to the community of life (Principle 2.1) based on "the intrinsic values of nature and of all living beings" (Principle 2.2). However, on a strict statutory interpretation of the Hague Principles, it would be difficult to adopt such a position. Firstly, the only mention of "future generations" in the Hague Principles is in the Background Note. As previously expressed, our responsibilities for "each other, for future generations, for all living beings, and the Earth" suggest that the category "future generations" is inextricably bound to "each other" – namely, the present generation of human beings. If the drafters of the Hague Principles intended to extend the "future generations" to include the future generations of "living non-humans," they would have expressly done so.

Secondly, and as Behrens concedes, we (especially in the Western tradition) are unable to coherently account for moral duties for future persons, much less the future of other life-forms (Behrens 2012: 179–180). Lastly – and using the phrase of Behrens – what makes a particular species a "morally considerable" entity? Would it be fair to say that future generations of elephants are "morally considerable entities" and that the future generations of dragonflies are not? Environmental pragmatists confront this dilemma by arguing that an eco-centric value system, together with a long-term focus on our human intergenerational responsibilities, is enough to ensure the future protection of the natural environment, including the life of non-human species (Behrens 2012: 180). To put it another way, if we act as trustees for non-human species (e.g., wildlife, whales, and birds), indirectly we would be looking after their future generations, since the possibility of these species becoming extinct would become remote.

Earth Trusteeship for the Future

In 1997, former UN Secretary-General Kofi Annan proposed a "new concept of trusteeship" and that the defunct UN Trusteeship Council (one of the principal organs of the UN, whose purpose was to supervise the administration of trust territories under the International Trusteeship System) be converted to a forum where member states of the UN exercise their collective trusteeship for the integrity of the global commons – such as the oceans, atmosphere, and outer space (Laurenti 1997; UN Res A/51/950). A similar call was later made in 2001 by Maurice Strong, a co-chair of the Earth Charter Commission, who wrote about the need for a "refurbished Trusteeship Council" whose purpose would be to concentrate on issues that are global in nature and important (Strong 2001). Twenty years later, António Guterres, Secretary-General of the UN, issued a report titled *Our Common Agenda*, where he proposed, inter alia, a "repurposed Trusteeship Council" (Our Common Agenda 2021). Implicit in this proposal is that a "repurposed Trusteeship Council" would revive and/or replace the Trusteeship Council, one of the principal organs of the UN. Arising out of this report, the Secretary-General established a High-Level Advisory Board (HLAB) to "identify global public goods and other areas of common interest where governance improvements are most needed" (Our Common Agenda 2021).

In its 2023 report *A Breakthrough for People and Planet*, the HLAB wrote: "All of us will benefit collectively if we treat a healthy planet as a global public good" (United Nations 2023a). The HLAB added: "Our responsibilities to current and future generations can only be met if we act in trusteeship for the planet." The acknowledgment of "trusteeship for the planet" in the HLAB report suggests that Earth trusteeship could add some value to global governance (and perhaps gives a glimmer of hope for a new and improved 2.0 Trusteeship Council). The jury is still out on this issue since the HLAB report stands to be debated at the UN Summit of the Future, to be held in New York in 2024.

The concept of Earth trusteeship is also "waiting in the wings" to assist in the International Court of Justice Advisory Opinion (ICJAO) on climate change, initiated by the Pacific island state of Vanuatu. In March 2023, Vanuatu successfully led a global coalition of 132 co-sponsoring states that adopted a UN General Assembly resolution to proceed to the ICJ for an Advisory Opinion on climate change (UNGA Res. 77/276). The legal question to be put to the ICJ consists of two parts. The first part (part [a]) asks:

> What are the obligations of States under international law to ensure the protection of the climate system and other parts of the environment from anthropogenic emissions of greenhouse gases for States and for present and future generations?

In response to the preceding question – and bearing in mind international law, judicial decisions, constitutional provisions, and the teachings of highly qualified jurists in this field – it is possible to mount an argument that states have an obligation to act as trustees for the Earth for both present and future generations. This would not be the first time that the ICJ will be considering the issue of trusteeship within an environmental context. For example, in the *Gabčíkovo-Nagymaros Project* (1997 ICJ Rep 7) decision, Vice-President Weeramantry, in his Separate Opinion, raised the consciousness of trusteeship in the ICJ when he referred to "the principle of trusteeship of Earth resources" as being "the first principle of modern environmental law." This does not mean that it will be all "smooth-sailing" for Earth trusteeship at the ICJ. One foreseeable challenge is that the Court may be reluctant to apply private law systems (such as a trust) as a category of "general principles of law recognized by civilized nations," as seen in Sir Arnold McNair's Separate Opinion in the ICJ's Advisory Opinion in *International Status of South-West Africa* (1950 ICJ Rep 128). Added to this, one could argue that it may be a bit premature for the Court to rule that an ecological trust, such as Earth trusteeship, is an "international custom" or a "general practice" that is accepted as law by states, as envisaged under the Statute of the ICJ (art 38(1)(b)). Again, here the issue remains too close to call, and the outcome depends on how persuasive and convincing the legal arguments would be before the Court.

Conclusion

The core message of the Hague Principles is that humans are members of the community of life, and this determines what relationship we should have with others. Rights do not exist in a vacuum, and any global legal system needs to be based on responsibilities, from which any rights must follow (Bosselmann 2018). As human beings, we inhabit one planet, which consists of many other life-forms. The Earth, therefore, must be seen as a harmonious system that supports all life. Thus, our main collective responsibility is to be trustees of the Earth, which includes other life-forms, nature, and Earth's ecological systems. The idea of protecting the integrity of ecological systems is not new in international environmental law. It must, however, now be formalized and conceptualized, in relation to the entire Earth, as a living system. Earth trusteeship requires trustees to act altruistically when caring for the Earth – a subject which does not belong to them. As trustees of the Earth, we hold the planet in trust for the benefit of others, who are the present generation, future generations (persons unborn), all living beings, and even the Earth itself as a living entity. Earth trusteeship is an ethical responsibility. It is also non-anthropocentric, because it requires selfless service to others, such as living non-humans and Earth's natural spaces.

As an ecological trust, Earth trusteeship has the potential to be that "new consciousness in law" and a future tool for global governance. To be an effective tool, Earth trusteeship needs to first be accepted by states. This is where the difficulty may lie. Even if the concept of trusteeship of the Earth is an ancient tradition rooted within Indigenous cultures and religion, the political idea of Earth trusteeship was birthed five years ago at the Peace Palace in the Hague. Earth trusteeship may therefore need time to blossom before UN member states and the ICJ consider it an international custom or general practice in law. Earth trusteeship, therefore, is a theoretical framework waiting to be fully unraveled. And the sooner we unravel and grapple its ecological contents, the better it will be for the future of the planet.

References List

Agreement under the United Nations Convention on the Law of the Sea on the Conservation and Sustainable Use of Marine Biological Diversity of Areas Beyond National Jurisdiction [High Seas Treaty] (opened for signature on 20 September 2023).

Behrens, K.G. (2012) "Moral obligations towards future generations in African thought," *Journal of Global Ethics*, 8(2–3), p. 179.

Birks, P. (2000) "The content of fiduciary obligation," *Israel Law Review*, 34, p. 3 (Lionel Cohen Lecture).

Bosselmann, K. (2013) "The concept of sustainable development," in: K. Bosselmann, D. Grinlinton, & P. Taylor (eds) *Environmental Law for a Sustainable Society*, 2nd ed., Auckland: New Zealand Centre for Environmental Law.

Bosselmann, K. (2017) *The Principle of Sustainability – Transforming Law and Governance*, 2nd ed., Oxon: Routledge.

Bosselmann, K. (10 December 2018) "Opening of earth trusteeship forum," *Peace Palace*, The Hague.
Bosselmann, K. (2019) *Can kaitiakitanga Save the Planet*? Available at: www.auckland.ac.nz/en/news/2019/12/20/can-kaitiakitanga-save-planet.html.
Bosselmann, K. (2020) "The role of earth trusteeship in earth governance," in: Laura Westra, Klaus Bosselmann, & Matteo Fermeglia (eds) *Ecological Integrity in Science and Law*, Cham: Springer.
Bosselmann, K. (2021) "The framework of ecological law," in: B.H. Desai (ed) *Our Earth Matters – Pathways to a Better Common Environmental Future*, Amsterdam: IOS Press BV.
Bosselmann, K. (1 August 2023) "Earth trusteeship can help us solve climate change," *TEDxPorto*. Available at: www.youtube.com/watch?v=ZnaxYJbirTI.
Bosselmann, K. & Taylor, P. (2019) "Promoting global ethics: The Earth Trusteeship Initiative," in: P. Burdon, K. Bosselmann, & K. Engel (eds) *The Crisis in Global Ethics and the Future of Global Governance – Fulfilling the Promise of the Earth Charter*, Gloucestershire: Edward Elgar Publishing Limited.
Cicovacki, P. (2009) *Albert Schweitzer's Ethical Vision (A SourceBook)*, New York: Oxford University Press Inc.
Constitution of the Independent State of Papua New Guinea 1975 (Rev).
Constitution of the Kingdom of Bhutan 2008.
Convention on Biological Diversity (opened for signature 5 June 1992, entered into force 29 December 1993).
Convention on the Conservation of Antarctic Marine Living Resources (opened for signature 20 May 1980, entered into force 7 April 1982).
The Earth Charter (2000). Available at: https://earthcharter.org/library/the-earth-charter-text/.
Fuel Retailers Association of Southern Africa v Director-General Environmental Management CCT 67/06, 7 June 2007 at 102.
Gabčíkovo-Nagymaros Project (Hungary/Slovakia) (1997) ICJ Rep 7 at 102 and 108.
Garner, B.A. (ed) (2019) *Black's Law Dictionary*, St Paul: Thomas Reuters.
The Hague Principles for a Universal Declaration on Responsibilities for Human Rights and Earth Trusteeship [The Hague Principles] (December, 2018).
International Status of South-West Africa (Advisory Opinion) (1950) ICJ Reports 128 at 148.
Johnston, D. (1988) *The Roman Law of Trusts*, Oxford: Oxford University Press.
Kim, R.E. & Bosselmann, K. (2015) "Operationalizing sustainable development: Ecological integrity as a grundnorm of international law," *Review of European, Comparative & International Environmental Law*, 24(2).
Laurenti, J. (1997) "Renewing the United Nations – A critical assessment of the secretary-general's "track two" reform program," *Global Policy Forum*. Available at: www.globalpolicy.org.
Miglani v State of Uttarakhand (2017) (PIL) No. 140 of 2015 (HC) at 22.
Moffat, G. (2009) *Trusts Law*, 5th ed., Cambridge: Cambridge University Press.
Report of the UN Conference on Environment and Development UN Doc A/CONF.151/26 (Vol I) (12 August 1992) (Rio Declaration).
Report of the UN Conference on the Human Environment UN Doc A/CONF.48/14/Rev 1 (5–16 June 1972) (Stockholm Declaration).
Sobion, J. (2023) "Earth trusteeship: A theoretical framework waiting to be unravelled," in: J. Sobion & H. van Willenswaard (eds) *Reflections on Earth Trusteeship – Mother Earth and a New 21st-Century Governance Paradigm*, Nonthaburi: INI Books.

Strong, M. (2001) "Reforming the United Nations," *The Futurist* 35(5), World Future Society. Available at: www.mauricestrong.net/index.php?option=com_content&view=article&id=160&Itemid=85.
Taylor, P. (2020) "Opportunity within failure: Can the global pact for the environment learn from responsibility to protect," in: L. Westra, K. Bosselmann, & M. Fermeglia (eds) *Ecological Integrity in Science and Law*, Cham: Springer.
United Nations (1997) *Renewing the UN: A Programme for Reform – Report of the Secretary-General*, Res: A/51/950 at 85.
United Nations (2013) *Intergenerational Solidarity and the Needs of Future Generations – Report of the Secretary-General*, Res: A/68/100.
United Nations (2017) *Report of the Secretary-General – Harmony With Nature* (Res: A/72/175).
United Nations (2021) *Our Common Agenda – Report of the Secretary General*, New York: United Nations.
United Nations (2023a) *A Breakthrough for People and Planet: Effective and Inclusive Global Governance for Today and the Future*, New York: United Nations University. Available at: www://highleveladvisoryboard.org/breakthrough/.
United Nations (2023b) *Request for an Advisory Opinion of the International Court of Justice on the Obligations of States in Respect of Climate Change*, UNGA Res: 77/276.
United Nations Framework Convention on Climate Change (opened for signature 4 June 1992, entered into force 21 March 1994).
van Willenswaard, H. (2016) *The Wellbeing Society – A Radical Middle Path to Global Transformation*, Bangkok: Garden of Fruition.
van Willenswaard, H. (2020a) *Earth Trusteeship Working Group – Initial Concept Note for Discussion*, Summary draft, Right Livelihood College South East Asia campus, Bangkok.
van Willenswaard, H. (ed) (2020b) *Innovation for Life – A New Light on Right Livelihood*, Bangkok: School for Wellbeing Studies and Research.
Weeramantry, C.G. (2009) *Tread Lightly on the Earth – Religion, the Environment and the Human Future*, Pannipitiya: Stamford Lake (Pvt) Ltd.
Weiss, E.B. (1989) *In Fairness to Future Generations: International Law, Common Patrimony, and Intergenerational Equity*, Tokyo: United Nations University; New York: Transnational Publishers Inc.
Weiss, E.B. (2020) "The future of the planetary trust in a kaleidoscopic world," *Environmental Policy and Law*, 50, p. 449.
World Charter for Nature A/RES/37/7 (1982).

Chapter 25

Beyond Stockholm +50

Pushing for Earth Governance

Klaus Bosselmann

Introduction

The 1972 United Nations Conference on the Human Environment adopted the Stockholm Declaration on the Human Environment (Stockholm Declaration), which marked the beginning of international environmental law. Its first principle proclaimed the interrelatedness between humans and nature: "Man is both creature and moulder of his environment, which gives him physical sustenance and affords him the opportunity for intellectual, moral, social and spiritual growth." The next four principles describe various responsibilities that humans have toward the Earth.

Ten years later, the United Nations General Assembly adopted the World Charter for Nature, "aware that man is part of nature" and "convinced that every form of life is unique, warranting respect regardless of its worth to man" (Preamble). The first three principles state: "Nature shall be respected and its essential processes shall not be impaired," "the genetic viability of the Earth shall not be compromised," and "all areas of the Earth, both land and sea, shall be subject to these principles of conservation" (World Charter for Nature).

Fifteen years later, the Report of the World Commission on Environment and Development: Our Common Future (Brundtland Report) observed in its opening statement that the sight of our planet from space may have "a greater impact on thought than did the Copernican revolution of the 16th century," as this allowed us to "see and study the Earth as an organism whose health depends on the health of all its parts." Remarkably, these early documents of international environmental law highlighted the uniqueness and vulnerability of the Earth, the interconnectedness and intrinsic value of all life-forms, and a duty of caring for the Earth.

Twenty years later, the 1992 United Nations Conference on Environment and Development (Earth Summit) adopted the first global environmental treaties. They declared the "Earth's climate" and "biological diversity" a common concern of humankind but reaffirmed the sovereign rights of states to exploit their own resources. In this way, any global responsibilities

DOI: 10.4324/9781003440871-29

appeared as derivative and secondary, reflecting the lowest common denominator among states. As a result, environmental treaties have always focused on specific fragments of the Earth, with obligations dictated by feasibility rather than by urgency, and further compromised by signatory states deciding for themselves how they should be implemented. Despite the Brundtland Report's call "From One Earth to One World," states have done very little to narrow the gap.

Fifty years after Stockholm, it is only too obvious that international environmental law has failed. Some trees have been saved, but the forest is disappearing (Bosselmann 2010). The reason is that states have kept their proclaimed responsibilities toward the Earth non-committal and not binding. What may be less obvious is that this carelessness is not just a lack of political will but a repercussion of habitual state sovereignty. States and their representatives, some more than others, have used sovereignty as a shield against global action.

The biggest stumbling block has been the resistance against "new kinds of institutional arrangements and novel procedures of decision-making, on both national and international levels." The person who wrote this was Maurice Strong, initiator and Secretary-General of the 1972 Stockholm Conference. A year after Stockholm, he observed that the state's "responsibility to ensure that activities within their jurisdiction or control do not cause damage to the environment of other States or of areas beyond national jurisdiction" (Principle 21, Stockholm Declaration) is "a far more revolutionary matter than has yet been generally acknowledged or comprehended" (Strong 1973). Principle 21 reappeared as Principle 2 in the Rio Declaration, adopted at the 1992 Earth Summit, again under the leadership of Maurice Strong as Secretary-General. Probably the most significant provision of both Declarations is the no-harm rule now widely recognized as a principle of customary international law, whereby a state is duty-bound to prevent environmental harm to other states. The crux is that the no-harm rule does not include areas beyond national jurisdiction, the global commons, or the Earth as a whole. A conventional understanding of state sovereignty would not allow for this. An obligation *erga omnes* to safeguard the Earth does not exist. It may be emerging, though, considering the many initiatives, declarations, and agreements that have since been produced by civil society, professional bodies, and international organizations (Robinson 2018).

Maurice Strong himself was instrumental in the creation of the Earth Charter that the 1987 Brundtland Report had originally called for but was ultimately rejected by states during the 1992 Earth Summit. As a cross-cultural, cross-religion consensus of global values, the 2000 Earth Charter gained recognition among governmental agencies, several states, UNESCO, IUCN, and other global organizations. Its key principles are "Respect and Care for the Community of Life" (Part I) and "Ecological Integrity" (Part II).

Both principles directly address what international environmental law was and is in urgent need of, that is, an overall objective and obligation to protect the integrity of the Earth's ecological system. Its ethical core is our belonging to the community of life rather than long-standing Western anthropocentrism that has shaped traditional, ineffective environmental law. The Earth is not an assembly of natural resources for human consumption but a living whole that humans are part of and need to respect. Fundamentally, this means that humans – and their political institutions – have responsibilities toward non-human members of the community of life, in other words, toward the Earth community as a whole.

The following chapter will explore notions around state responsibilities to show the current weakness of international law, but also practical ways to strengthen it. There are important developments toward advancing state sovereignty and building a system of Earth governance. One critical element here is the cooperation between civil society and the United Nations.

State Responsibilities Toward the Earth

Article 2 of the 1992 Rio Declaration on Environment and Development (Rio Declaration) says: "States have, in accordance with the Charter of the United Nations and the principles of international law, the sovereign right to exploit their own resources pursuant to their own environmental and developmental policies." Like a private owner of land, the state has the undisturbed right to exploit its territory. Crucially, the state has no obligation to protect it or protect any areas outside national boundaries (e.g., oceans and the atmosphere).

On the other hand, the second half of Article 2 says that states have "the responsibility to ensure that activities within their jurisdiction or control do not cause damage to the environment of other States or of areas beyond the limits of national jurisdiction." Furthermore, Principle 7 of the Rio Declaration reads: "States shall co-operate in a spirit of global partnership to conserve, protect and restore the health and integrity of the Earth's ecosystem." So is there a legal obligation of states to protect the global environment and integrity of the Earth's ecosystem, after all?

The answer is no. The current system of international law does not require the sovereign state to protect the natural environment within or beyond national boundaries. It only expects states to consider – but not necessarily avoid – disastrous environmental consequences of their actions. There is huge discretion involved here, a degree of moral responsibility, but no legally enforceable obligations whatsoever. Only negotiated treaties and fundamental principles of international law could change that, but so far, all treaties have been too weak and fragile – not to mention the lack of enforcement – to urge states into the logic of common responsibility for the Earth. For this to happen, we need a deliberate, bold move toward trusteeship for the Earth.

There is an ever-growing ecological movement that has found its legal expression in Earth jurisprudence, Earth law, and ecological law (Bosselmann & Taylor 2017). Some recent developments give us a sense of just how significant this legal movement has been. In 2016, some 100 professors of environmental law adopted a manifesto called "From Environmental Law to Ecological Law" (Oslo Manifesto) at the IUCN Academy of Environmental Law Colloquium in Oslo, Norway. The Oslo Manifesto (2016) has since been endorsed by hundreds of environmental lawyers and environmental law organizations from around the world and has led to the establishment of the Ecological Law and Governance Association (ELGA) in 2017 (Ecological Law and Governance Association 2017). ELGA is a global network of lawyers, universities, and environmental activists that coordinates initiatives for transforming law and governance.

One of these initiatives is the Earth Trusteeship Initiative (ETI), established on 10 December 2018 in the Peace Palace in The Hague, Netherlands (Earth Trusteeship Initiative 2018). This day marked the 70th anniversary of the adoption of the Universal Declaration of Human Rights. With the support and endorsement of many human rights, Indigenous rights, and environmental organizations, the ETI launched the "Hague Principles for a Universal Declaration on Responsibilities for Human Rights and Earth Trusteeship" (Hague Principles) (The Hague Principles 2018). The Hague Principles aim for a coherent concept of human rights and responsibilities that states have with respect to the protection of human rights and to the protection of our common home, Earth.

The ethical underpinnings of these responsibilities are not new but need to be articulated in the context of our current predicament, as the following passage from a 2017 address to the UN General Assembly explained:

> The ethics of stewardship or guardianship for the community of life is one of the most foundational concepts in the history of humanity. It is inherent in the teachings of the world's religions and the traditions of indigenous peoples and is, an integral part of humanity's cultural heritage. Yet, our political and legal institutions have not taken Earth ethics to heart. The Earth as an integrated whole may be featuring in images, in science and in ethics, but does not feature in law. Earth and the areas outside national jurisdictions (the global commons) are considered as res nullius, a legal nullity without inherent rights. Not that Earth cares about such rights. It is we humans who must choose to care about them. If we keep ignoring them, then basically we are saying that the Earth system doesn't really matter. We take it for granted – like sunshine and rain – and of no relevance to the system of law that governs society and states. Given that the ethics of earth stewardship are widely accepted today we should be ready for taking the next step: Earth trusteeship.
>
> Earth trusteeship is the essence of what Earth jurisprudence is advocating, but, more importantly, it has also been advocated in key international environmental documents. Earth trusteeship is the institutionalization of the duty to protect the integrity of ecological systems.

> This duty is expressed in no less than 25 international agreements – from the 1982 World Charter for Nature right through to the 2015 Paris Climate Agreement! (Kim & Bosselmann 2015). To act on this duty, "states need to cooperate in the spirit of global partnership" as, for example, Principle 7 of the 1992 Rio Declaration says (Bosselmann 2017: 2, 3).

The legal argument for Earth trusteeship can be firmly based on ethics common to all cultures and fundamental obligations of states expressed in many international agreements. The challenge ahead is to convince governments that the step to Earth trusteeship is not only necessary but also actually possible and not too difficult to take.

Ultimately, the legitimacy of the sovereign state rests on its core task to act for, and in the interest of, its citizens. This requires consent with the governed (Ashcraft, citing John Locke 1991: 524). Governmental duties can therefore be understood as fiduciary obligations toward citizens (Fox-Decent 2012). Such fiduciary obligations are recognized typically in public law, exist in common law and civil law (although in varying forms and degrees), and are also known in international law (Blumm & Guthrie 2012; Brown Weiss 1984). The fiduciary function of the state can also be described as a trusteeship function (Finn 1995).

Sovereignty and Trusteeship

How, then, can state sovereignty be reconciled with trusteeship? *Prima facie*, both seem to have different purposes, yet they are part of the same basic function of the state, that is, to serve the citizens it depends on and is accountable for.

Furthermore, global commons governance brings sovereignty and trusteeship close together (Stec 2010). As has been noted, the traditional concept of sovereignty is less compelling today than it was in the past because of a "glaring misfit between the scope of the sovereign's authority and the sphere of the affected stakeholders" (Benvenisti 2013: 301). This "glaring misfit" engenders inefficient, undemocratic, and unjust outcomes for under- or unrepresented affected stakeholders. Non-citizens, future generations, and the natural environment all fall into such a category of "affected stakeholders." To overcome this misfit, states need to increasingly perform trusteeship functions.

The state gains its legitimacy exclusively from the people who create it. While the legality of a state depends on recognition by other states, once in existence, a state can only ever legitimize its continued existence through the ongoing trust by its people. The core idea of the modern democratic state is that it acts through its people, by its people, and for its people. This implies a fiduciary relationship between people and state and is arguably the only legitimate basis for political authority in the English Civil War, the American

Revolution, and then again confirmed in the French Revolution (Reisman 1990: 867). It is echoed in constitutional documents, such as the 1776 Pennsylvania Declaration of Rights: "[A]ll power being . . . derived from the people; therefore, all officers of government, whether legislative or executive, are their trustees and servants, and at all times accountable to them." John Locke had famously asserted that legislative power is "only a fiduciary power to act for certain ends" and that "there remains still in the people a supreme power to remove or alter the legislative, when they find the legislative act contrary to the trust reposed in them" (Locke 1690).

Likewise, Immanuel Kant drew the moral basis of fiduciary obligations from the duty-bound relationship between parents and children (Criddle & Fox-Decent 2009: 352). Kant claimed that children have an innate and legal right to their parents' care. In a similar sense, he believed that state legitimacy was the result of a contract that is necessarily created between people to form a Rousseauian *volonté général* (general will). Through this process, Kant claimed, we jointly authorize the state, who, in turn, acts in the form of trusteeship governance.

That state sovereignty is fundamentally a trust relationship cannot be dismissed as a mere ideal. Trusts and the implicit fiduciary relationships can be traced back to Middle Eastern origins and Roman and Germanic law. They are also inherent in the teachings of the world's religions and are prevalent in non-Western cultures (Criddle & Fox-Decent 2009: 378, 379). In fundamental terms, trust relationship is also anchored in the Universal Declaration of Human Rights. Article 21(3) states that "the will of the people shall be the basis of the authority of government."

At its most simplistic, the state's legitimacy to govern is based on its ability to serve the will of the people, usually described as the common interest. Aristotle saw the purpose of the state as "for the common good." John Locke also hinted at such a purpose. But of course, who defines common good, and what does it include? According to Locke's definition, the *common good* was what arose from there being surplus produce that could be sold in the marketplace.

Following Eyal Benvenisti, we can conceive of three normative arguments for state trusteeship. Firstly, sovereignty should be viewed as a vehicle for the exercise of personal and collective self-determination (Benvenisti 2013: 301). Collective self-determination embodies the freedom of a group to pursue its interests, further its political status, and "freely dispose of [its] natural wealth and resources" (International Covenant 1966, Art. 1) or, in other words, protect and preserve them. Secondly, sovereign states are agents of humanity as a whole (Benvenisti 2013: 305), as all human beings are holders of rights not because states grant them but because they are entitlements of free-born, equal human beings. The legitimacy of a state ultimately depends on its ability to honor and respect human rights, hence the trusteeship function of the state with respect to humanity. Thirdly, the right to own natural resources ("territorial sovereignty") is intrinsically linked with the responsibility to protect them. Any disjuncture would jeopardize the sustained use by citizens, hence the need for state trusteeship of natural resources. In essence, the legitimacy of the state of the 21st century rests on its ability to function as a trustee of human rights and the natural environment.

Toward Earth Governance

Earth system science, political science, and jurisprudence have evidenced that the integrity of the Earth's ecosystem must be protected and restored (Rockström et al. 2023). It is critical, therefore, to integrate this duty into evolving policies and institutions of global environmental governance. Conceptually, Earth governance (Bosselmann 2015) requires a shift from thinking about the environment from a state-centered perspective to an Earth-centered perspective. This involves the creation of Earth trusteeship mechanisms at local, national, and global levels.

Following the framework of the Earth Charter, the Hague Principles conceive rights and responsibilities as embedded in the Earth community. They define responsibilities for the Earth, the community of life, and human rights (The Hague Principles 2018). They are designed around the need for trusteeship responsibilities that people and their political institutions have for the protection of human rights and the Earth system. In this regard, both citizens and states have responsibilities as trustees of the Earth (Bosselmann 2023: 118). Trusteeship – whether public, private, or hybrid – implies a legal mandate to govern, yet bound by the duty to benefit those who cannot speak and act for themselves (in contrast to self-interest implicit in private or corporate ownership and the national interest served by sovereign states). Earth trusteeship would apply this concept to global governance, ensuring care for the Earth system, or Earth community, for the benefit of future generations.

After the adoption of the Hague Principles in 2018, a period of further studies and exchanges in various fora followed. Five years after its launch, 2023 initiated a new period of public dialogue and tentative application of Earth trusteeship to concrete areas of concern, as well as the reassessment of the principles in light of new developments.

Among the documents to guide education, academia, and activism is the "Declaration on Education for Earth Trusteeship," which was adopted in 2020 by the laureates of the Right Livelihood Award:

> It is vitally important to alert all Earth citizens to their responsibilities as trustees for current and future generations of life on our planet home. Education for Earth Citizenship is necessary for all people of all cultures and religions, across all generations. This aspect tends to be neglected in educational curricula throughout the world.

> Trusteeship implies protection and care of the commons, the environment, conservation of the rights and interests of future generations, protection of our cultural heritage, and the protection of human rights and dignity worldwide.
>
> An inter-generational, inter-disciplinary and cross-cultural approach needs to be inspired with the thought that every individual can make some contribution to this cause, through Right Livelihood: ethical living to achieve Earth Democracy.
>
> False ideas of ownership of natural resources as opposed to the Principles of Earth Trusteeship; individual rights as opposed to community rights, social duties, and trusteeship of the commons; present profit at the cost of future damage – all these tend to dominate modern society.
>
> It is vitally important to take educational measures on Earth Trusteeship, including strengthening and spreading learning centres, widely organized through the support of Right Livelihood Laureates and others. We can correct the negative trends and make a lasting, transformative contribution worldwide to develop a shared global Right Livelihood consciousness through Principles of Earth Trusteeship leading to the preservation of our humanity and our planet.
>
> (Gunasekera et al. 2023: 25)

Since 2020, manifold actions and reflections on Earth trusteeship occurred in the form of conferences, workshops, and writings. They have been summarized in a comprehensive book launched in 2023 titled *Reflections on Earth Trusteeship: Mother Earth and a new 21st-century governance paradigm* (Sobion & Willenswaard). Of particular importance is the cooperation between civil society organizations and the United Nations in preparation of the Summit of the Future in September 2024.

The 2023 Report of the High-Level Advisory Board on Effective Multilateralism (HLAB) summarized its call for effective and inclusive global governance in this way: "Our responsibilities to current and future generations can only be met if we act in trusteeship for the planet" (High-Level Advisory Board 2023: 23). The report makes specific recommendations on a number of issues toward institutional change. They will be discussed at the 2024 Summit of the Future.

In preparation of its report, the HLAB invited contributions from a range of civil society networks. They included recommendations for integrating Earth into the system of global governance. An extract of these recommendations reads as follows:

> The High-Level Advisory Board on Effective Multilateralism (HLAB) has a unique opportunity to further develop Our Common Agenda's idea of a repurposed Trusteeship Council, especially in the light of some resistance from a number of Member States. Under international law, there is

at present no legally binding duty of States to protect the global environment, i.e. the integrity of the Earth's ecosystem. This makes it necessary, and indeed urgent, to investigate the merits of trusteeship responsibilities that Member States may have towards future generations and the Earth. A repurposed Trusteeship Council would remain a UN principal organ and focus on intergenerational equity and the global environment. It could adopt a new multilateral governance model – beyond the traditional state-centric framework – to protect and restore the global commons and Earth's ecological integrity beyond the sovereignty and reach of individual states. The process towards institutional reform of this nature takes commitment and time, however it is important to start the process and also explore new multi-sectoral coalitions at international and national levels to take up common trusteeship challenges. Not taking this opportunity now, could cause a situation where any efforts towards effective multilateralism and protection of global public goods will be too late.

There are, at least, three options for further investigating Earth governance.

(1) One is the process of further implementing Agenda 2030 and achieving its Sustainable Development Goals. From an Earth system perspective, the seventeen SDGs, while all relevant, need to foremost ensure that humanity stays within the limits of planetary boundaries. The economy- and society-related SDGs will not be achieved without ensuring that the environment-related SDGs including Water (6), Climate (13), Life Below Water (14) and Life on Land (15) are realized. These need to be safeguarded, arguably through trusteeship governance, to make any prospect for sustainable development realistic and achievable. The HLAB should highlight the need for this "strong sustainability" approach in its report and ask the High-Level Political Forum (HLPF) for a systematic review of the implementation of Agenda 2030.

(2) Another opportunity is the Summit of the Future in September 2024 where the merits of repurposing of the Trusteeship Council – and/or an Earth Trusteeship Dialogue programme in conjunction with the appointment of a Special Civil Society Envoy – could be discussed including its mandate, governance structure and possible amendments to the UN Charter (Preamble and Chapter XIII – Trusteeship Council – to include intergenerational equity and future generations). To be successful, the HLAB's report needs to accompany the preparations for the Summit's outcome document "Pact for the Future" and also inform the preparatory ministerial meeting to be held in September 2023.

(3) At the Summit, the establishment of a UN Special Envoy for Future Generations is likely to be decided on to help ensure that future generations inherit a healthy planet. To have sufficient impact, the Envoy needs to be in a position of responsibility and access with a clearly

defined mandate and with the support from a – well-resourced – UN Future Generations Office. The terms of reference for both, the Envoy and the Office should include responsibility and capacity to pursue dialogue on trusteeship including a repurposed Trusteeship Council.

> These three options are not mutually exclusive and should be pursued simultaneously to encourage cross-sectoral discussions within the UN, but also within civil society. Critically, the UN needs to open up to citizens and civil society. No recommendations of the HLAB, no matter how ambitious, will be realized without multi-sector collaboration involving governments, the corporate sector and civil society as equal consensus building partners.
>
> (Bosselmann 2022)

The time following the 2024 Summit of the Future will be of critical importance to build partnerships between civil society organizations, the UN, and individual member states. The common goal must be to follow the Hague Principles and define our trusteeship responsibilities with respect to the Earth.

Reference List

Ashcraft, R. (ed) (1991) *John Locke: Critical Assessments*, London: Routledge.

Benvenisti, E. (2013) "Sovereigns as trustees of humanity: On the accountability of states to foreign stakeholders," *AJIL*, 107(2), pp. 295–333.

Blumm, M. & Guthrie, R. (2012) "Internationalizing the public trust doctrine," *UC Davis Law Rev*, 45, pp. 741–808.

Bosselmann, K. (2010) "Losing the forest for the trees," *Sustainability*, 2(8), pp. 2424–2448.

Bosselmann, K. (2015) *Earth Governance: Trusteeship of the Global Commons*, Cheltenham: Edward Elgar.

Bosselmann, K. (2017) "The next step: Earth trusteeship," *Address to the United Nations General Assembly*, Earth Day 2017. Available at: http://files.harmonywithnatureun.org/uploads/upload96.pdf.

Bosselmann, K. (2022) *Recommendations to the High-Level Advisory Board on Effective Multilateralism Concerning the Environment as a Global Public Good.* Available at: https://highleveladvisoryboard.org/download/integrating-earth-into-global-governance/.

Bosselmann, K. (2023) "The hague principles: Responsibilities and rights concerning humans and the earth," in: J. Sobion & H. Willenswaard (eds) *Reflections on Earth Trusteeship*, pp. 115–127, Nonthaburi/Thailand: INI Books.

Bosselmann, K. & Taylor, P. (eds) (2017) *Ecological Approaches to Environmental Law*, Cheltenham: Edward Elgar.

Brown Weiss, E. (1984) "The planetary trust: Conservation and intergenerational equity," *Ecology Law Quarterly*, 11, pp. 495–581.

Criddle, E. & Fox-Decent, E. (2009) "A fiduciary theory of jus cogens," *Yale Journal of International Law*, 34, pp. 331–387.

Earth Trusteeship Initiative (2018). Available at: www.earthtrusteeship.world/.

Ecological Law and Governance Association (2017). Available at: www.elga.world/.

Finn, P. (1995) "The forgotten 'trust': The people and the state," in: M. Cope (ed) *Equity: Issues and Trends*, pp. 131–151, Leichhardt/NSW: The Federation Press.
Fox-Decent, E. (2012) *Sovereignty's Promise: The State as a Fiduciary*, Oxford: Oxford University Press.
Gunasekera, N., Ura, K., & Sobion, J. (2023) "Introduction," in: J. Sobion & H. Willenswaard (eds) *Reflections on Earth Trusteeship*, pp. 23–26, Nonthaburi/Thailand: INI Books.
The Hague Principles (2018). Available at: www.earthtrusteeship.world/the-hague-principles-for-a-universal-declaration-on-human-responsibilities-and-earth-trusteeship/.
High-Level Advisory Board on Effective Multilateralism (2023) *A Breakthrough for People and Planet: Effective and Inclusive Global Governance for Today and the Future*, New York: United Nations University.
International Covenant on Civil and Political Rights (ICCPR) (1966), 999 UNTS 171.
Kim, R. & Bosselmann, K. (2015) "Operationalizing sustainable development: Ecological integrity as a grundnorm in international law," *Review of European, Comparative and International Environmental Law*, 24(2), pp. 194–208.
Locke, J. (1690) *Second Treatise of Government*, Ch. 13–15. Available at: https://english.hku.hk/staff/kjohnson/PDF/LockeJohnSECONDTREATISE1690.pdf.
Oslo Manifesto (2016) *From Environmental to Ecological Law*. Available at: www.elga.world/oslo-manifesto/.
Reisman, W. (1990) "Sovereignty and human rights in contemporary international law," *AJIL*, 84, pp. 866–876.
Robinson, N. (2018) *Environmental Law: Is an Obligation Erga Omnes Emerging?* Gland: IUCN. Available at: www.iucn.org/sites/default/files/content/documents/2018/environmental_law_is_an_obligation_erga_omnes_emerging_interamcthradvisoryopinionjune2018.pdf.
Rockström, J., Kotzé, L., Milutinović, L., Biermann, Brovkin, V., Donges, J., Ebbesson, J., French, D., Gupta, J., Kim, R., Lenton, T., Lenzi, D., Nakicenovic, N., Neumann, B., Schuppert, F., Winkelmann, R., Bosselmann, K., Folke, C., Lucht, Schlosberg, D., Schlosser, P., Steffen, W., & Richardson, K. (2023) "Redefining the Global Commons in the Anthropocene," *Proceedings of the National Academy of Sciences*. Available at: www.pnas.org/.
Sobion, J. & Willenswaard, H. (eds) (2023) *Reflections on Earth Trusteeship: Mother Earth and a New 21st-Century Governance Paradigm*, Nonthaburi/Thailand: INI Books.
Stec, S. (2010) "Humanitarian limits to sovereignty: Common concern and common heritage approaches to natural resources and environment," *International and Comparative Law Review*, 12, pp. 384–385.
Strong, M. (1973) "One year after Stockholm," *Foreign Affairs Magazine*, 51(4), pp. 690–707.
World Charter for Nature (1982) *UN GA RES 37/7*. Available at: https://uploads-ssl.webflow.com/60c9c12fa8875207255f7f22/61095204ba2bee73192ff55d_World%20Charter%20for%20Nature%20(1982).pdf.

Index

Note: Page numbers in *italic* refer to figures; those in **bold** refer to tables.

FSC
www.fsc.org
MIX
Papier | Fördert
gute Waldnutzung
FSC® C083411

Kolibri 360 GmbH
Ferdinand-Jühlke-Straße 7
99095 Erfurt, Deutschland
produktsicherheit@kolibri360.de